Power Electronics
Design, Testing and Simulation
Laboratory Manual

KR Varmah
Professor
Department of Electrical and Electronics Engineering
Muthoot Institute of Technology and Science
Kochi, India

Former
Professor and Head
Department of Electrical and Electronics Engineering
Manipal Institute of Technology
Manipal University, Manipal, India
E-mail: krvarmah@gmail.com

Ginnes K John
Assistant Professor
Department of Electrical and Electronics Engineering
Rajagiri School of Engineering and Technology
Kakkanad, Kochi, India
E-mail: ginneskjohn@gmail.com

Chikku Abraham
Associate Professor
Department of Electrical and Electronics Engineering
Muthoot Institute of Technology and Science
Kochi, India
E-mail: chikkuabrahamkalathil@gmail.com

CBS

CBS Publishers & Distributors Pvt Ltd

New Delhi • Bengaluru • Chennai • Kochi • Kolkata • Mumbai
Bhopal • Bhubaneswar • Hyderabad • Jharkhand • Nagpur • Patna
• Pune • Uttarakhand • Dhaka (Bangladesh) • Kathmandu (Nepal)

Power Electronics
Design, Testing and Simulation
Laboratory Manual

ISBN: 978-93-86310-89-7

First Edition: 2017
Reprint: 2020

Published by Satish Kumar Jain and produced by Varun Jain for

CBS Publishers & Distributors Pvt Ltd
4819/XI Prahlad Street, 24 Ansari Road, Daryaganj, New Delhi 110 002, India.
Ph: 23289259, 23266861, 23266867   Fax: 011-23243014   Website: www.cbspd.com
e-mail: delhi@cbspd.com; cbspubs@airtelmail.in.

Corporate Office: 204 FIE, Industrial Area, Patparganj, Delhi 110 092
Ph: 011-4934 4934   Fax: 011-4934 4935   e-mail: publishing@cbspd.com;
                                                                 publicity@cbspd.com

Branches

- **Bengaluru:** Seema House 2975, 17th Cross, K.R. Road,
  Banasankari 2nd Stage, Bengaluru 560 070, Karnataka, India
  Ph: +91-80-26771678/79   Fax: +91-80-26771680   e-mail: bangalore@cbspd.com
- **Chennai:** 7, Subbaraya Street, Shenoy Nagar, Chennai 600 030, Tamil Nadu, India
  Ph: +91-44-26260666, 26208620   Fax: +91-44-42032115   e-mail: chennai@cbspd.com
- **Kochi:** 42/1325, 1326, Power House Road, Opp KSEB, Kochi 682 018, Kerala, India
  Ph: +91-484-4059061-65   Fax: +91-484-4059065   e-mail: kochi@cbspd.com
- **Kolkata:** 6/B, Ground Floor, Rameswar Shaw Road, Kolkata-700014 (West Bengal), India
  Ph: +91-33-2289-1126, 2289-1127, 2289-1128   e-mail: kolkata@cbspd.com
- **Mumbai:** 83-C, Dr E Moses Road, Worli, Mumbai-400018, Maharashtra, India
  Ph: +91-22-24902340/41   Fax: +91-22-24902342   e-mail: mumbai@cbspd.com

Representatives

Bhopal	0-8319310552	Bhubaneswar	0-9911037372	Hyderabad	0-9885175004
Jharkhand	0-9811541605	Nagpur	0-9421945513	Patna	0-9334159340
Pune	0-9623451994	Uttarakhand	0-9716462459	Dhaka	01912-003485
Kathmandu (Nepal)	977-9818742655			(Bangladesh)	

Printed at Glorious Printers, Daryaganj, Delhi, India

to

Our Parents and Gurus

Preface

Rapid developments in the area of power electronic devices and circuits have revolutionized the field of research and development of electronic appliances, industrial drives, power grid, etc. At present, power electronics is a subject of study in many branches of engineering in almost all universities of India and abroad. It is important for the students of electrical and electronics engineering and other related disciplines to have a thorough knowledge of the fundamental aspects of power electronics.

Needless to say that gaining practical knowledge reinforces theoretical aspects and this is the importance of performing experiments in a laboratory. However, to conduct an experiment in a laboratory, proper guidance is necessary and it is expected that this manual will definitely serve the desired purpose. This is written in a very simple and lucid manner. The book is designed for self-study, so that students can learn the fundamental aspects clearly and perform the experiments in a laboratory without much difficulty. This manual is prepared to cater to the needs of students of both undergraduate and graduate engineering degree courses.

Many years of classroom teaching and practical experience in power electronics laboratory in guiding students to conduct experiments have helped us greatly in planning and preparing this manual. The procedures to conduct experiments are presented in simple steps with a systematic approach. First, experiments to study the characteristics of power electronic devices are discussed followed by the design and testing of thyristor triggering and commutation circuits and power electronic converters. Systematic procedures for the simulation of power electronic circuits and converters using MATLAB/Simulink are presented. Finally, MATLAB programs to simulate a few power electronic circuits are also given at the end.

Accordingly, this manual is divided into five sections. *Section A* gives a brief introduction to a few power electronic devices like thyristor, power BJT, triac, power MOSFET and IGBT. The fabrication details, principle of operation and their performance characteristics are explained in brief. This section also includes step-by-step procedures to conduct the experiments and to obtain the device characteristics. *Section B* discusses the design and testing of certain power electronic circuits like thyristor triggering circuits, thyristor commutation circuits, rectifiers, AC voltage controllers and DC–DC choppers. The design equations and design procedures are also presented. *Section C* provides the procedures for simulation of power electronic circuits, discussed in Section B, using MATLAB and Simulink. In addition, simulation of single phase and three phase inverters are also discussed. Also, a simulation study of power quality, aspects of power electronics circuits is also elaborated in detail. In *Section D* a few MATLAB programs to simulate power electronic circuits such as triggering circuits, rectifier circuits, AC voltage regulator, DC–DC chopper and inverter are given. *Section E* is supplemented with *viva voce* and sample questions which can be a ready-reckoner for students preparing for the examination.

The objective of performing power electronic experiments is to make the students familiar with the characteristics of power electronic devices, design and testing of certain power electronic circuits and understand the working of a few power electronic converters. We feel that high voltage power circuits are not essential for this and they can be avoided. This saves a lot of electric power.

Instructions and procedures are given in such a way that the students are urged to make circuit connections using a breadboard as per the circuit diagrams. This method is adopted to gain certain advantages:

1. Only low voltage and low power circuits are utilized for the sake of convenience and students' safety.

2. Only a small amount of electric power is utilized in performing the experiments.

3. Even though modules of power electronic circuits are available in the market, some of them may be very costly. Normally, students are unaware of these devices and components that are used in the module. To them, it is like a "black box" and when asked to 'play' with certain switches and buttons they are totally unaware of what really happens. We feel that this is a definite disadvantage. On the other hand, if the students are encouraged to wire-up low power electronic circuits using breadboards, they can be trained in a systematic way to make proper connections of components to form the required power electronic circuits. In addition, they will gain a firsthand knowledge about the devices and components that make the circuits and how they are interconnected, and also practical knowledge about the formation of power electronic circuits. We expect that this will make the experiments interesting and will instil much enthusiasm and confidence in students.

Because of safety measures, which are explained in this manual, a safe working voltage of 24 V RMS is adopted in performing the experiments. Hence, in almost all experiments a 230/12-0-12 V transformer is used. All experiments are laboratory tested. We sincerely hope that the students will enjoy conducting experiments in the laboratory by reading and understanding the contents of this manual, with utmost attention.

We hereby sincerely acknowledge with gratitude the permission granted to us to use the licensed version of 2010 MATLAB® and Simulink® in the power electronics laboratory of Rajagiri School of Engineering and Technology, Kochi. We hereby place on records the immense help rendered to us by the staff of electrical and electronics department of Rajagiri School of Engineering and Technology, Kochi.

We would like to place on record the sincere and dedicated efforts of Mr Ananthakrishnan (Isa Publishers) to have this manual published. We thank all those who have directly or indirectly helped us in executing this work.

Constructive suggestions for the improvement of this book will also be highly appreciated. We also convey our gratitude to the entire staff of CBS Publishers & Distributors for making all possible efforts in the publication of this book.

KR Varmah
Ginnes K John
Chikku Abraham

General Instructions to Use the Manual

Design and Testing

1. It is expected that the students should know how to use breadboard, DSO/CRO and other electronic equipment and components which are used in the lab.
2. To observe the waveform of load current or device current, connect a 1 Ω resistor in series with the load or the device and observe the voltage profile across it.
3. A step down transformer of 230 V/24 V or 230 V/12–0–12 V is used for doing experiments with AC circuits explained in this manual. Since experiments can be conducted with low voltage and low power, breadboards are used to make the circuit connections.
4. Source voltage of 12 V or 24 V DC voltage is normally used for performing DC circuit experiments. For this purpose DC voltage regulators are used which are commonly available in the lab.
5. The data sheets of power semiconductor devices can be downloaded using internet. Search the device using its IC number. For example C106 D. The relevant data of ICs are given wherever it is required while designing.

Circuit Simulation Using MATLAB

1. Basic knowledge in MATLAB/Simulink and MATLAB programming is a pre-requisite. The MATLAB *Simulink* and *Program* "Help" is very useful to understand the basics of simulation. Also, there are a number of books and online materials are available. Introduction to the basics of simulations is out of the scope of this book.
2. Simulation speed can be adjusted by selecting a proper solver in the Simulink model window. Select *simulation > configuration parameters* in the Simulink model file and from the *solver options*, different solver can be chosen.
3. In Simulink, 10^{-3} is entered as $e - 3$. Similarly 10^{-6} is entered as $e - 6$.
   For example:
   8 mH is entered as $8e - 3$
   100 µF is entered as $100e - 6$
   1 millisecond is entered as $1e - 3$
   1 kΩ is entered as $1e3$
4. If a thyristor is to be turned ON using a pulse generator, a pulse width (% of period) of 5 is only required, since it is a latching device. For a MOSFET pulse width should be its ON time. For example, for a MOSFET, pulse width (% of period) is 75 if its duty ratio is 0.75.

A Few Words of Caution

1. While working with electrical and electronic circuits, personal safety is of paramount importance.
2. Incorrect connections of electronic devices and components and other gadgets, made inadvertently, can cause serious hazards.
3. For making any change in connections already made, it is always safe to switch OFF the supply and remove the mains' plug.
4. For carrying out any adjustment, use only one hand with the other hand kept safely away and free from any metal contact.
5. Ensure that the components selected are of correct values and the measuring instruments are set to proper ranges.
6. Make sure that the DSO is set properly to make the necessary measurements.
7. Study the experiments correctly and gain confidence to conduct the experiments alone.

Contents

Section A: Characteristics of Devices

Section B: Circuit Design and Testing

Section C: Circuit Simulation Using MATLAB/Simulink

Section D: Circuit Simulation Using MATLAB Program

Section E

Section A

Characteristics of Devices

Experiment A1

Static Characteristics of a Thyristor

A1.1 OBJECTIVE

a. To obtain the minimum gate voltage and minimum gate current
b. To plot the static v-i characteristics
c. To measure the latching current
d. To measure the holding current

A1.2 THEORY

A thyristor is a four layer PNPN device as shown in the figure. When the device is forward biased, the junction J_1 is forward biased, J_2 is reverse biased and the junction J_3 is forward biased. The device is said to be in the forward blocking state or off state. However a leakage current flows because of a positive voltage gradient across junction J_2. Minority carriers from P_2 region will migrate to N_1 region and this constitutes the leakage current. When forward voltage is gradually increased the leakage current increases accordingly. At a particular value of the forward voltage avalanche breakdown of the reverse biased junction J_2 will occur due to large reverse current. This value of the forward voltage is known as "forward break over voltage". After the avalanche breakdown there will be free movement of charge carriers across the entire three junctions and a large anode to cathode current flows. This current is limited only by the external impedance in the circuit. The forward voltage blocking capability of the device depends on the thickness of the junction J_2.

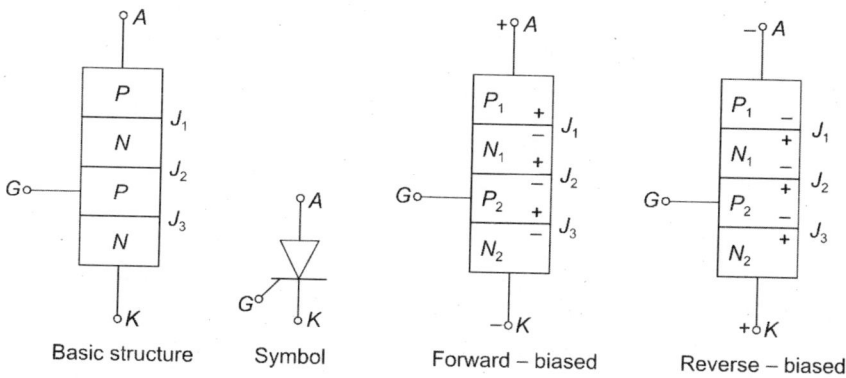

| Basic structure | Symbol | Forward – biased | Reverse – biased |

3

When the thyristor is reverse biased with anode negative with respect to cathode, it can be seen that the junction J_1 and J_3 are reverse biased and the junction J_2 is forward biased. However a reverse leakage current flows. If the reverse voltage is gradually increased, the reverse biased junction J_1 and J_3 will breakdown and large reverse current flows. This occurs at a particular value of the reverse voltage and this is known as *reverse break down voltage.*

To turn ON the device, a small positive voltage is applied to the gate with respect to the cathode, when the device forward biased. A small gate current flows and this is because of migration of electrons from N_2 to P_2 region. Minority carriers in P_2 increase because of large voltage gradient V_{AK}. Some of these electrons in the P_2 region reach the Junction J_2 thus increasing the minority carriers near this junction. This will increase the reverse leakage current across junction J_2 and ultimately cause this junction to break down. The device is said to be triggered into the conduction state and this is called "forward ON state". Once the device is turned in to conduction mode the gate current has no control over the device. Even when the gate supply is removed, the device will continue to remain in the forward conduction mode. The device is said to be latched in to conduction. Hence the device is said to be a latching device.

Latching current is the minimum value of the anode current which it must attain during turn ON process to remain in the forward ON state, even when the gate signal is removed. Latching current is the minimum anode current required for the transition from forward blocking state to forward ON state.

Holding current is the minimum value of the anode current in the forward conduction mode, below which the device is turned off. Holding current is the minimum value of anode current for the device to make a transition from forward ON state to forward blocking state.

A thyristor is turned ON by triggering it with a gate current. It is turned OFF by itself when the anode current is less that the holding current. Hence it is called a semi-controlled device. To remain in conduction, the anode current must be greater than the holding current. For this reason it may be called a current controlled device.

Forward Characteristics

The forward characteristic of the device shows the variation of the anode current with increase in anode-cathode voltage, with the gate current held constant at a particular value.

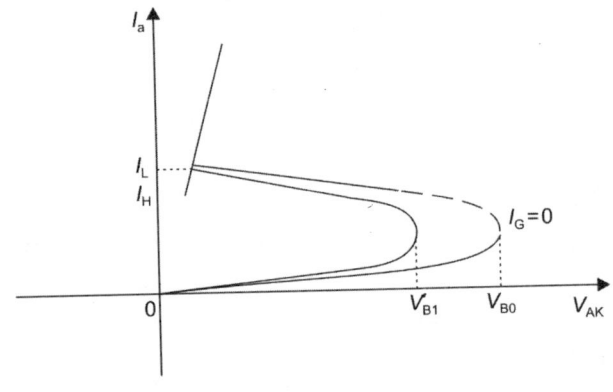

A1.3 CIRCUIT DIAGRAM

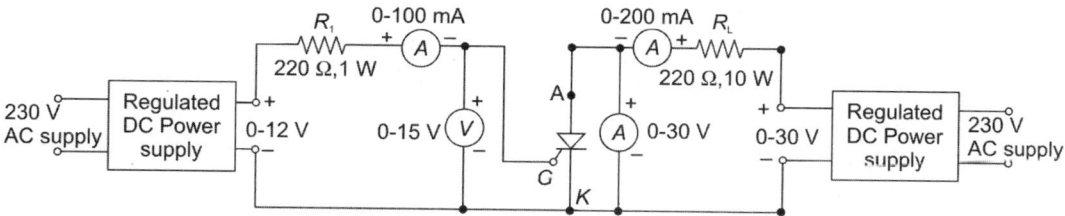

A1.4 DEVICE SPECIFICATION

Thyristor   TYN 616   600 V, 16A
Maximum gate current = 60 mA
Minimum gate current < 15 mA

A1.5 CIRCUIT DESIGN

Anode supply voltage = 30 V
Select the load resistance R_L = 220 Ω, 10 W

Maximum load current $I_L = \dfrac{30}{220} = 0.136$ A

Gate supply voltage = 12 V
Maximum gate current = 60 mA

Resistance in the gate circuit $R_1 = \dfrac{12}{0.06} = 200\,Ω$ Select R_1 = 220 Ω, 1 W

A1.6 EQUIPMENT AND COMPONENTS REQUIRED

Sl No	Component	Specification	Quantity
1.	Regulated power supply	0 – 12 V DC	1
		0 – 30 V DC	1
2.	Resistors	220 Ω, 10 W	1
		220 Ω, 1 W	1
3.	Ammeters	0 – 100 mA DC	1
		0 – 200 mA DC	1
4.	Voltmeters	0 – 15 V DC	1
		0 – 30 V DC	1
5.	Thyristor	TYN 616	1
6.	Breadboard		

A1.7 PROCEDURE

A. To Measure Minimum Gate Voltage and Minimum Gate Current

1. Make the connections as per the circuit diagram.
2. Keeping the output voltages of the voltage regulators at minimum values, switch ON the power supplies.
3. Adjust the anode supply voltage to around 24 V

4. Adjust the gate supply voltage to make the thyristor just turn into conduction. This can be viewed by observing the ammeter in the load circuit.
5. Note down the readings of voltmeter and ammeter in the gate circuit. These values are the $V_{g\,min}$ and $I_{g\,min}$ respectively for the thyristor.
6. Reduce the supply voltages to minimum values and switch off the power supplies.
7. Steps 2 to 6 may be repeated to arrive at the minimum values of gate voltage and the gate current.

B. To Plot the Static v–i Characteristics

1. With the output voltages of the voltage regulators set to minimum values, switch ON the power supply.
2. Adjust the gate current to a value higher than $I_{g\,min}$.
3. Vary the anode voltage gradually in steps and at each step note down the values of anode voltage and anode current. At some value of anode voltage the thyristor is turned into conduction mode from the forward blocking state.
4. Take two or more sets of readings of anode voltage and anode current.
5. After reducing the output voltages of the voltage regulators to minimum values, switch off the power supply.
6. Repeat the above steps for two more values of gate current.
7. Reduce the supply voltages to minimum values and switch off the power supply.

C. To Measure the Latching Current

1. With the output voltages of the voltage regulators set to minimum values, switch ON the power supplies.
2. Adjust the gate current to a value greater than $I_{g\,min}$.
3. Adjust the anode voltage to turn the thyristor into conduction mode.
4. Switch off gate supply. If the thyristor is still in the conduction mode, switch ON the gate supply. Adjust the anode current to a lower value by reducing the anode voltage and then switch off the gate supply once again. If the thyristor is still in the conduction mode, switch ON the gate supply and decrease the anode current further. This process is repeated to obtain the minimum value of anode current at which the thyristor remains in the conduction mode, even when the gate supply is removed. This minimum value of anode current is the latching current of the thyristor.
5. Reduce the output voltages of the voltage regulators to minimum values and switch off the supply voltages.

D. To Measure the Holding Current

1. With the output voltages of the voltage regulators set to minimum values, switch ON the power supply.
2. Adjust the gate current to a value greater than $I_{g\,min}$.
3. Adjust the anode voltage just to bring the thyristor in to the conduction mode.
4. Switch off the gate supply.
5. Reduce the anode voltage gradually to decrease the anode current. At a certain minimum anode current, the thyristor is suddenly turned off.

6. Repeat steps 2, 3, 4 and 5 to observe the minimum value of anode current at which the thyristor is turned off. This minimum value of the anode current is the holding current of the device.
7. Switch off the power supply.

A1.8 OBSERVATIONS

Minimum gate voltage Minimum gate current		
Trial	$V_{g\ min}$	$I_{g\ min}$

v-i Characteristics					
$i_{g1} =$		$i_{g2} =$		$i_{g3} =$	
v_a	i_a	v_a	i_a	v_a	i_a

Latching current		Holding current	
Trial	I_L	Trial	I_h

Minimum gate voltage =   Minimum gate current =
Latching current =   Holding current =

A1.9 GRAPH: v–i CHARACTERISTICS

Plot the *v–i* characteristics on a graph sheet.

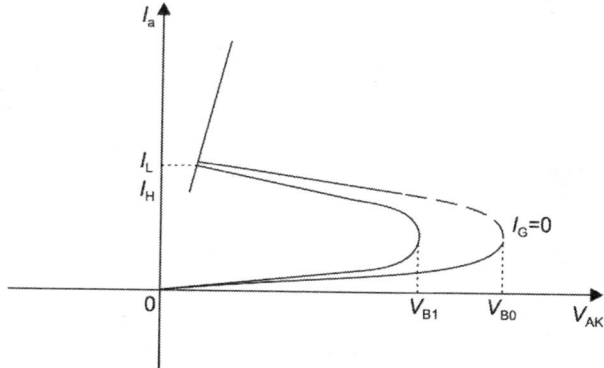

A2.1 OBJECTIVE

To plot the v–i characteristics of a triac

A2.2 THEORY

A triac is basically a bidirectional device which can conduct in either direction. A triac is electrically equivalent to two thyristors connected in antiparallel. The basic structure and the circuit symbol are shown in the figure.

The terminals are marked as MT_1 (main terminal 1), MT_2 (main terminal 2) and G (gate). With no signal applied to the gate, the triac can block both the half cycles of an ac voltage, if the peak value of the supply voltage is less than both the forward and reverse break over voltage of the device. A triac can be turned ON in each half cycle of the applied ac voltage by applying a small positive or negative voltage to the gate with respect to the terminal MT_1.

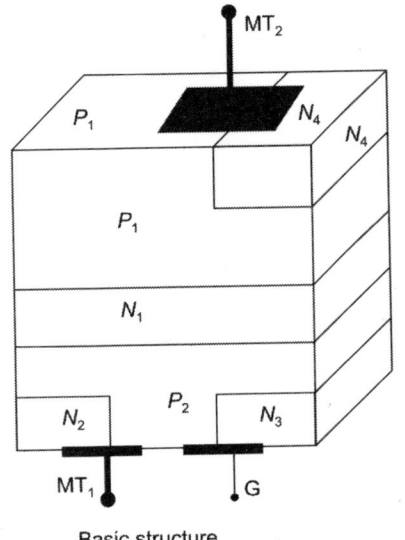

Basic structure

Symbol

Thyristor Equivalence

The device can have four different modes of operation depending on the polarity of the voltage applied between MT_1 and MT_2 and the polarity of Gate with respect to MT_1.

1. MT_2 is positive with respect to MT_1 and Gate is positive with respect to MT_1
2. MT_2 is positive with respect to MT_1 and Gate is negative with respect to MT_1
3. MT_2 is negative with respect to MT_1 and Gate is positive with respect to MT_1
4. MT_2 is negative with respect to MT_1 and Gate is negative with respect to MT_1

The first two modes can be used for positive half cycles and the other two for negative half cycles of the AC supply voltage. But the first and fourth modes are generally used for practical circuits. First mode exhibits forward characteristics and the fourth mode exhibits the reverse characteristics.

In mode I, when MT_2 is positive with respect to MT_1, the junctions P_1–N_1 and P_2–N_2 are forward biased and the junction N_1–P_2 is reverse biased. Consider that a small positive voltage is applied at the gate with respect to MT_1. A gate current flows through P_2 to N_2 layers. This injects enough electrons into the P_2 layer and this causes the reverse biased junction N_1–P_2 to break down. Thus the triac starts conducting through P_1–N_1–P_2–N_2 layers and is turned ON. The device is said to operate in the first quadrant.

In mode IV, when MT_2 is made negative with respect to MT_1, the junctions N_4–P_1 and N_1–P_2 are forward biased whereas the junction P_1–N_1 is reverse biased. Consider that a small negative voltage is applied at the gate with respect to MT_1. A gate current flows through P_2 to N_3 layers. This will cause the reverse biased junction P_1–N_1 to break down and the triac starts conducting through P_2–N_1–P_1–N_4 layers. The device is said to operate in the fourth quadrant.

Forward Characteristics

The forward characteristic of a triac shows the variation of the current through the device from MT_2 to MT_1, when a positive voltage applied at MT_2 with respect to MT_1 is gradually increased, for a particular value of positive gate current.

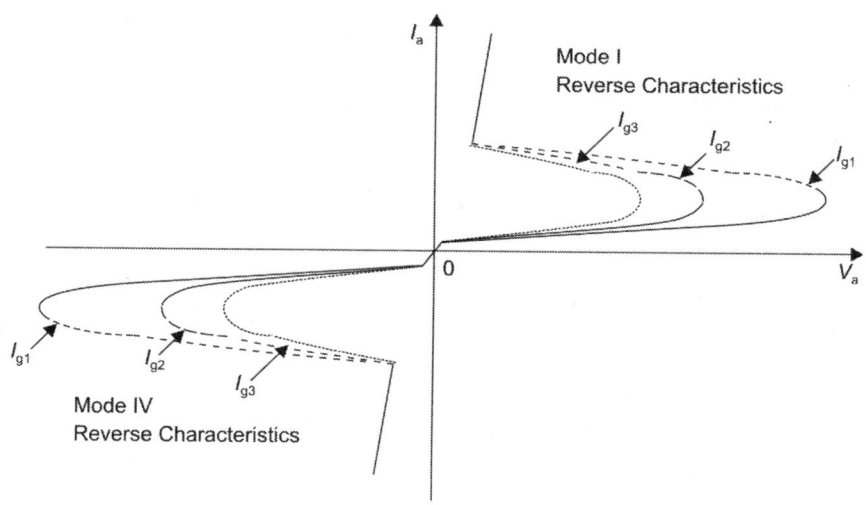

Reverse Characteristics

The reverse characteristic of a triac shows the variation of the current through the device from MT_1 to MT_2, when a positive voltage applied at MT_1 with respect to MT_2 is gradually increased, for a particular value of negative gate current.

A2.3 CIRCUIT DIAGRAM

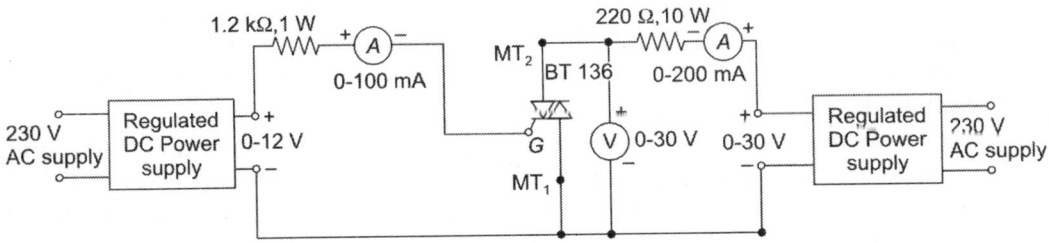

A2.4 DEVICE SPECIFICATION

Triac   BT 136   500 V   4 A
Gate trigger current = 10 mA

A2.5 CIRCUIT DESIGN

Anode supply voltage = 30 V
Select the load resistance R_L = 220 Ω, 10 W

Maximum load current = $\dfrac{30}{220}$ = 0.136 A

Gate supply voltage = 12 V

Resistance in the gate circuit $R_1 = \dfrac{12}{0.01} = 1200\ \Omega$

Select R_1 = 1.2 k Ω, 1W

A2.6 EQUIPMENT AND COMPONENTS REQUIRED

Sl No	Component	Specification	Quantity
1.	Regulated power supply	0 – 12 V DC	1
		0 – 30 V DC	1
2.	Regulated power supply	220 Ω, 10 W	1
		1.2 k Ω, 1 W	1
3.	Ammeters	0 – 200 mA DC	1
		0 – 100 mA DC	1
4.	Voltmeters	0 – 30 V DC	1
5.	Thyristor	BT 136	1
6.	Breadboard		

A2.7 PROCEDURE

A. Forward Characteristics

1. Make the connections as per the circuit diagram.
2. With minimum output voltages at the voltage regulators, switch ON the power supplies.
3. Adjust the gate current to a particular value.
4. Vary the MT_2 to MT_1 voltage gradually in steps and at each step note down values of anode voltage and anode current. At some stage the triac is turned into conduction mode.
5. Take two more sets of readings by varying the MT_2 to MT_1 voltage.
6. Switch off the power supplies.
7. Repeat the above procedure for two more values of gate current.
8. Decrease the supply voltages to minimum values and switch off the power supplies.

B. Reverse Characteristics

The following changes are made to the circuit

1. Interchange the connections to the positive and negative terminals of the gate supply. This will make the gate negative with respect to MT_1.
2. Make MT_2 negative with respect to MT_1, by interchanging the connections.
3. Interchange the polarity of all meters.
4. Follow the procedure as given in forward characteristics.

A2.8 OBSERVATIONS

Forward Characteristics						Reverse Characteristics					
$i_{g1} =$		$i_{g2} =$		$i_{g3} =$		$i_{g1} =$		$i_{g2} =$		$i_{g3} =$	
v_a	i_a	v_a	i_a	v_a	i_a	v_a	i_a	v_a	i_a	v_a	i_a

A2.9 GRAPH

Plot the forward and the reverse characteristics on the same graph sheet.

Experiment A3

Static Characteristics of a BJT

A3.1 OBJECTIVE

To plot the input and output characteristics of a BJT.

A3.2 THEORY

A power bipolar junction transistor (BJT) is a semiconductor device which is fabricated to carry high power when compared to a conventional small signal transistor. Even though both n-p-n and p-n-p types are available, an n-p-n power BJT is more commonly used in power electronic circuits. Because of higher mobility of electrons when compared to holes, an n-p-n device can be fabricated on a smaller silicon area to provide the same performance as an equivalent p-n-p device. A BJT has a higher voltage blocking capability while in the off state and a large current carrying capacity during conduction. The term bipolar indicates that the current flow in the device is due to both the holes and electrons.

Basic Structure

The basic triple diffused structure and the symbol of a BJT is shown in the figure. A BJT has three terminals namely the collector, the emitter and the base. The n-type layer in the emitter region is heavily doped with high impurity concentration. A p-type layer is diffused over this n^+ layer to which the base terminal is connected. The base region is moderately doped. A moderately doped n-layer is diffused over the

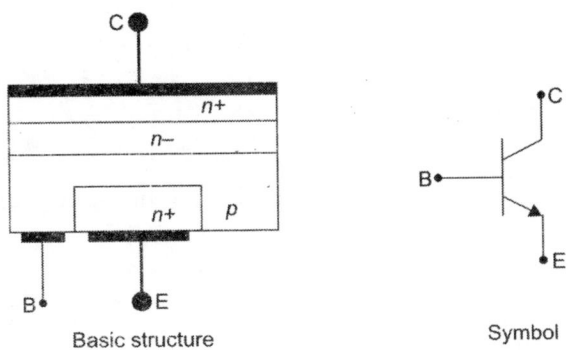

Basic structure   Symbol

p-layer and on to this a heavily doped n-layer is fabricated through a third diffusion process. The collector terminal is connected to this region. The thickness of this heavily doped region near the collector region determines the breakdown voltage of the transistor.

Principle of Operation

A BJT when used in switching applications is normally connected in the common emitter configuration. This is because this configuration offers higher current gain and high input impedance. Consider that a positive voltage is applied at the collector terminal with respect to the emitter. When a positive voltage is applied at the base terminal with respect to the emitter, the base-emitter junction is forward biased. The emitter acts as a source of mobile charge carriers and hence the name emitter. For an NPN transistor the emitter region has n^+ layer and the mobile charge carriers are electrons. These electrons moving from n^+ region cross the emitter-base junction and then reach the base–collector junction. These electrons are swept into the collector region because the collector–base junction is reverse-biased. Consequently the electrons reaching n^+ collector region are collected by the metal electrode attached to n^+ and hence the name collector. This constitutes the collector current.

Some of the electrons crossing the emitter-base junction and reaching the p-region re-combine with holes there. A few reach the base terminal. The electrons that do not reach the collector are responsible for the base current. Hence the collector current is less than the emitter current.

Important characteristics of a BJT are the input characteristics and the output characteristics.

Input Characteristics

This characteristic of BJT shows the variation of the base current with the base-emitter voltage, keeping the collector-emitter voltage a constant. For different values of the collector-emitter voltage, a family of such characteristics can be drawn.

Output Characteristics

This characteristic of BJT shows the variation of the collector current with the collector-emitter voltage, keeping the base current a constant. For different values of base current, a family of such characteristics can be drawn.

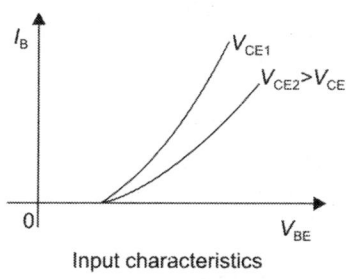

Input characteristics

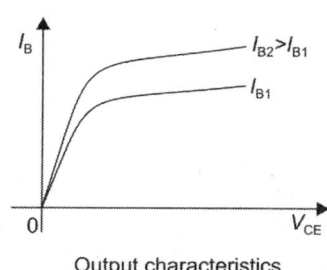

Output characteristics

A3.3 CIRCUIT DIAGRAM

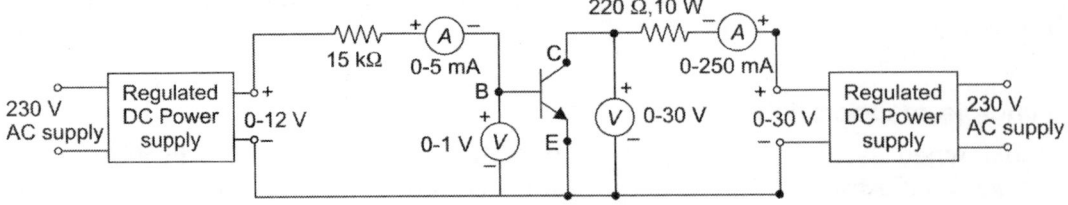

A3.4 DEVICE SPECIFICATION

BJT   2N3055 HV3   100 V   12 A   115 W   β = 20 to 100

A3.5 CIRCUIT DESIGN

Anode supply voltage = 30 V

Select the load resistance R_L = 220 Ω, 10 W

Load current = $\dfrac{30}{220}$ = 0.136 A

Collector current = 0.136 A

Selecting β = 100, base current $I_B = \dfrac{0.136}{100}$ = 0.00136 A = 1.36 mA

Assume that the base current is limited to I_B = 1 mA

Base circuit supply voltage = 12 V

Resistance in the base circuit $R_1 = \dfrac{12}{0.001}$ = 12 kΩ   Select R_1 = 15 kΩ

Maximum base to emitter voltage = 0.8 V

A3.6 EQUIPMENT AND COMPONENTS REQUIRED

Sl No	Component	Specification	Quantity
1.	Regulated power supply	0 – 12 V DC	1
		0 – 30 V DC	1
2.	Resistors	220 Ω, 10 W	1
		15 k Ω, 0.5 W	1
3.	Ammeters	0 – 250 mA DC	1
		0 – 5 mA DC	1
4.	Voltmeters	0 – 1 V DC	1
		0 – 30 V DC	1
5.	BJT	2N3055	1
6.	Breadboard		

A3.7 PROCEDURE

A. Input Characteristics

1. Make the connections as per the circuit diagram.
2. With minimum output voltages at the voltage regulators, switch ON the power supplies.
3. Keep the collector-emitter voltage at zero value, i.e. $V_{CE} = 0$.
4. Vary the base-emitter voltage in steps and measure the values of base current.
5. Keep the collector-emitter voltage at a particular value, say $V_{CE} = 2$ V.
6. Repeat step 4.
7. Bring the regulator voltages to minimum values and then switch off the supply.

B. Output Characteristics

1. Make the connections as per the circuit diagram.
2. With minimum output voltage at the voltage regulators, switch ON the power supplies.
3. Keep the collector-emitter voltage to a minimum value.
4. Vary the base-emitter voltage to drive some base current, say $I_B = 50$ μA.
5. Vary the collector-emitter voltage in steps and at each step observe the collector current. Note that the base current should be kept constant at the pre-set value, at each step.
6. Repeat step 5 for a different value of base current, say $I_B = 100$ μA.
7. Bring the regulator voltages to minimum values and then switch off the supply.

A3.8 OBSERVATIONS

Input Characteristics					
$V_{CE} = 0$		$V_{CE} = 2$V		$V_{CE} = 4$V	
v_{BE}	i_{BE}	v_{BE}	i_{BE}	v_{BE}	i_{BE}

Output Characteristics					
$i_B = 50$ μA		$i_B = 100$ μA		$i_B = 150$ μA	
v_{CE}	i_C	v_{CE}	i_C	v_{CE}	i_C

Minimum value of base-emitter voltage =

A3.9 GRAPH

Plot the input and output characteristics on a graph sheet.

Experiment A4

Static Characteristics of a MOSFET

A4.1 OBJECTIVE

To study the static characteristics of a power MOSFET

A4.2 THEORY

A metal oxide semiconductor field effect transistor (MOSFET) is a unipolar device and its operation depends on the flow of the majority carries. It is a voltage controlled device and it acts as a fast acting switch. It has very high input impedance and has a positive temperature coefficient of resistance.

There are two types of enhancement MOSFETs. They are:
1. n-channel enhancement MOSFET and
2. p-channel enhancement MOSFET.

The n-channel enhancement power MOSFET is more generally used because of higher mobility of electrons.

Basic Structure

The device symbol and basic structure of an n-channel power MOSFET are shown in the figure. It has three terminals: the drain (D), the source (S) and the gate (G). The source and the drain terminals are power terminals.

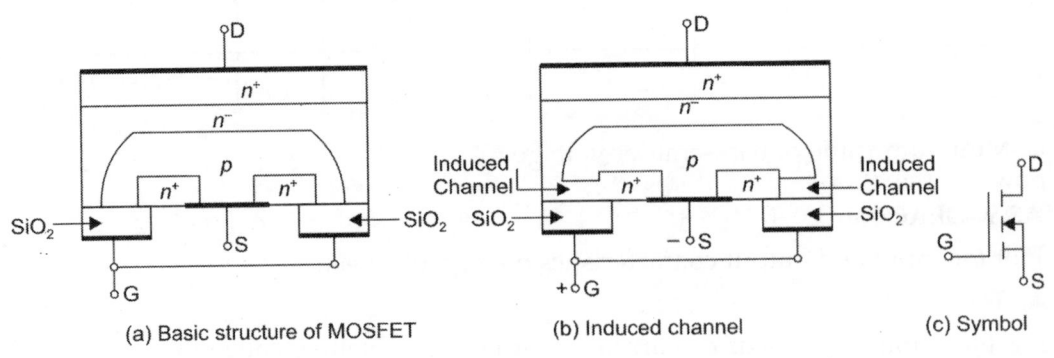

(a) Basic structure of MOSFET   (b) Induced channel   (c) Symbol

On a heavily doped n substrate (n^+), a high resistivity or lightly doped epitaxial n⁻ layer (n^-) is grown. A metal contact is made to the n^+ layer which forms the drain terminal. The thickness and resistivity of this epitaxial n^- layer decides the voltage blocking capability of the device. A relatively lightly doped p-region is then diffused into the epitaxial layer and into this region two heavily doped n^+ regions are diffused. An insulating layer of Silicon dioxide (SiO_2) is grown on the surface. This insulating layer is etched to provide metallic terminals for the source. The two n^+ regions make contact with the source terminals. Another metal contact is made to the SiO_2 layer which forms the gate terminal.

Principle of Operation

With the gate circuit open, no current flows from the drain to the source because of one n^+ p junction, which is reverse biased. A low control voltage is applied across the gate and the source, to turn ON the device.

When a positive supply is given to the gate with respect to the source, electrons are pulled from the n^+ region to the p-region near the gate. This forms an n-channel between the source n^+ and drain n^- regions as shown in the figure. This channel provides a path for the flow of electrons from the source to the drain. In effect a current flows from the drain to the source through this channel. If the positive gate voltage is not of sufficient magnitude to create a channel, the current cannot flow. If the gate supply is made more positive, the n-channel becomes deeper and therefore more current flows from the drain to the source. The n-channel is enhanced by the increase of gate to source voltage and hence the name n-channel enhancement MOSFET. The ON state resistance of the device depends on this n-channel. The input impedance is high and the gate current is extremely small. The device has a very high gain between the output power and the gate power.

It is important to note that in a MOSFET, the current conduction is by majority carriers only. There are no minority charge carriers. Hence there is no problem of storage of minority carriers, the presence of which affects the switching characteristics. Because of this the turn ON and turn OFF times are extremely small of the order of less than one microsecond and the device can be used in applications that require very fast switching.

The internal structure is such that there is an inbuilt body diode, which forms an integral part of the device. Hence it can carry full current in the reverse direction. Thus the device acts as a controlled device in the forward direction and as an uncontrolled device in the reverse direction.

The gate current required in a MOSFET is extremely small because of the fact that the gate circuit impedance is extremely large of the order of about 10^9 ohm. The main disadvantage of n-channel enhancement MOSFET is that the conducting n-channel between the drain and the source has large ON–state resistance. This leads to high power dissipation in the n-channel while the device conducts.

Two important characteristics of a power MOSFET are (1) The Transfer characteristics and (2) The output characteristics.

A. The Transfer Characteristics

It is the variation of the drain current I_d with the gate- source voltage V_{GS}, keeping the drain-source voltage V_{DS} a constant. From this characteristic the trans-conductance can be determined. It is defined as the ratio of change in drain current for a given change in gate to source voltage, when the drain to source voltage is kept constant.

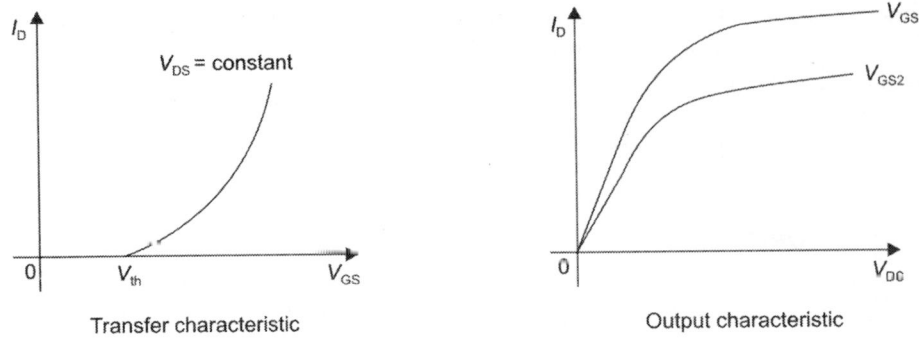

| Transfer characteristic | Output characteristic |

B. The Output Characteristics

This is the variation of the drain current with the variation of drain to source voltage, keeping the gate to source voltage constant. From this characteristic, the dynamic drain resistance can be determined.

A4.3 CIRCUIT DIAGRAM

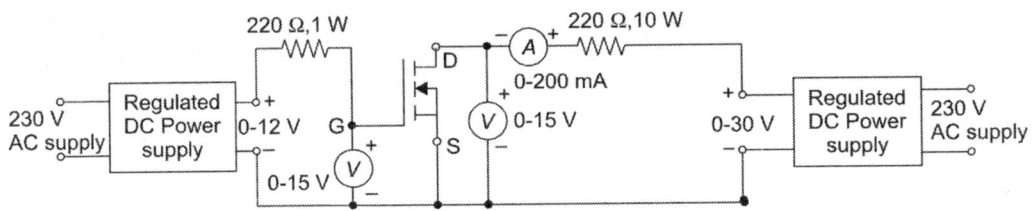

A4.4 DEVICE SPECIFICATION

MOSFET   IRF740   400 V   10 A
Gate-source voltage: 3 V to 15 V

A4.5 CIRCUIT DESIGN

Supply voltage = 30 V
Select load resistance R_L = 220 Ω, 10 W

Load current = $\dfrac{30}{220}$ = 0.136 A1

Gate supply voltage = 12 V
Maximum gate current = 60 mA

Resistance in the gate circuit $R_1 = \dfrac{12}{0.06} = 200\ \Omega$

Select R_1 = 220 Ω, 1 W

A4.6 EQUIPMENT AND COMPONENTS REQUIRED

Sl No	Component	Specification	Quantity
1.	Regulated power supply	0 – 12 V DC	1
		0 – 30 V DC	1
2.	Resistors	220 Ω, 10 W	1
		220 Ω, 1 W	1
3.	Ammeters	0 – 200 mA DC	1
4.	Voltmeters	0 – 15 V DC	1
		0 – 30 V DC	1
5.	MOSFET	IRF740	1
6.	Breadboard		

A4.7 PROCEDURE

A. Transfer Characteristics

1. Connections are made as shown in the diagram.
2. With the output voltages at the voltage regulators set to minimum values switch on the power supplies.
3. Adjust the power supply to keep the drain-to-source voltage at a constant value, say 5 V
4. Adjust the gate to source supply to vary the gate voltage. Up to a particular value of V_{GS} the drain current will remain zero. This voltage is known as the threshold voltage. After this value of voltage, the drain current starts increasing from zero value to higher values.
5. For different values of gate to source voltage note down the corresponding values of drain currents. For every set of readings the drain-to-source voltage must be kept constant.
6. Reduce the supply voltages to minimum values and switch off the power supplies.

B. Output Characteristics

Refer to the same connection diagram.

1. With the output voltages at the voltage regulators set to minimum values, switch ON the DC power supplies
2. Adjust the gate-source voltage to a value a little more than the threshold voltage.
3. Adjust the drain-source voltage gradually in steps and note down the values of the drain current. At each step make sure that the gate-to-source voltage is kept constant.
4. Repeat the experiment for two more values of gate-to-source voltage.
5. Reduce the supply voltages to minimum values and switch off the power supplies.

A4.8 OBSERVATIONS

Transfer Characteristics					
$V_{DS} =$		$V_{DS} =$		$V_{DS} =$	
v_{GS}	i_D	v_{GS}	i_D	v_{GS}	i_D

Output Characteristics					
$V_{GS} =$		$V_{GS} =$		$V_{GS} =$	
v_{DS}	i_D	v_{DS}	i_D	v_{DS}	i_D

Threshold Voltage =

A4.9 GRAPH

Plot the transfer characteristics and the output characteristics.

Experiment A5

Static Characteristics of an IGBT

A5.1 OBJECTIVE

To study the static characteristics of an IGBT

A5.2 THEORY

An Insulated Gate Bipolar Transistor combines the advantages of BJTs and MOSFETs. An IGBT has high impedance gate like that of a MOSFET which requires only a small gate current and a low ON- state power loss while in conduction as that of a BJT. An IGBT is free from the problem of second breakdown as in a BJT. The symbol and the basic structure of an n-channel IGBT is shown in the figure.

The device has three terminals namely the collector, the emitter and the gate. The collector and the emitter terminals are power terminals. A low control voltage is applied across the gate and the emitter, to turn ON the device.

Basic Structure

The structure of IGBT is very much similar to that of a MOSFET. An n-channel IGBT is fabricated on a p^+ substrate. This region constitutes the collector of the IGBT. Over this an additional n^+ layer is formed which is the buffer layer. This is followed by a lightly doped n^- layer. Next to this is relatively large p-layer. Two n^+ regions are diffused into this p-layer as shown in the figure. An insulating layer of Silicon dioxide (SiO_2) is made on the surface. This insulating layer is etched to provide metallic terminals for the emitter. The two n^+ regions make contact with the emitter terminals. Another metal contact is made to the SiO_2 layer which forms the gate terminal.

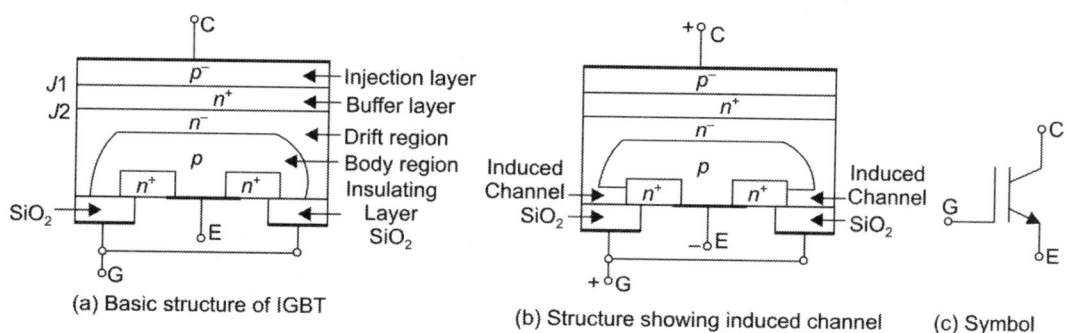

(a) Basic structure of IGBT   (b) Structure showing induced channel   (c) Symbol

Principle of Operation

When a positive voltage is applied at the collector terminal with respect to the emitter, in the absence of a gate-to-emitter voltage, the device cannot conduct because the junction J_2 remains reverse biased. When a positive voltage is applied at the gate with respect to the emitter, electrons are pulled from the n^+ region to the gate terminal. Consequently, an n-type channel is created between the n^+ region and the n^- region through the p region as shown in figure. This n-channel short circuits the n^- region with the emitter n^+ region. The flow of electrons, from the n^+ emitter region through the channel to the n^- drift region, acts as the base drive current for the PNP transistor. Hence electrons flow from the emitter to the collector. Also this base current induces injection of holes from the p^+ collector region to the n-drift region. This high level of injection of holes increases the conductivity of the drift region. Eventually a forward current is established. Thus the current flow from the collector to the emitter is due to the flow of both the electrons and holes and so the IGBT is a bipolar device. Its symbol is shown in figure.

The n^+ buffer layer improves the turn OFF speed by reducing the quantity of minority charge carrier injection and by raising the re-combination rate during switching transition. The n^+ buffer layer also lowers the ON-state voltage drop and improves the forward voltage blocking capability of the device. But it reduces the reverse voltage blocking capability.

The structure of IGBT shows that there is a PNPN structure between the collector and the emitter. Note that $p^+ n^- p$ constitute a pnp transistor with p^+ as emitter, n^- as base and p as collector. Also $n^- p n^+$ form an NPN transistor with n^- as collector p as base and n^+ as emitter.

E	p^+	n^+
B	n^-	p
C	p	n^-

Similar to MOSFET, IGBT has a threshold value for the gate-emitter voltage below which the device is not turned ON. There are two characteristics of an IGBT. They are (1) the transfer characteristics and (2) the output characteristics.

A. Transfer Characteristics

This shows the variation of collector current with respect to the variation of gate to emitter voltage, with the collector to emitter voltage held constant.

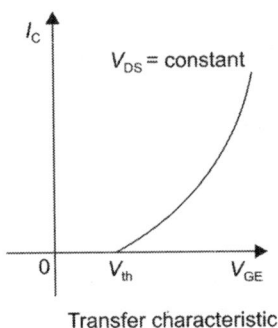

Transfer characteristic

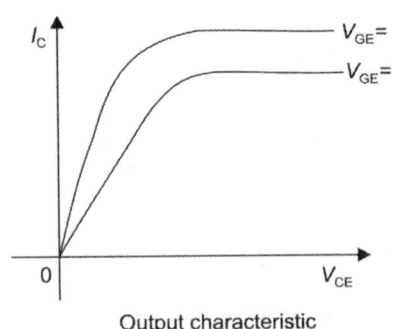

Output characteristic

B. Output Characteristics

This shows the variation of the collector current with collector-emitter voltage for a constant value of gate-emitter voltage.

A5.3 CIRCUIT DIAGRAM

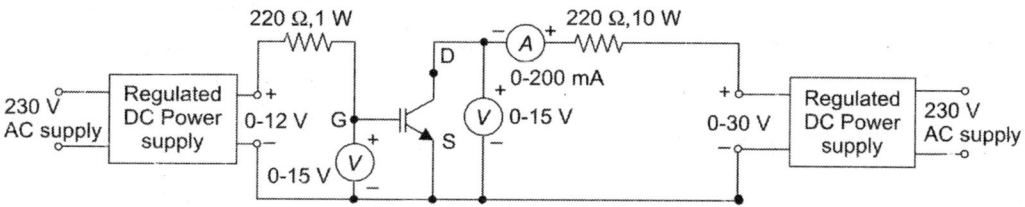

A5.4 DEVICE SPECIFICATION

IGBT   CT60   900 V   60 A   200 W

Gate to emitter voltage 3.5 V to 20 V

A5.5 CIRCUIT DESIGN

Collector to emitter supply voltage = 30 V

Select the load resistance R_L = 220 Ω, 10 W

$$\text{Load current} = \frac{30}{220} = 0.136 \text{ A}$$

Gate supply voltage = 12 V

Maximum gate current = 60 mA

$$\text{Resistance in the gate circuit } R_1 = \frac{12}{0.06} \quad 200 \text{ Ω} \qquad \text{Select } R_1 = 220 \text{ Ω, 1 W}$$

A5.6 EQUIPMENT AND COMPONENTS REQUIRED

Sl No	Component	Specification	Quantity
1.	Regulated power supply	0 – 12 V DC	1
		0 – 30 V DC	1
2.	Resistors	220 Ω, 10 W	1
		220 Ω, 1 W	1
3.	Ammeters	0 – 200 mA DC	1
4.	Voltmeters	0 – 15 V DC	2
5.	IGBT	CT60	1
6.	Breadboard		

A5.7 PROCEDURE

A. Transfer Characteristics

1. Connections are made as per the circuit diagram.
2. With the output voltages at the voltage regulators set to minimum values, switch ON the power supplies.

3. Adjust the power supply to keep the collector- emitter voltage at a constant value, say 5 V.
4. Gradually increase the gate–emitter voltage and find the value of the voltage up to which the collector current remains zero. This voltage is the threshold voltage.
5. Gradually increase the gate-emitter voltage in steps and note down the corresponding value of the collector current. At each step ensure that the collector-emitter voltage is kept constant at the initially set value.
6. Reduce the supply voltages to minimum values and switch off the power supplies.

B. Output Characteristics

Refer to the same connection diagram.
1. With the output voltages at the voltage regulators set to minimum values, switch ON the power supplies.
2. Adjust the gate to emitter voltage at a constant value, slightly more than the threshold value.
3. Vary the collector–to-emitter voltage gradually in steps and note down the values of the corresponding collector current. In each step ensure that the gate-to-emitter voltage is held constant at the initial value.
4. Repeat the experiment for two more different values of gate to emitter voltage.
5. Reduce the supply voltages to minimum values and switch OFF the power supplies.

A5.8 OBSERVATIONS

Transfer Characteristics						Output Characteristics					
$V_{CE}=$		$V_{CE}=$		$V_{CE}=$		$V_{GE}=$		$V_{GE}=$		$V_{GE}=$	
v_{GE}	i_C	v_{GE}	i_C	v_{GE}	i_C	v_{CE}	i_C	v_{CE}	i_C	v_{CE}	i_C

Gate – emitter Threshold voltage =

A5.9 GRAPH

Plot the transfer characteristics and the output characteristics on a graph sheet.

B

Circuit Design and Testing

Experiment B1

Design and Testing of Resistance-Triggering Circuit for a Thyristor

B1.1 OBJECTIVE

To design and study the working of an R triggering circuit for triggering the gate of a thyristor.

B1.2 THEORY

A thyristor can be switched from the forward blocking state to the forward conduction state in a variety of ways.

1. **Forward voltage triggering:** Applying a forward voltage greater than the forward break over voltage, a thyristor can be brought into conduction mode from the forward blocking state without applying a gate voltage. This method is likely to be destructive for the device and so it is never employed.

2. **dV/dt triggering:** If the anode to cathode voltage applied has a high value greater than the rating of the thyristor, the reverse biased junction J_2 breaks down and the device is brought into conduction.

3. **Temperature triggering:** During the forward blocking state most of the applied voltage appears across the junction J_2. There can be a leakage current across this junction. The voltage across the junction J_2 may raise the temperature at this junction which in turns causes increase in leakage current. The leakage current across junction J_2 rises to such a value as to cause the breakdown of this junction. The device is thus brought into forward conduction state, even without the gate supply.

4. **Light triggering:** Light rays are directed into the P layer near the junction J3. This generates enough charge carriers in this layer. Electrons will drift towards junction J_2 and finally J_2 breaks down. The thyristor which is forward biased is thus turned into conduction mode. Such a thyristor is called light activated SCR (LASCR).

5. **Gate voltage triggering:** This is the most commonly adopted method of triggering a thyristor. With the device forward biased, a small voltage is applied at the gate with respect to cathode. A gate current flows. This generates enough charge carriers in the P layer near J_3 junction. Electrons drift towards Junction J_2 because of high voltage gradient (across the anode to cathode) and the junction, which is otherwise reverse biased, breaks down. The thyristor is now switched into forward ON state.

A step DC voltage, a slowly rising DC voltage or a rectified positive half-wave voltage can be used to trigger a thyrsitor. In DC voltage triggering, a dc supply is applied to the gate with respect to the cathode. The gate voltage must be greater than the gate trigger level which is the minimum voltage required for firing a thyristor and bringing it into forward ON state. The disadvantage is that the gate current is maintained even when the thyristor is turned ON and this leads to more gate circuit loss.

The gate voltage can also be in the form of pulses. This method is called pulse triggering. Pulse triggering is the method of triggering the gate with periodic DC pulses. The pulse width is generally less than $100\ \mu$ sec. Firing pulses are periodic in nature. In pulse triggering method, one circuit generates a pulse or a spike. It is then amplified by a driver circuit and then applied to the gate of the thyristor using an isolation circuit. Since the gate voltage is in the form of pulses it generates less ohmic loss in the gate circuit.

A varying voltage is applied across the gate and the cathode. When this voltage reaches the gate trigger voltage the thyristor is triggered. Pulse triggering method offers many advantages over other methods. Some of the advantages are low gate power consumption improved reliability, fast triggering and easiness to adjust the firing angle.

In pulse triggering it must be ensured that the magnitude of the pulse must be sufficiently high to trigger the gate of the thyristor so that it is brought into conduction. Higher the magnitude lesser is the turn-ON time of the device. Also the pulse width must be sufficiently large to allow the anode current to exceed the latching current; then only the device is latched into conduction.

B1.3 R-TRIGGERING

The circuit used for triggering a thyristor at the gate terminal is known as triggering circuit or firing circuit. R-triggering circuit is a very simple and economical circuit; even though it is not used commercially.

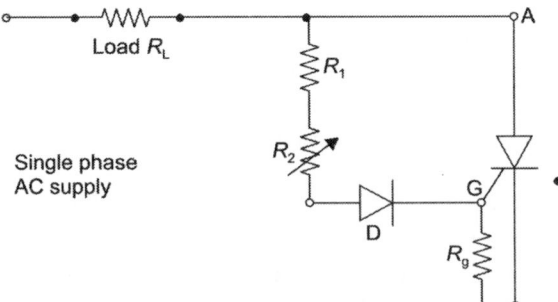

In the circuit shown, by varying the resistance the gate current and the gate voltage can be adjusted. When the gate voltage reaches the minimum gate voltage required to turn ON the thyristor, known as the gate trigger voltage, thyristor is turned ON. R_1 is a fixed resistance to limit the gate current to a value less than the maximum permissible value of the gate current, when the thyristor is in the off state. R_2 is a variable pot.

When the thyristor is in the off state, the gate voltage is given by,

$$v_{GK} = V_m \sin \omega t - i_g (R_L + R_1 + R_2) - V_D$$

where R_L is the load resistance.

The gate current is given by,

$$i_g = \frac{V_m \sin \omega t - V_D - V_{GK}}{(R_1 + R_2 + R_L)}$$

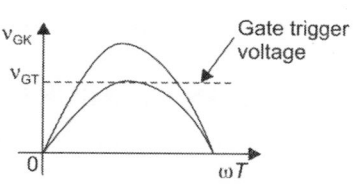

The gate voltage is sinusoidal and this voltage is in phase with the supply voltage. When R_2 is decreased, the gate voltage V_{GK} increases. Let V_{GT} is gate trigger voltage required to trigger the thyristor. When V_{GK} reaches the V_{GT} level, the thyristor is turned ON.

The diode prevents the gate current during negative half cycle of the AC supply voltage. The gate voltage is thus in the form of DC pulses. The amplitude of this dc pulse can be controlled by varying R_2. If the maximum value of the gate voltage V_{GK} is less than V_{GT}, the thyristor cannot be fired. If the maximum value of the gate voltage V_{GK} is equal to V_{GT}, the firing angle is 90° as shown in the figure. If the maximum value of the gate voltage V_{GK} is greater than V_{GT}, the firing angle is less than 90°. Hence in R-triggering scheme, the maximum value of firing angle is limited to 90°. The minimum firing angle is little above 0°.

The resistor R_g is a gate stabilizing resistor. It limits the gate voltage to be less than the permissible value V_{Gm}, when the thyristor is in the OFF-state. In the worst case when $R_2 = 0$,

$$\frac{R_g}{R_L + R_1 + R_g} V_m \sin \alpha \le V_{Gm} \qquad R_g \le \frac{V_{GT}(R_1 + R_L)}{V_m - V_{GT}}$$

It can be seen from the waveform that at the instant of triggering, the gate voltage is

$$V_{GT} = (V_m \sin \alpha)\frac{R_g}{(R_L + R_1 + R_2 + R_g)}$$

$$\sin \alpha = \frac{V_{GT}}{V_m} \frac{(R_L + R_1 + R_2 + R_g)}{R_g}$$

and α depends on R_2 and the range of α is $0° < \alpha < 90°$.

B1.4 CIRCUIT DIAGRAM

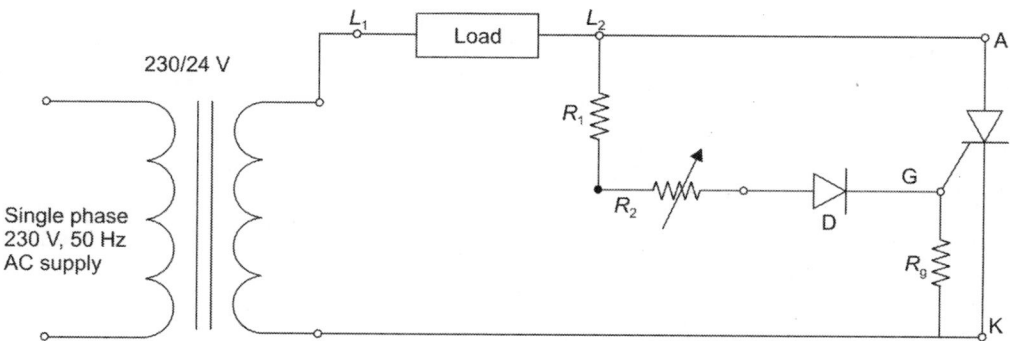

B1.5 DEVICE SPECIFICATION

Thyristor   C106 D

Minimum gate current = 6 mA   Minimum gate voltage = 0.8 V

B1.6 DESIGN EQUATIONS

Select a load resistance R_L. Depending on the power rating of the load resistor, determine the permissible load current. This is the current through the thyristor.

$$\text{Allowable thyristor current} = I_{TH} = \sqrt{\frac{\text{Power rating of the load}}{\text{Load resistance}}}$$

Select a normal gate current I_g. At the instant of triggering, let the minimum and the maximum triggering angles be 5° and 90° respectively. In the off state, the voltage across the thyristor is $v_T = V_m \sin \alpha$. This is the voltage across the gate circuit.

At the instant of minimum triggering angle, with $R_2 = 0$ in the worst case

$$V_m \sin 5° = I_g(R_L + R_1) + V_D + V_{GT}$$

From this R_1 can be calculated.

At the instant of maximum triggering angle

$$V_m \sin 90° = I_g(R_L + R_1 + R_2) + V_D + V_{GT}$$

From this R_2 can be calculated.

The value of R_g is so selected that the voltage across it is not greater than V_{GT}.

$$(V_m \sin \alpha)\frac{R_g}{(R_L + R_1 + R_2 + R_g)} < V_{GT}$$

$$R_g < \frac{V_{GT}(R_1 + R_L)}{V_m \sin \alpha - V_{GT}} \text{ with } \alpha = 5° \text{ and } R_2 = 0$$

B1.7 R-TRIGGERING CIRCUIT DESIGN

Take the input voltage as 24 V (rms value)

Select the load resistance $R_L = 100\ \Omega$, 10 W

$$\text{Maximum load current} \quad I_{TH} = \sqrt{\frac{10}{100}} = 0.316 \text{ A}$$

At the instant of minimum triggering angle, with $R_2 = 0$ in the worst case

$$V_m \sin 50 = I_g(R_L + R_1) + V_D + V_{GT}$$

Choose a gate current $I_g = 6$ mA, $V_{GT} = 0.8$ V and $V_D = 0.7$ V

Solving this $R_1 = 143\ \Omega$. Select $R_1 = 150\ \Omega$

At the instant of maximum triggering angle

$$V_m \sin 90° = I_g(R_L + R_1 + R_2) + V_D + V_{GT}$$

$$V_m \sin 90° = 0.006\ (100 + 150 + R_2) + 0.7 + 0.8$$

Solving this $R_2 = 5167 \, \Omega$ Select $R_2 = 5 \, k\Omega$ and $R_1 = 330 \, \Omega$

$$Rg = \frac{V_{GT}(R_1 + R_L)}{V_m \sin\alpha - V_{GT}} = \frac{0.8(330 + 100)}{34\sin 5° - 0.8} = 159 \, \Omega$$

Choose $R_g = 100 \, \Omega$

B1.8 EQUIPMENT AND COMPONENTS REQUIRED

Sl No	Component	Specification	Quantity
1.	Single phase transformer	230/24 V, 50 Hz	1
2.	Resistors	330 Ω, 0.25 W	1
		100 Ω, 0.25 W	1
		100 Ω, 100 W	1
3.	Potentiometer	0 – 5 kΩ	1
4.	Thyristor	C106 D	1
5.	Diode	IN4007	1
6.	DSO/CRO		
7.	Breadboard		

B1.9 PROCEDURE

1. Connections are made as per the connection diagram.
2. Keep the resistor R_2 at the maximum value.
3. Switch ON the AC mains.
4. Observe the source voltage waveform using probe connected to channel-II of the DSO.
5. Decrease R_2 so that the thyristor is just turned ON. This corresponds to a firing angle of nearly 90°.
6. Observe the gate pulses using the probe connected to channel-I and the waveform of voltage across the thyristor using the probe connected to channel-II of the DSO.
7. Disconnect both the probes connected to the channels of the DSO.
8. Without altering the settings of R_2, observe the waveform of voltage across the load using the probe connected to channel-II.
9. Disconnect the probes from the channel.
10. Decrease R_2 further to decrease the thyristor firing angle to a value less than 90°.
11. Repeat steps 6, 7, 8 and 9.
12. Switch OFF the AC supply.

Precaution: Make the voltage measurements one by one, to avoid the common ground problem of the scope.

B1.10 OBSERVATION

Observe the following wave forms on the DSO and draw them on graph sheet.
1. Source voltage
2. Load voltage
3. Thyristor voltages for the following cases:
   (a) $\alpha = 30°$ and (b) $\alpha = 60°$

Sample waveforms are shown below:

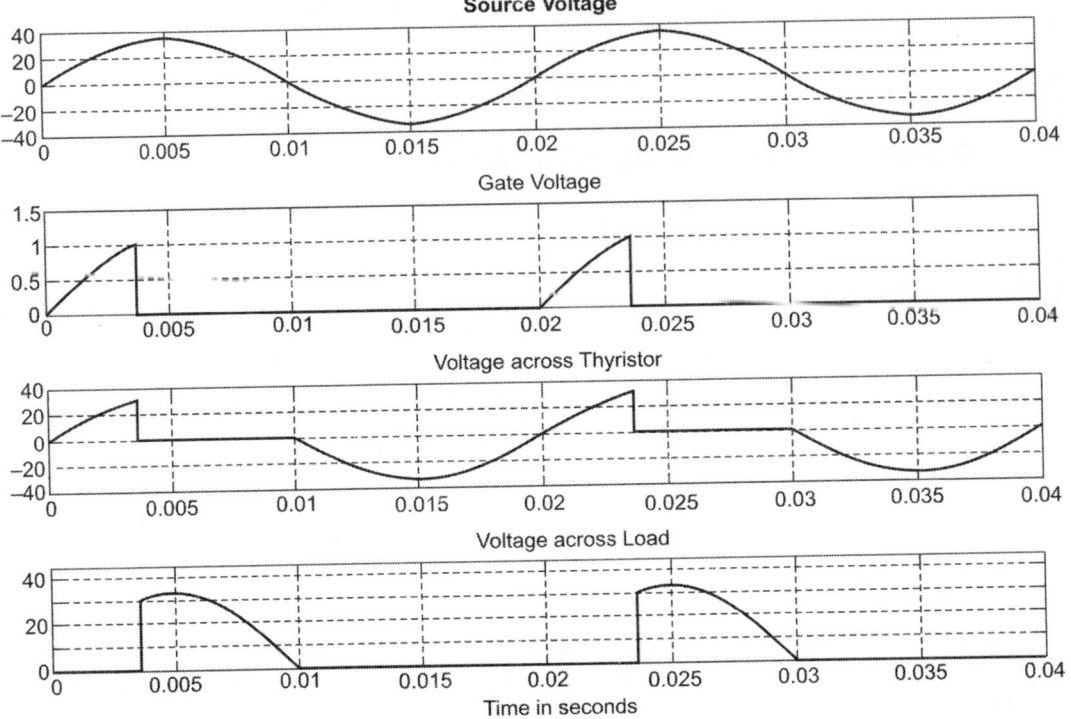

Waveforms R-Triggering $\alpha = 60°$

Experiment B2

Design and Testing of RC Triggering Circuit for a Thyristor

B2.1 OBJECTIVE

To design and study the working of an RC triggering circuit for triggering the gate of a thyristor.

B2.2 THEORY

In the circuit shown, during the negative half cycle of the supply voltage, the capacitor charges through R_1 and R_2 with V_{ab} negative. The resistance R_1 is a fixed resistance included in the circuit to limit the gate current when R_2 is decreased to a minimum. During the positive half cycle when the supply voltage rises, the capacitor starts charging through R_1 and R_2. When the capacitor voltage reaches a value equal to the gate trigger voltage, it discharges producing a pulse. This pulse appears across the gate of the thyristor and it gets fired. Now the capacitor discharges and the voltage across it becomes almost zero.

During the negative cycle of the ac supply, the capacitor gets charged through R_1 and R_2 to maximum reverse voltage and remains at this value till the supply voltage reaches zero value. During the next positive half cycle the capacitor charges through

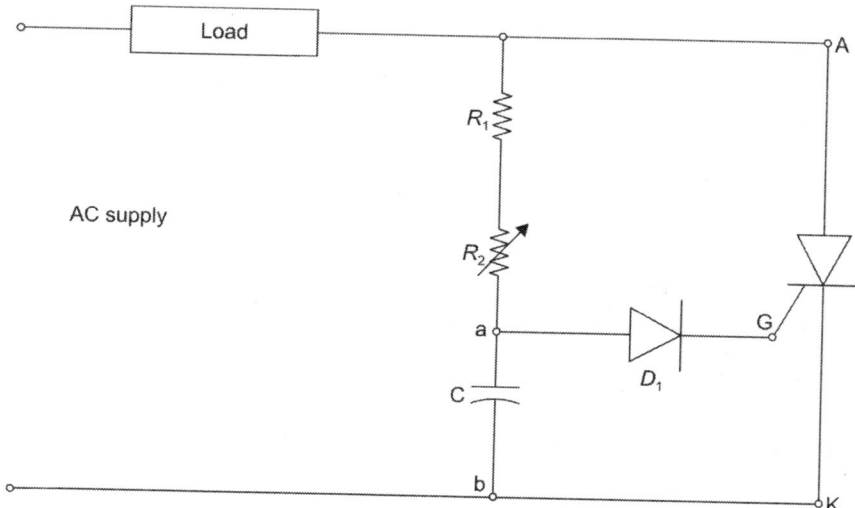

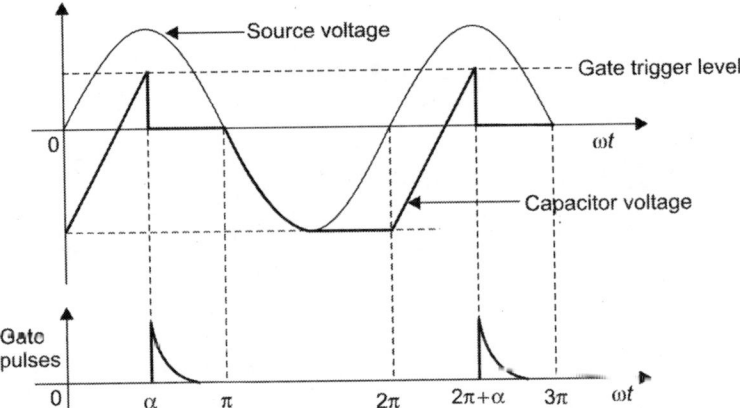

R_1 and R_2 until its voltage reaches the gate trigger voltage. At this instant it discharges producing a pulse sufficient to trigger the thyristor. The cycle is thus repeated. The diode D prevents a negative voltage appearing at the gate. The charging and discharging of the capacitor is shown in the figure.

The charging time of the capacitor depends on the time constant $(R_1 + R_2)C$. When R_2 is high, the time taken by the capacitor to get charged to the gate trigger voltage level is large. This in turn provides a large firing angle α, nearly equal to 180°. When R_2 is minimum, the time constant is less and so the capacitor quickly charges to the gate trigger voltage and this makes the firing angle minimum and closely near to zero value. Hence theoretically the triggering angle can be varied from 0° to 180°. An approximate equation for time constant is $RC > 0.65T$ where $T = 20$ ms for 50 Hz.

B2.3 RC TRIGGERING CIRCUIT FOR HALF-WAVE RECTIFIER

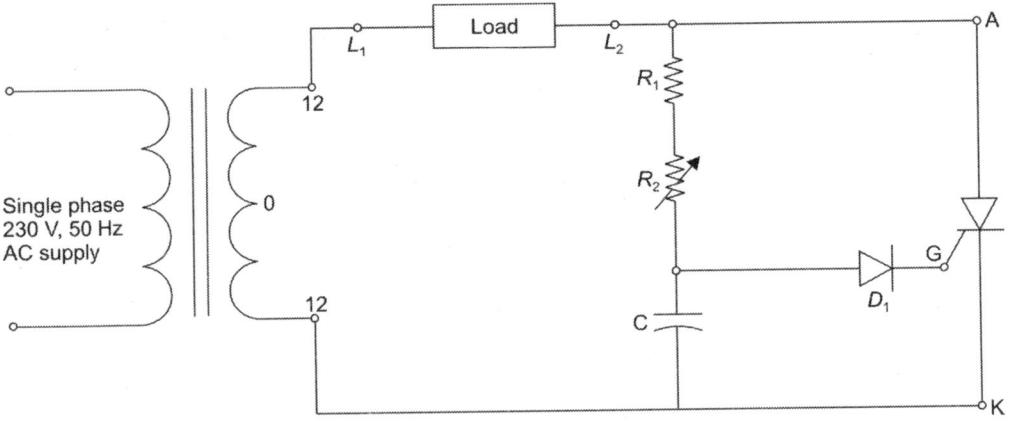

B2.4 DESIGN OF RC TRIGGERING CIRCUIT (HALF-WAVE)

Select the load resistance $R_L = 100\ \Omega$, 10 W

Select $R_1 = 220\ \Omega$

To get the maximum firing angle very close to 180°, a high value of R_2 is to be selected.

Choose $R_2 = 100\ \text{k}\Omega$

$RC \geq 0.65$ T where $T = 20$ m sec for 50 Hz supply

$$C \geq \frac{0.65T}{(R_1 + R_2)} = \frac{0.65 \times 20 \times 10^{-3}}{100.22 \times 10^3} = 0.1297 \ \mu F$$

Select $\quad C = 0.22 \ \mu F$

B2.5 EQUIPMENT AND COMPONENTS REQUIRED

Sl No	Component	Specification	Quantity
1.	Single phase transformer	230/24 V, 50 Hz	1
2.	Resistors	220 Ω, 0.25 W	1
		100 Ω, 10 W	1
3.	Potentiometer	0 – 100 kΩ	1
4.	Capacitor	0.22 μF	1
5.	Thyristor	C106 D	1
6.	Diode	1N4007	1
7.	DSO/CRO		
8.	Breadboard		

B2.6 PROCEDURE

1. Make the connections per the connection diagram.
2. Keep the resistor R_2 at the maximum value.
3. Switch ON the AC mains.
4. Observe the source voltage waveform using the probe connected to channel-II of the DSO.
5. Decrease R_2 so that the thyristor is just turned ON. This corresponds to a firing angle of nearly 180°. Adjust R_2 to decrease the firing angle to value about 120°.
6. Connect the positive lead of probe, connected to channel–I of the DSO, to the gate terminal G and the common terminal to K of the thyristor and observe the gate triggering pulses.
7. Connect the positive lead of the probe connected to channel–II of the DSO to the upper terminal of the capacitor and negative lead to K of the thyristor and observe the waveform of voltage across the capacitor.
8. Connect the positive lead of the probe, connected to channel–II of the DSO, to the terminal A of the thyristor and negative lead to K of the thyristor and observe the waveform of voltage across the thyristor.
9. Disconnect both the channel connections of the DSO.
10. Without altering the settings of R_2, connect the positive lead of the probe, connected to channel–II of the DSO, to the terminal L_1 of the load and negative lead to L_2 of the load and observe the waveform of voltage across the load.
11. Disconnect the probe, connected to channel-II of the DSO, from the circuit.
12. Decrease R_2 further to decrease the thyristor firing angle to a value less than 90°.
13. Repeat steps 6, 7, 8, 9 and 10.
14. Disconnect both the channel connections and switch OFF the AC supply.

B2.7 OBSERVATION

Observe and draw the following wave forms for the cases α < 90° and (b) α > 90°

1. Source Voltage
2. Gate pulse
3. Capacitor voltage
4. Thyristor voltage and
5. Load voltage

Sample waveforms are given below for different triggering angles.

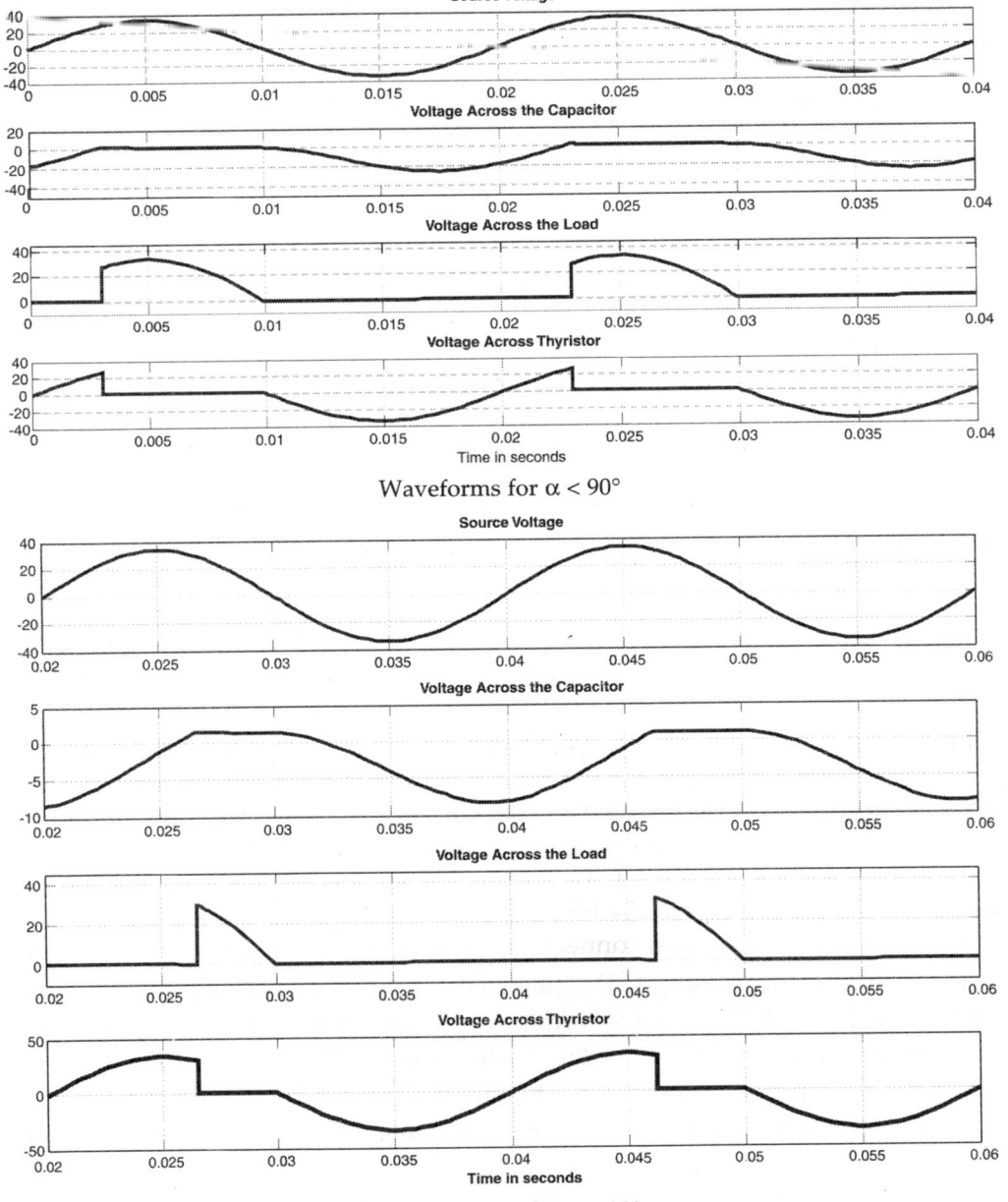

Waveforms for α < 90°

Waveforms for α < 90°

B2.8 RC TRIGGERING FOR FULL-WAVE RECTIFIER

Experiment is similar to that of the *RC* firing circuit for half-wave rectifier except that the supply is taken from a single phase full-wave diode bridge rectifier. During every half cycle of the ac supply, the anode to cathode voltage of the thyristor is positive and the capacitor is charged through the resistors to the level of the gate triggering voltage. At this instant the capacitor discharges producing a pulse that appears across the thyristor and hence it is triggered in to conduction.

The charging of the capacitor depends on the time constant *RC*. The value of *RC* is given by the empirical relation, $RC \geq 25\,T$ where $T = 20$ m sec for $f = 50$ Hz supply.

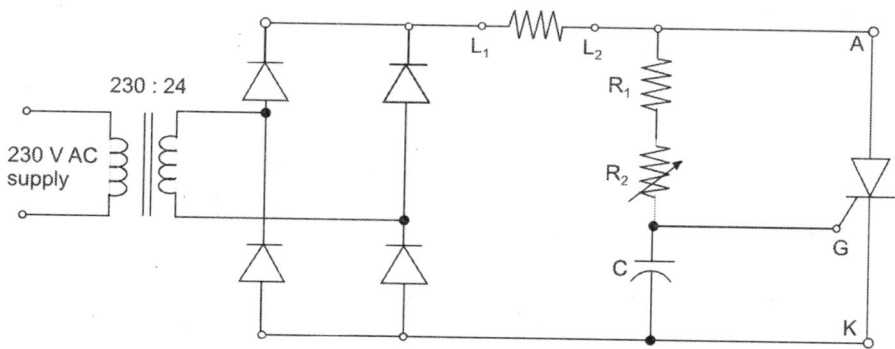

B2.9 DESIGN OF RC TRIGGERING CIRCUIT (FULL-WAVE)

$$RC \geq 25\,T \quad T = \frac{1}{f} = \frac{1}{50} = 20 \text{ m sec}$$

To get the maximum firing angle very close to 180°, a high value of R_2 is to be selected. Select $R_1 = 220\,\Omega$ and $R_2 = 500\,k\Omega$

$$C \geq \frac{25T}{(R_1 + R_2)} = \frac{25 \times 20 \times 10^{-3}}{500.22 \times 10^3} = 1\,\mu F$$

Select $C = 1.0\,\mu F$

Select load resistance = 100 Ω, 10 W

B2.10 EQUIPMENT AND COMPONENTS REQUIRED

Sl No	Component	Specification	Quantity
1.	Single phase transformer	230/24 V, 50 Hz	1
2.	Resistors	220 Ω, 0.25 W	1
		100 Ω, 10 W	1
3.	Potentiometer	0 – 500 kΩ	1
4.	Capacitor	1 μF	1
5.	Thyristor	C106 D	1
6.	Diode	1N4007	4
7.	DSO/CRO		
8.	Breadboard		

B2.11 PROCEDURE

Experimental procedure is similar to that given for the RC triggering circuit for half-wave rectifier.

B2.12 OBSERVATION

Observe and draw the following wave forms
1. Source voltage
2. Rectifier output voltage
3. Load voltage and
4. Thyristor voltages for the following cases
    (a) α = 20°   (b) α = 45°

Sample waveforms are given below:

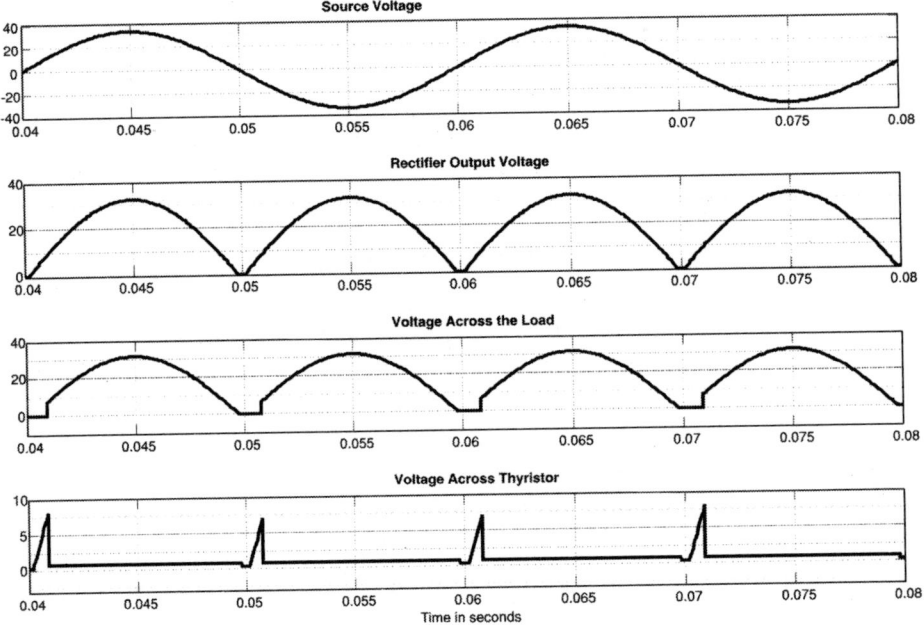

Experiment B3

Design and Testing of a Triggering Circuit Using UJT

B3.1 OBJECTIVE

To study the working of a pulse trigger circuit for a thyristor using UJT

B3.2 THEORY

Even though many pulse triggering circuits are available, UJT trigger circuits are generally used because of simplicity in operation, good reliability and low cost. A UJT has satisfactory operation over a wide range of temperatures. It has high input impedance and hence low current requirement for triggering.

The unijunction transistor consists of a bar of n-type Silicon with terminals at each end; Base-1 and Base-2. P-type silicon is alloyed to the opposite side of the n-type bar. A terminal connection to this p-type material serves as the emitter.

When a voltage V_{BB} is applied across the base terminals B_2 and B_1, the voltage gets divided across the bar so that a portion of the voltage V_{BB} appears across emitter $-B_1$ junction. This voltage is independent of temperature or current and depends only on the internal resistance R_{B_1} and R_{B_2}. This emitter $-B_1$ junction voltage is equal to

$$V_{BB}\left(\frac{R_{B_1}}{R_{B_1} + R_{B_2}}\right) = \eta V_{BB},$$

where $\eta = \dfrac{R_{B_1}}{R_{B_1} + R_{B_2}}$ is called the intrinsic stand-off ratio; $0 < \eta < 1$

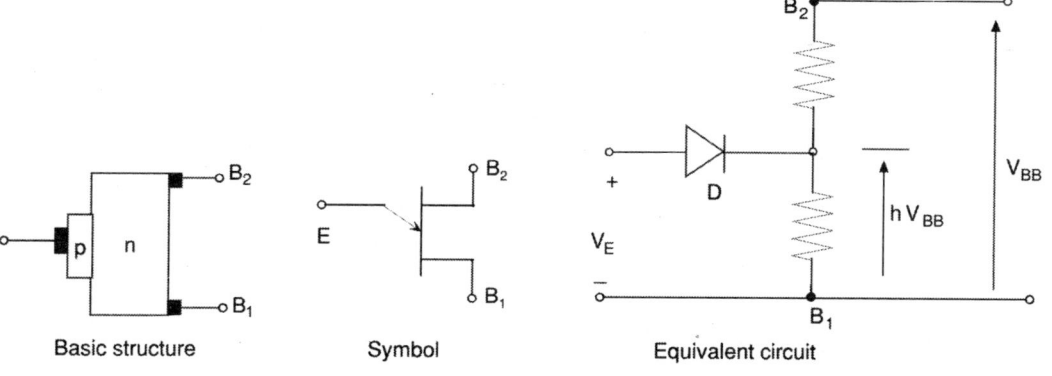

Basic structure   Symbol   Equivalent circuit

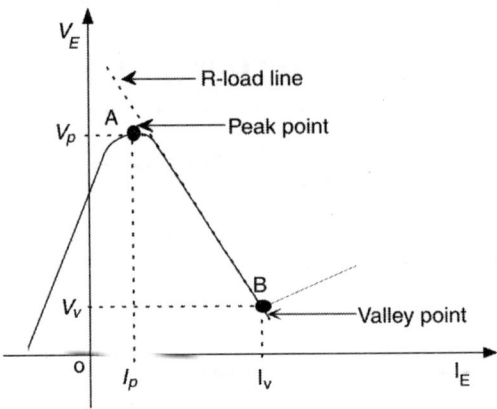

So long as the voltage applied to the emitter V_E is less than ηV_{BB}, the emitter B_1 junction is reverse biased and only a very small leakage current is present; that is I_e is negative. When V_E is made equal to $(\eta V_{BB} + V_D)$ where V_D is the forward voltage drop, the Emitter $-B_1$ junction is forward biased, the device conducts and the emitter current is positive. This is represented by the point A in the device characteristics. The point A is called the peak point. When the B_1 junction is forward biased holes are injected towards Base-1 region. Because of the increased number of charge carriers in the Base-1 region, the resistance R_{B1} decreases so that emitter to Base-1 voltage drops and the current increases. The device thus exhibits negative resistance characteristics. When the current increases further, a stage may reach when the Base-1 region gets saturated with the charge carriers and the resistance R_{B1} will not decrease further. This is represented by the point B in the characteristic and is called the valley point. A further increase in the emitter current will be followed by a rise in emitter voltage.

In the circuit shown UJT is used as a relaxation oscillator. This circuit can produce sharp pulses at constant frequency.

The resistances R_T and R_g are external resistances connected to the base terminals B_2 and B_1 respectively. The capacitor C charges through the resistance R at a rate decided by the time constant RC. When the capacitor voltage reaches the peak point voltage V_p, the UJT gets triggered and it conducts. The capacitor discharges through the resistance R_g producing a pulse across it. This pulse across R_g is used to trigger the gate of thyristor. The capacitor voltage (Emitter $-B_1$ voltage) drops down to valley voltage

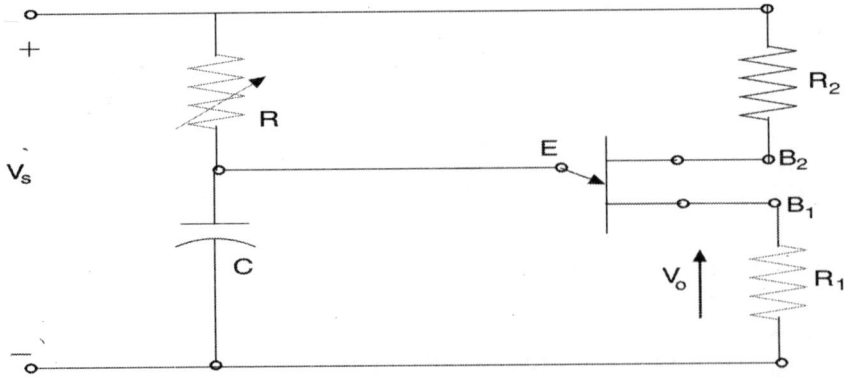

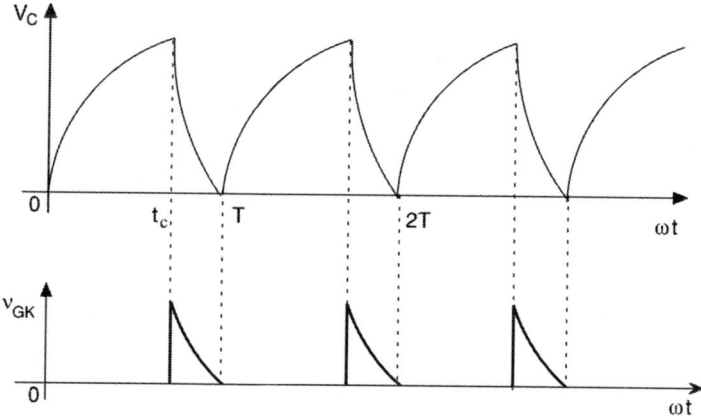

V_V. At this point, the device is reverse-biased and is blocked. The output is zero until the capacitor is once again charged to the peak point voltage V_p. This process will repeat and a train of pulses is generated at the output terminals.

The emitter $-B_1$ voltage varies depending on the rate at which the capacitor is charged. The resistance R should be such that the load line intersects the device characteristics only at the negative resistance region. If the load line intersects at any other region of the characteristic, the resulting operation is stable and the circuit will not produce oscillations. Thus the resistance R has maximum and minimum values.

$$R_{max} = \frac{V_s - V_p}{I_p}$$

and

$$R_{min} = \frac{V_s - V_v}{I_v}$$

The device conducts when the capacitor charges to the peak point voltage V_p exponentially.

$$V_E = V_V + V_s\,(1 - e^{-t/RC}) \qquad \qquad ...(1)$$

Also

$$V_E = V_P = \eta V_{BB} + V_D = \eta V_s + V_D \qquad \qquad ...(2)$$

Assuming $V_v = V_D$, $\eta V_s = V_s\,(1 - e^{-t/RC})$, where V_s is the supply voltage.

Hence, $\eta = 1 - e^{-t/RC}$

If t_c is the time taken by the capacitor to charge to the peak point voltage,

$$\eta = 1 - e^{-t_c/RC}$$

Neglecting the time taken the capacitor to discharge, $t_c = T$, the period of oscillations. R_T improves thermal stability and may be determined from the empirical relation,

The value of R_g is chosen to be sufficiently small so that the voltage drop across it due to normal leakage current should not trigger the thyristor. For isolation R_g may be replaced by a pulse transformer. The secondary of the pulse transformer can be connected across the gate and K of the thyristor. Triggering pulses are induced at the secondary of a pulse transformer to fire the gate of the thyristor.

B3.3 CIRCUIT DIAGRAM

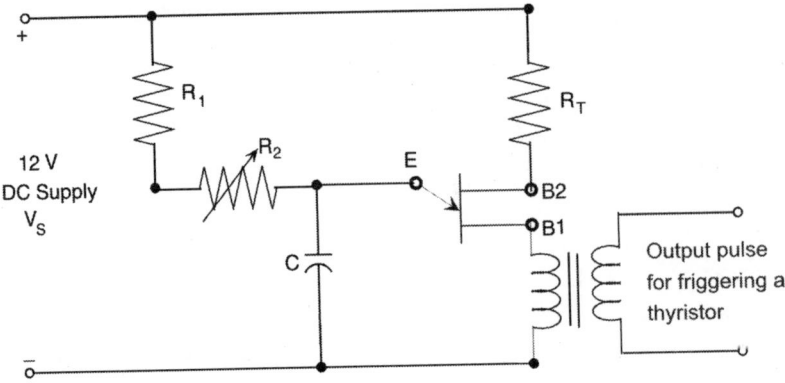

B3.4 DEVICE'S SPECIFICATIONS

Select UJT 2N2646

Intrinsic stand-off ratio	$\eta = 0.65$
Base to base voltage	$V_{BB} = 12$ V
Peak point current	$I_P = 100$ μA
Valley point current	$I_V = 4$ mA

B3.5 DESIGN EQUATIONS FOR TRIGGERING CIRCUIT

$$V_P = \eta V_{BB} + V_D$$

$$R_{max} = \frac{V_{BB} - V_p}{I_p}$$

$$T = -RC \, l_n(1-\eta) \quad C = \frac{-T}{Rl_n(1-\eta)}$$

An empirical formula to calculate R_T is

$$R_T = \frac{10^4}{\eta V_{BB}} \, ohm$$

B3.6 TRIGGERING CIRCUIT DESIGN

Peak point voltage

$$V_P = \eta V_{BB} + V_D = 0.65 \times 12 + 0.7 = 8.5 \text{ V}$$

$$R_{max} = \frac{V_{BB} - V_p}{I_p} = \frac{12 - 8.5}{100 \times 10^{-6}} = 35 \text{ kW}$$

$$R = R_1 + R_2$$

Select $R_1 = 220 \, \Omega$ and $R_2 = 50$ kΩ pot

The period for charging and discharging the capacitor is taken to be 10 ms.

$$C = \frac{-T}{Rl_n(1-\eta)} = -\frac{10 \times 10^{-3}}{50 \times 10^3 \ln(1-0.65)} \quad \text{Select } C = 0.22 \text{ μF}$$

$$R_T = \frac{10^4}{\eta V_{BB}} = \frac{10^4}{0.65 \times 12} = 1282 \text{ W} \quad \text{Select } R_T = 1.2 \text{ k}\Omega$$

Select a load resistance of 100 Ω, 20 W

B3.7 EQUIPMENTS AND COMPONENTS

Sl No	Component	Specification		Quantity
1.	DC source	12 V		1
2.	Resistors	R_1	220 Ω, 0.25 W	1
		R_T	1.2 kΩ, 0.25 W	1
		R_L	100 Ω, 20 W	1
3.	Potentiometer	R_2	50 kΩ	1
4.	Capacitor	C	0.22 μF	1
5.	Pulse transformer		1:1:1	1
6.	Thyristor	C106 D		1
7.	DSO/CRO			
8.	Breadboard			

B3.8 PROCEDURE

1. Connections are made as per the circuit diagram
2. Keep R_2 at the maximum value for a large triggering angle.
3. Set the DSO ready for measurement, after necessary adjustments.
4. Switch ON the DC supply.
5. Connect the leads of the probe, connected to channel-I of the DSO, across the secondary of the pulse transformer to observe the triggering pulses generated.
6. Decrease the value of R_2 and repeat step 5.
7. Switch OFF the supply.

B3.9 OBSERVATION

Observe the triggering pulses in the DSO and draw the waveform in a graph sheet.

Design and Testing of a Line Synchronized Triggering Circuit Using UJT

B4.1 OBJECTIVE

To study the working of a line synchronised UJT triggering circuit for a thyristor.

B4.2 THEORY

A relaxation oscillator using UJT can be used for triggering the gate of a thyristor. When such a relaxation oscillator circuit is used for triggering a thyristor operating with alternating current, some method must be adopted to synchronize its oscillations with the frequency of the ac supply. For this, the supply voltage for the relaxation oscillator is taken from an ac supply through a diode rectifier and a zener diode. The diode rectifier can either be a half-wave rectifier or a full-wave rectifier. The zener diode is so selected that it clips the voltage at V_{BB}. This voltage becomes zero at the zero crossing points of the ac supply, i.e at $\omega t = 0$, π, 2π,.... Since the zener diode voltage goes to zero at the end of each half cycle, the synchronization of the trigger circuit with ac supply voltage is achieved. The capacitor voltage also becomes zero at the zero crossing points of the ac supply. Hence line synchronization with zero crossing points can be achieved.

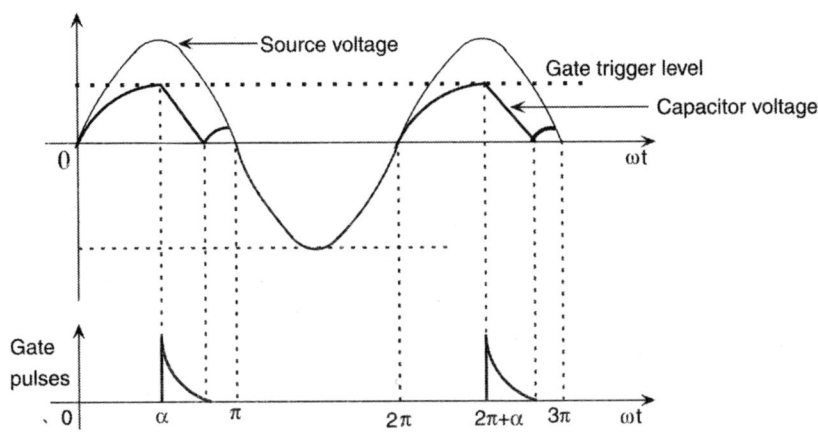

Triggering pulses for a half-wave rectifier

In case of half-wave rectifier, the zener diode clips the voltage to the level of V_{BB} during every positive half cycle. This voltage is applied to the capacitor through R and the charging rate is decided by RC. When the capacitor voltage reaches the level of $V_p = \eta V_{BB}$, it discharges through UJT and a pulse appears across the primary of a pulse transformer. The pulse induced across the secondary is used to trigger the gate of a thyristor. The capacitor starts charging in the next positive half cycle. The charging rate can be controlled by varying the resistor R and this method is called ramp control.

If the value of the R is reduced, the charging time of the capacitor gets reduced and hence the capacitor can charge and discharge more than once during a half cycle. As the first pulse will trigger the thyristor, other successive pulses generated during the same half cycle are redundant.

In case of full-wave rectifier, the zener diode clips the voltage to the level of V_{BB} during every half cycle. The circuit operation is the same except that trigger pulses are generated during every half cycle. These are shown in figure given below.

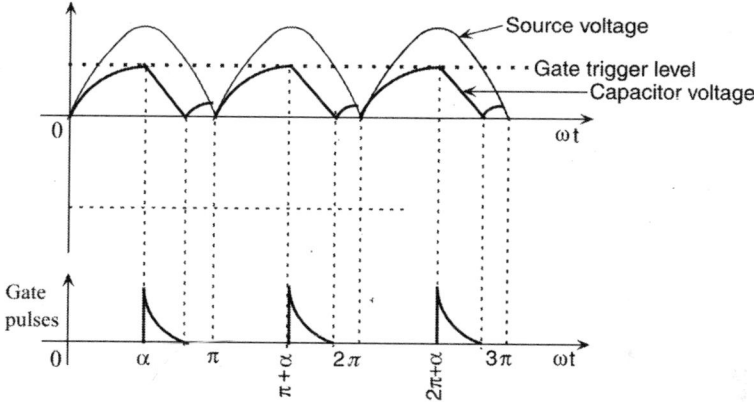

Triggering pulses for a full-wave rectifier

B4.3 UJT TRIGGERING CIRCUIT FOR HALF-WAVE RECTIFIER

The resistance R_S is included in the circuit to limit the peak Zener current and the peak emitter current of the UJT. A pulse transformer is used for transmission of generated gate triggering pulses from a low power control circuit to a high power circuit.

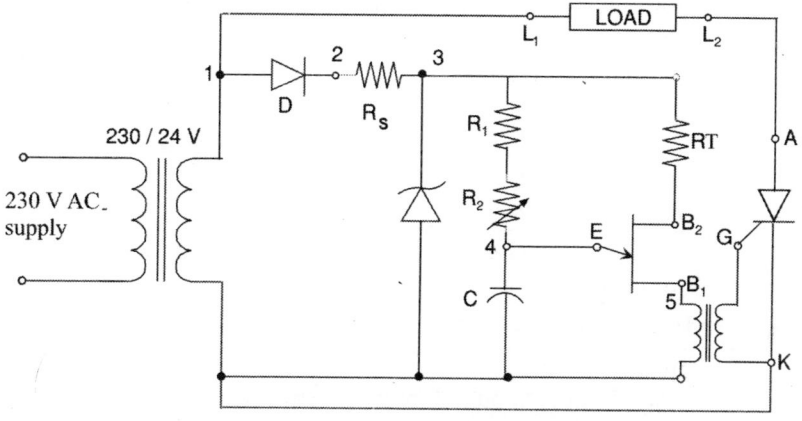

B4.4 DEVICE SPECIFICATIONS

UJT   2N 2646
Base to base voltage V_{BB}
Intrinsic stand-off ratio η
Zener diode
Rating of the zener diode
Thyristor   C106D

B4.5 DESIGN EQUATIONS FOR TRIGGERING CIRCUIT

Maximum value of the rectifier output voltage V_{rmax}
Zener diode voltage rating V_Z
Zener power rating W_z

Maximum zener current $I_z = \dfrac{W_z}{V_z}$

$$R_S = \frac{V_{rmax} - V_z}{I_z}$$

Select $\alpha_{min} = 10°$ and $\alpha_{max} = 170°$

$$t_{min} = \frac{10\pi}{180\omega} \qquad t_{max} = \frac{170\pi}{180\omega}$$

$$R_{min} = \frac{-t_{min}}{C \ln(1-\eta)} \qquad R_{max} = \frac{-t_{max}}{C \ln(1-\eta)}$$

An empirical formula to calculate R_T is

$$R_T = \frac{10^4}{\eta V_{BB}}$$

Select a load resistance for a chosen value of allowable thyristor current.

B4.6 TRIGGERING CIRCUIT DESIGN

Let $V_{BB} = 12$ V, $\eta = 0.65$
Maximum value of the rectifier output voltage, $V_{rmax} = 34$V
Voltage across the zener diode $V_Z = 12$ V
Rating of the Zener diode 0.5 W, 12 V

Zener current $I_z = \dfrac{0.5}{12} = 0.417$ A

$$R_S = \frac{V_{rmax} - V_z}{I_z} = \frac{34-12}{0.0417} = 527 \ \Omega$$

Select $R_s = 560 \ \Omega$, 1 W
Select $\alpha_{min} = 10°$ and $amax = 170°$

$$t_{min} = \frac{10\pi}{180\omega} = \frac{10\pi}{180 \times 2\pi 50} = 0.555 \ \text{m sec}$$

$$t_{max} = \frac{170\pi}{180\omega} = \frac{170\pi}{180 \times 2\pi 50} = 9.444 \text{ m sec}$$

Select the capacitor $C = 0.22 \ \mu F$

$$R_{min} = \frac{-t_{min}}{C \ln(1-\eta)} = \frac{-0.555 \times 10^{-3}}{0.22 \times 10^{-6} \times \ln(1-0.65)} = 2.4 \text{ k}\Omega$$

$$R_{max} = \frac{-t_{max}}{C \ln(1-\eta)} = \frac{-9.444 \times 10^{-3}}{0.22 \times 10^{-6} \times \ln(1-0.65)} = 40.89 \text{ k}\Omega$$

$$R = R_1 + R_2 \quad \text{Select } R_2 = 50 \text{ k}\Omega \text{ pot}$$

The resistance R_1 is selected as 220 Ω. The resistance R_1 is a fixed resistance included in the circuit to limit the gate current when R_2 is decreased to a minimum.

$$R_T = \frac{10^4}{\eta V_{BB}} = \frac{10^4}{0.65 \times 12} = 1282 \ \Omega \quad \text{Select } R_T = 1.2 \text{ k}\Omega$$

Select load resistance $R_L = 100 \ \Omega$, 10 W

B4.7 EQUIPMENT AND COMPONENTS

Sl No	Component	Specification	Quantity
1.	Single phase transformer	230/24 V, 50 Hz	1
2.	Resistors	R_1 220 Ω, 0.25 W	1
		R_T 1.2 kΩ, 0.25 W	1
		R_S 560 Ω, 1 W	1
		$R_L = 100 \ \Omega$, 10 W	1
3.	Potentiometer	$R_2 = 50$ kΩ	1
4.	Capacitor	0.22 μF	1
5.	Thyristor	C106D	1
6.	Diode	IN4007	1
7.	Zener diode	1N5242BT, 0.5 W, 12 V	1
8.	Pulse transformer	1:1:1	1
9.	DSO/CRO		1
10.	Breadboard		

B4.8 PROCEDURE

1. Connections are made as per the circuit diagram.
2. Keep the variable pot R_2 at the maximum value for a large triggering angle.
3. Set the DSO ready for measurement, after necessary adjustments.
4. Make observation of the voltage waveform at the different points marked on the firing circuits, using DSO and plot them on a graph sheet.
5. Observe the wave forms of voltage across the load and the voltage across the thyristor. For this connect, using probes, the positive terminal of the channel-I of DSO to the point L_1 of the load and common negative terminal to the point L_2 or anode A of the thyristor. The positive terminal of the channel II is connected to the cathode K of the thyristor and the negative terminal to the anode A. Invert channel II by pressing the "*invert*" button. Plot the wave form on the same graph sheet.
6. Vary the value of R_2 and repeat steps 4 and 5.

B4.9 OBSERVATIONS

Observe the following wave forms and plot on a graph sheet.

1. Supply voltage
2. Rectified voltage
3. Zener voltage
4. Capacitor voltage
5. Output pulse of UJT
6. Load voltage
7. Thyristor voltage

B4.10 UJT TRIGGERING CIRCUIT FOR FULL-WAVE RECTIFIER

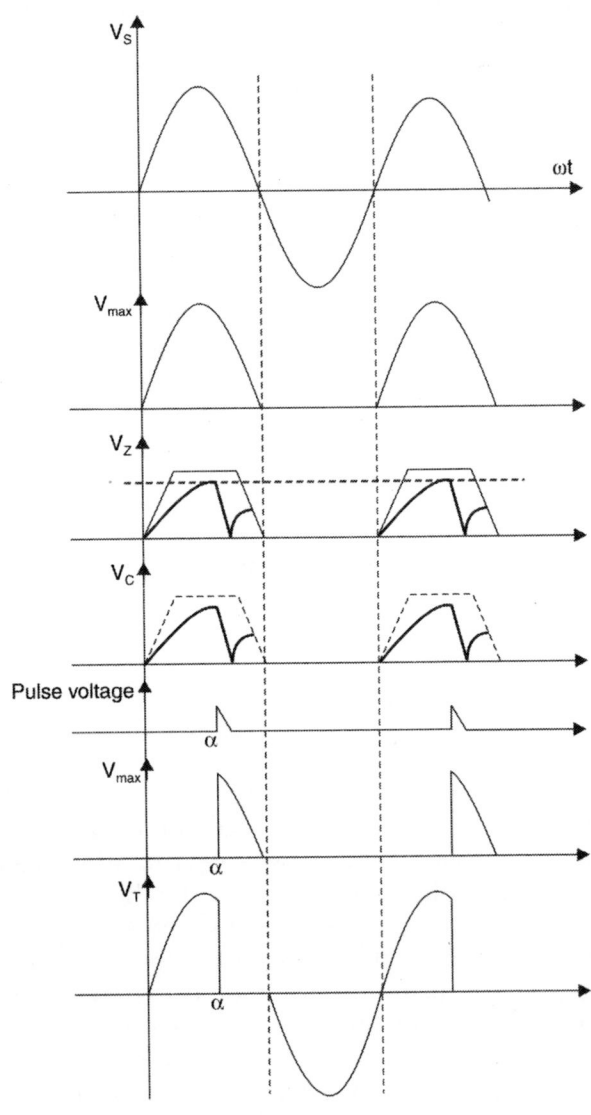

Waveforms–UJT triggering for half-wave rectifier

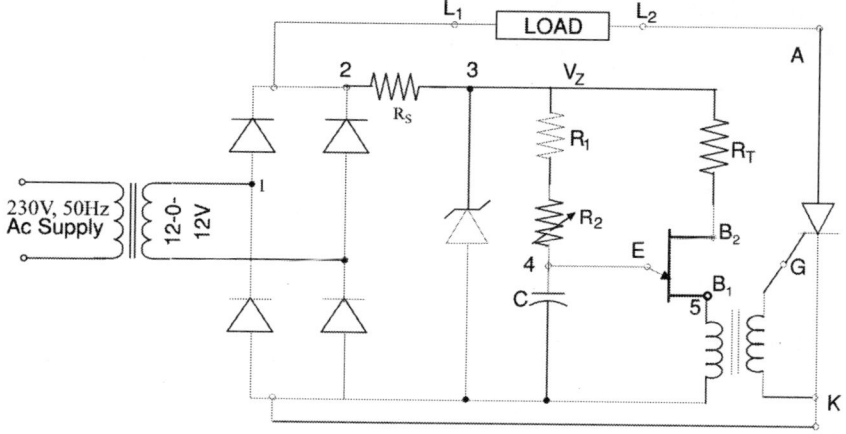

B4.11 DESIGN OF TRIGGERING CIRCUIT FOR A FULL-WAVE RECTIFIER

Design procedure is the same as that in the case of trigger circuit for half-wave rectifier.

B4.12 EQUIPMENT AND COMPONENTS—FULL-WAVE RECTIFIER

Sl No	Component	Specification	Quantity
1.	Single phase transformer	230/24 V, 50 Hz	1
2.	Resistors	R_1 220 Ω, 0.25 W	1
		R_T 1.2 kΩ, 0.25 W	1
		R_S 560 Ω, 1 W	1
		R_L = 100 Ω, 20 W	1
3.	Potentiometer	R_2 = 50 kΩ	1
4.	Capacitor	0.22 µF	1
5.	Thyristor	C106D	1
6.	Diode	IN4007	1
7.	Zener diode	1N5242BT, 0.5 W, 12 V	1
8.	Pulse transformer	1:1:1	1
9.	DSO/CRO		
10.	Breadboard		

B4.13 PROCEDURE

The procedure is in the same steps as given in the case of triggering circuit for half-wave rectifier.

B4.14 OBSERVATIONS

Observe the following waveforms.
1. Supply voltage
2. Rectified voltage
3. Zener voltage
4. Capacitor voltage

5. Output pulse of UJT
6. Load voltage
7. Thyristor voltage.

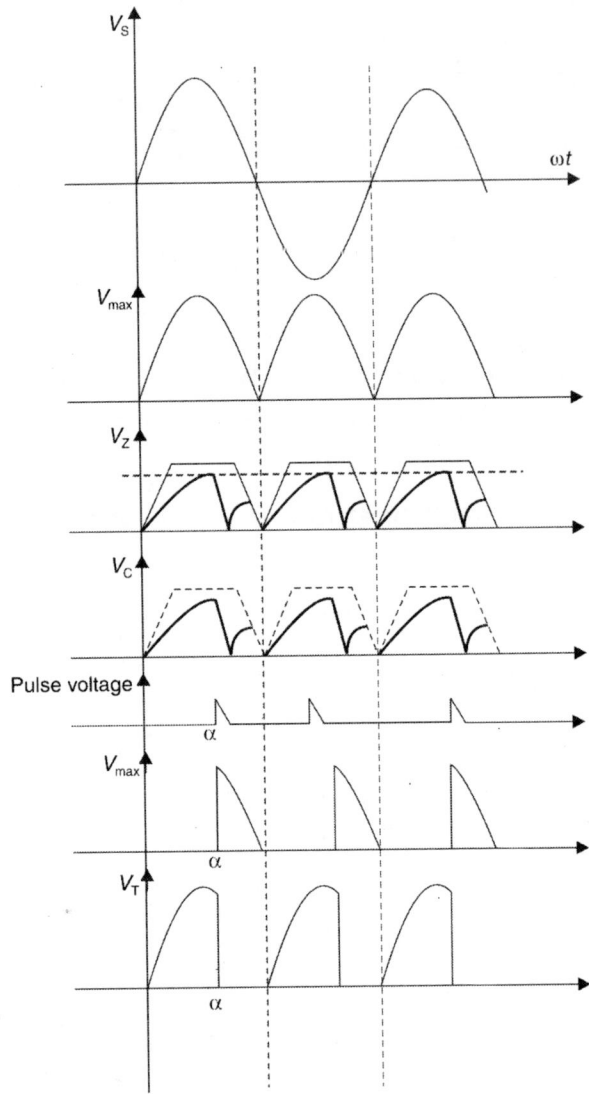

Waveforms- UJT triggering for full wave rectifier

Experiment B5

Design and Testing of a Single Phase Half-wave Rectifier

B5.1 OBJECTIVE

To study the performance of a single phase full-wave rectifier under the following loads:

1. Purely resistive load
2. RL load
3. RL load with freewheeling Diode

B5.2 THEORY

A converter that converts an AC voltage to a DC voltage is called a rectifier. An AC voltage can be converted to a DC voltage by means of a diode. However, a diode rectifier provides only a fixed DC voltage at the output and so it is known as an uncontrolled rectifier. To obtain a controlled output voltage, controllable power semiconductor switches like BJTs, MOSFETs, IGBTs and thyristors can be used. A converter using such devices is called a controlled rectifier. A thyristor is turned ON and OFF to control the period of its operation by which the output voltage is controlled. A thyristor used for this purpose is turned ON by applying a very short pulse at its gate terminal and is turned OFF by natural or line commutation.

The time delay at which the thyristor is triggered is called the *firing angle* or *triggering angle* or *delay angle*. This is measured from the instant from where the anode voltage starts going positive. By varying the triggering angle the output voltage of a rectifier can be controlled. This method of control is known as the phase angle control and the converter is called a phase-controlled converter.

In single-phase rectifiers, the input voltage to the rectifier is from a single-phase source and it is assumed to be purely sinusoidal. Single-phase rectifiers are classified into two, namely half-wave controlled rectifier and full-wave controlled rectifier. In a half-wave controlled rectifier, there is only one output pulse to a period of the input sinusoidal voltage and is sometimes referred to as "one pulse rectifier". In a single-phase full-wave rectifier, there are two output pulses, one each in the positive and negative half cycles of the supply voltage.

B5.3 HALF-WAVE RECTIFIER WITH RESISTIVE LOAD

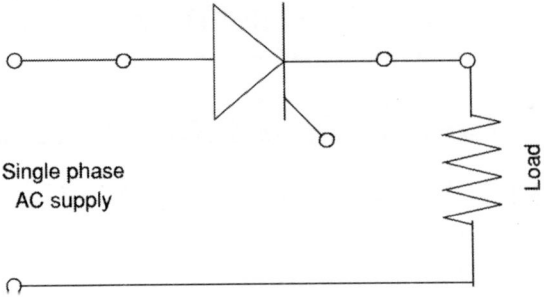

Working

The input voltage is an AC sinusoidal voltage and can be expressed as,

$$v_s(t) = V_m \sin \omega t$$

1. During the positive half cycle of the input voltage, the thyristor has anode voltage positive with respect to cathode and is said to be forward biased. At the instant $\omega t = \alpha$, the thyristor is triggered and it starts conducting. The load current flows and it is given by,

$$i(t) = \frac{V_m}{R} \sin \omega t$$

The waveform of load current is similar to the input voltage because the current is in phase with the input voltage.

2. At $\omega t = \pi$, the anode voltage is zero and therefore the current is zero. Hence at $\omega t = \pi$, the thyristor is turned off.

3. For $\omega t > \pi$, the anode to cathode voltage is negative and so the thyristor is reverse biased. Thus it blocks the supply voltage during the negative half cycle and the load voltage remains zero.

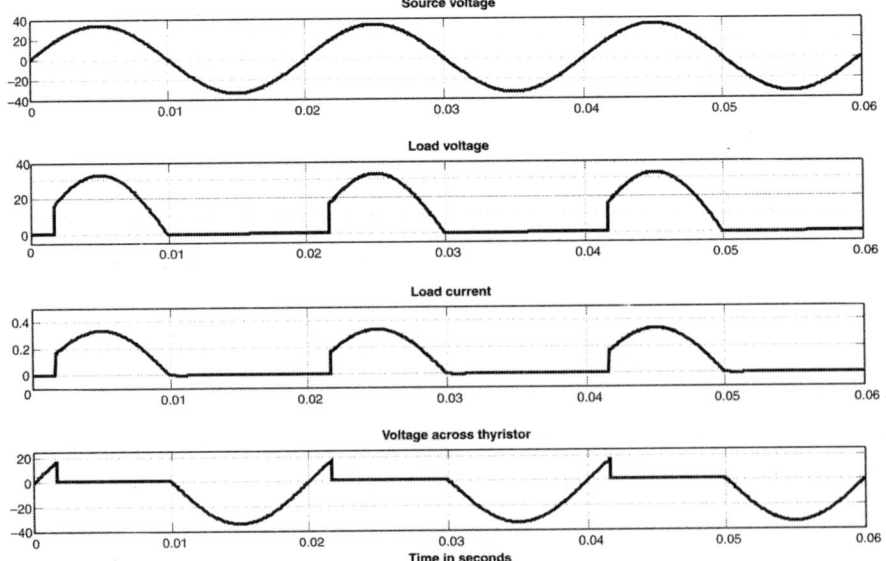

Waveform: Half-wave rectifier with R Load

4. After one cycle, for $\omega t > 2\pi$, the input voltage is positive, the thyristor is forward biased and it is turned ON once again at $\omega t = 2\pi + \alpha$, by triggering its gate. The thyristor conducts and the load current flows until when $\omega t = 3\pi$, where the load current becomes zero and the device is turned OFF. The cycle is thus repeated.

Note:

1. In each positive half cycle the thyristor conducts only for a period corresponding to $(\pi - \alpha)$ and the conducting time is given by,

$$t_c = \frac{\pi - \alpha}{\omega} \text{ sec}$$

where $\omega = 2\pi f$ is the angular frequency of the supply voltage and f is the supply frequency in H_Z.

2. The circuit turn off time is the time for which the thyristor remains in the reverse blocking state once it is turned off. In this case the thyristor remains in the reverse blocking state for a time duration of π radians. Hence the circuit turn off time is given by,

$$t_{off} = \frac{\pi}{\omega} \text{ sec.}$$

3. The average value of the load voltage is

$$V_{av} = \frac{1}{2\pi} \int_\alpha^\pi V_m \sin\theta \, d\theta = \frac{V_m}{2\pi}(1 + \cos\alpha)$$

At the output, there is only one pulse of load current and this appears during the positive half cycle of the input voltage. During the negative half cycle the load current is zero. Hence this rectifier is called a half-wave rectifier.

The output voltage and current have only one polarity and the region of converter operation is said to be in the first quadrant. Hence the converter is called a single quadrant converter.

The disadvantages of this converter are: (a) high ripple content and (b) low ripple frequency.

B5.4 HALF-WAVE RECTIFIER WITH RL LOAD

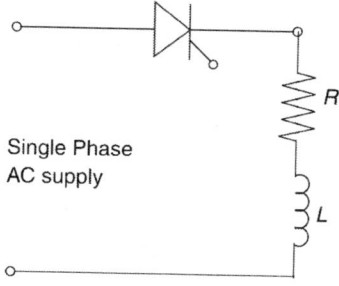

Single Phase
AC supply

Working

The load consists of a pure resistor of resistance R ohm and a pure inductor with inductance L Henry. The input voltage is assumed to be purely sinusoidal and is expressed as

$$v_s(t) = V_m \sin \omega t$$

1. The thyristor is triggered at the instant $\omega t = \alpha$. The output voltage follows the input voltage and is given by, $v_0(t) = V_m \sin \alpha$ and the output current is zero as the inductor current cannot change instantaneously. The output current gradually increases and reaches a maximum value. After the current reaching a maximum value it decreases. Hence (di/dt) is negative and so the voltage across the inductor $v_L = L(di/dt)$ is negative. In other words, when the current starts decreasing, the polarity of inductor voltage reverses and it aids the supply voltage in maintaining the load current. The output voltage is a maximum at $\omega t = \pi/2$ and decreases thereafter and becomes zero at $\omega t = \pi$.

2. At $\omega t = \pi$, even though the output voltage is zero, the output current does not become zero because of the inductance effect. For $\omega t > \pi$, the input voltage is negative. The load current flows for a little more time before it becomes zero, even when the output voltage is negative. This effect of the inductor is to maintain the load current in the same direction. In other words, the energy stored in the inductor is released to maintain the current through the load even when the output voltage is negative. This is called the inductance effect.

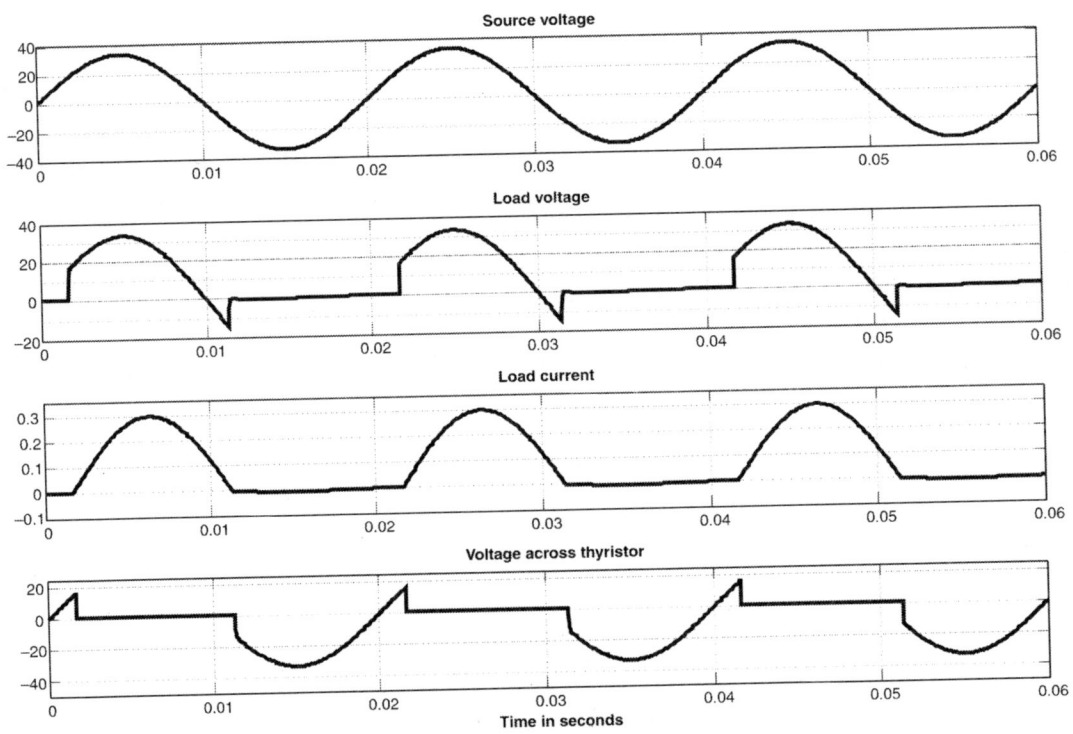

Waveform: Half-wave rectifier with RL load-DCM

3. After some time the inductor energy is completely dissipated in the resistor R and the current becomes zero at $\omega t = \beta$ as shown in figure. When the load current is zero the thyristor current is also zero. When the thyristor current is less than its holding current it is turned off. The angle β is known as the current extinction angle and is the instant at which the load current decreases to zero value. The angle $(\beta - \alpha)$ is called the conduction angle and it is the period during which the thyristor conducts.

4. From to $\omega t = \beta$ to $\omega t = 2\pi + \alpha$, the thyristor remains in the OFF state. For $\omega t > 2\pi$, the thyristor is forward biased. At $\omega t = 2\pi + \alpha$, when the thyristor is triggered, it is turned ON and the cycle is repeated. The waveforms of load voltage, load current and thyristor voltage are shown in figure with the supply voltage as reference.

Note:
1. From $\omega t = \alpha$ to $\omega t = \pi$, both the output voltage and the output current are positive and the power is positive. This power is consumed by the load.
2. From $\omega t = \pi$ to $\omega t = \beta$, the output voltage is negative whereas the output current is positive and the power is negative. During this period power flows from the load to the AC supply; the energy stored in the inductor is returned to the source.
3. Thus the net power consumed by the load is the difference between these two powers.
4. The voltage across the thyristor is the input voltage minus the voltage across the load.
5. From the waveform of thyristor voltage, it can be seen that the thyristor remains reverse biased from to $\omega t = \beta$ to $\omega t = 2\pi$. This period of time is called the circuit turn off time and is given by,

$$t_{off} = \frac{2\pi - \beta}{\omega}$$

6. The average value of the load voltage is

$$V_{av} = \frac{1}{2\pi} \int_{\alpha}^{\beta} V_m \sin\theta \, d\theta = \frac{V_m}{2\pi}(\cos\alpha - \cos\beta)$$

B5.5 DEMERITS OF HALF–WAVE RECTIFIER WITH RL LOAD

1. The output voltage has negative excursion which causes the average output voltage to decrease from what it would be for a resistive load, for the same triggering angle.
2. The presence of negative excursion of the output voltage increases the peak-to-peak value of the voltage on the DC side. This creates voltage "ripple" in the output voltage resulting in a much less smooth DC output voltage.
3. Increase in the ripple content in the load current.
4. The load current is discontinuous.
5. During the period when the output voltage is negative, power flows from the load side to the AC source. This is a reactive power flow due to the energy stored in the inductor.
6. The reactive power flow from the DC load side to the AC source decreases the power factor.

B5.6 HALF–WAVE RECTIFIER WITH RL LOAD AND FREEWHEELING DIODE

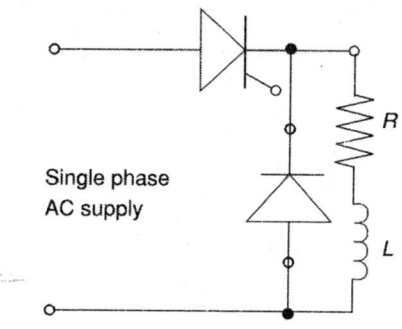

Working—Continuous Current Mode

In the case of a half-wave rectifier with RL load, the load current is discontinuous and is with high ripple content. The load current can be made continuous when a diode is connected across the load as shown in the figure. This diode is called a freewheeling diode or a fly wheeling diode.

1. The thyristor is triggered at $\omega t = \alpha$ and is turned ON. It conducts from $\omega t = \alpha$ to $\omega t = \pi$ during the first positive half cycle of the supply voltage. The output current gradually rises from zero and increases as output voltage increases. It reaches a maximum decided by the RL load and then decreases because by this time the supply voltage starts decreasing. At $\omega t = \pi$, the supply voltage is zero.

2. For $\omega t > \pi$, the supply voltage is negative. The diode now is forward biased and it conducts and the energy stored in the inductor is released to the load resistor through the freewheeling diode. The load is now short-circuited through the diode and the thyristor current is brought to zero at $\omega t = \pi$. When the thyristor current is less than its holding current it is turned off. The thyristor is thus turned off and a reverse voltage appears across it. The load current that otherwise flows through the thyristor, is transferred to the conducting diode and the thyristor is commutated. The freewheeling diode thus ensures the commutation of the thyristor at the end of the positive half cycle of the supply voltage and hence it is called a commutating diode.

3. Even though the thyristor is turned off at $\omega t = \pi$, the load current is not zero due the inductance effect. As the load current starts decreasing the polarity of the voltage across the inductor quickly reverses and the energy stored in it during the positive half cycle of the supply voltage is released through the freewheeling diode. This is known as the freewheeling action of the diode and hence the name freewheeling diode. Thus, the load current continues to flow even though the voltage across the load is zero. The load current is the same as the diode current, which is an exponentially decaying current.

4. For $\omega t > 2\pi$, the thyristor is once again forward biased. At $\omega t = 2\pi + \alpha$, the thyristor is fired again this cycle is repeated. After a few cycle of the input voltage, the load current attains a steady state level.

Note:

1. When the thyristor conducts, the voltage across the load is the input voltage.
3. When the freewheeling diode conducts, the voltage across the load is zero but the load current is not zero.
3. The voltage across the thyristor is the input voltage minus the load voltage.
4. If the value of the inductance is high, the energy stored in the inductor is high. Hence the load current will not decay down to zero during the freewheeling period and before the thyristor is fired again at $\omega t = 2\pi + \alpha$. This makes the load current continuous.
5. The average value of the load voltage is

$$V_{av} = \frac{1}{2\pi} \int_{\alpha}^{\pi} V_m \sin\theta \, d\theta = \frac{V_m}{2\pi}(1 + \cos\alpha)$$

When the Inductance of the Load is Low-Discontinuous Current Mode

When the L/R ratio of the load is low and the thyristor firing angle is large, the diode current may decay down to zero value before the thyristor is fired again. This is because of the energy stored in the inductor is not sufficient to maintain the load current during freewheeling until the thyristor is fired again at $\omega t = 2\pi + \alpha$. As a result, the load current becomes discontinuous.

The waveforms of load voltage, load current, the freewheeling current and the voltage across the thyristor are shown in figure, with the supply voltage as the reference.

The average value of the load voltage is

$$V_{av} = \frac{1}{2\pi} \int_{\alpha}^{\pi} V_m \sin\theta \, d\theta = \frac{V_m}{2\pi}(1 + \cos\alpha)$$

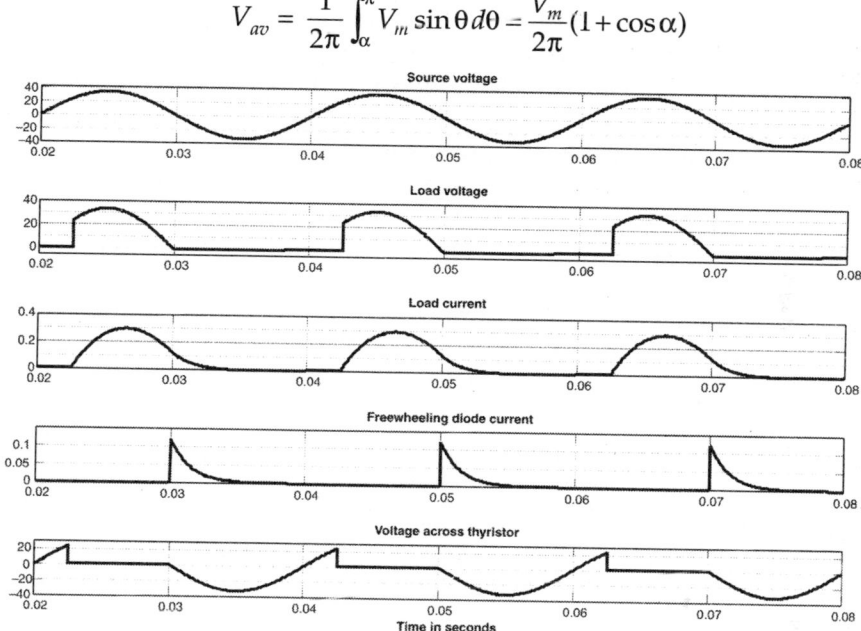

Waveform: Half wave rectifier with RL load and freewheeling diode (DCM)

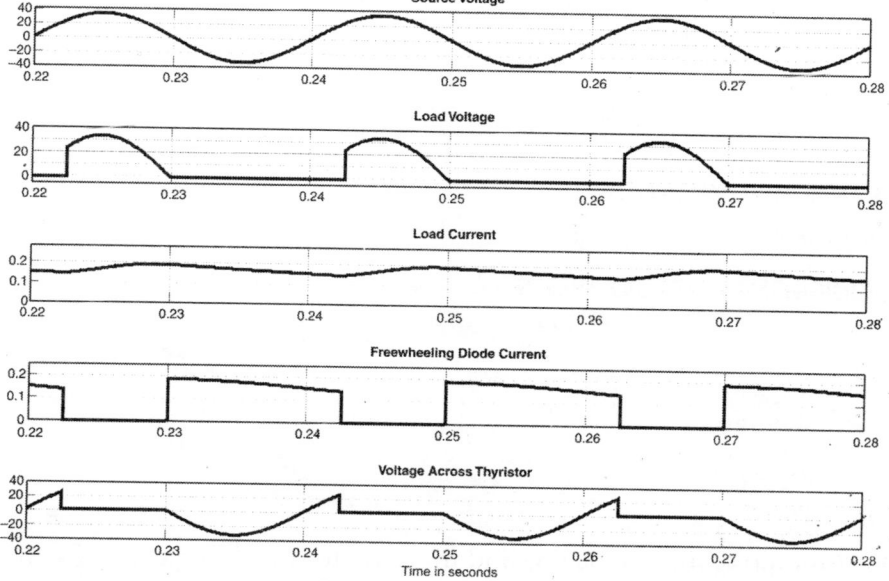

Waveform: Half wave rectifier with RL load and freewheeling diode (CCM)

B5.7 ADVANTAGES OF THE FREEWHEELING DIODE

1. Whenever the supply voltage becomes negative, the diode conducts and the negative voltage that might appear across the load is clipped.
2. Since the negative excursion of the output voltage is clipped, the average value of the output voltage is increased.
3. Because of the freewheeling action of the diode, there is no flow of reactive power from the load to the source so that the power factor is improved.
4. The current in the load is maintained even after the thyristor is turned off. This is because of the energy stored in the inductor forces a freewheeling current through the diode. This improves the discontinuity in the load current and the load current is made close to continuous or even continuous.
5. Ripple content in the load current is reduced.

B5.8 CIRCUIT DIAGRAM

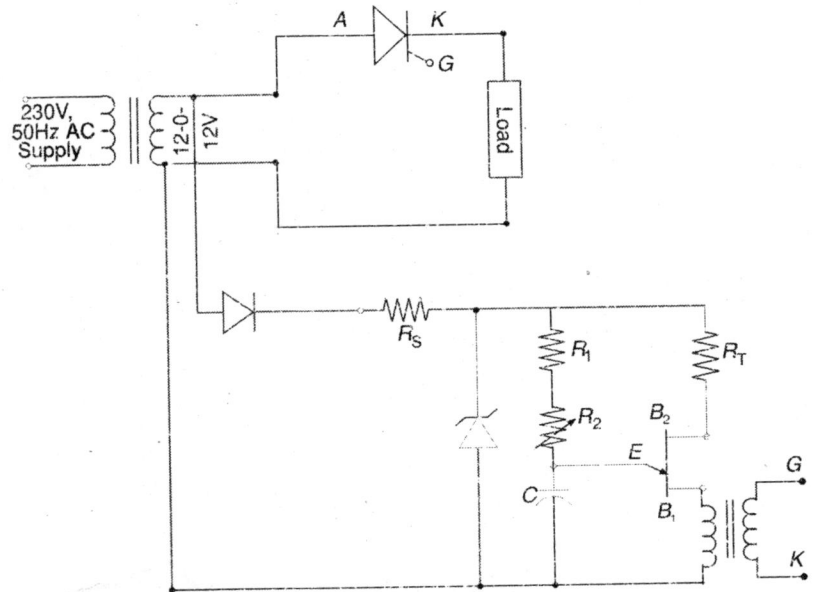

B5. 9 DEVICE SPECIFICATION

Thyristor C106D   Diode 1N4007

B5.10 EQUIPMENT AND COMPONENTS REQUIRED

Sl No	Component	Specification	Quantity
1.	Single phase transformer	230/24 V, 50 Hz	1
2.	Resistors	100 Ω, 20 W	1
3.	Inductor		1
4.	Thyristor	C106 D	1
5.	Diode	1N4007	1
6.	DSO/CRO		
7.	Breadboard		
8.	Thyristor triggering circuit using UJT		

B5.11 PROCEDURE

1. Make the connections as per the circuit diagram given in each case of the load.
2. Set the position of the pot in the triggering circuit at its maximum position and then switch ON the supply. Triggering circuit is the UJT triggering circuit used for a full-wave rectifier. For triggering circuit design, refer to section B4.
3. Adjust the firing angle to a value less than 90°.
4. Observe the following wave forms.
    (a) Voltage across the load.
    (b) Voltage across the thyristor.
    (c) Load current.
    (d) Thyristor current.
5. Adjust the firing angle to a value greater than 90°.
6. Repeat step 4.

B5.12 OBSERVATION

Draw the following wave forms, in each case of the load.
1. Voltage across the load
2. Voltage across the thyristor
3. Load current
4. Thyristor current.

Note:

1. To observe the waveform of thyristor current, connect a 1 Ω resistor in series with the thyristor and observe the voltage profile across it. Similarly to observe the waveform of load current, connect a 1 Ω resistor in series with the load and observe the voltage profile across it.
2. Care must be taken while connecting probes to both channels of the DSO simultaneously.
3. The waveforms are to be drawn with the supply voltage as the reference. While drawing the waveforms, it is assumed that when the device conducts the voltage across it is zero and the voltage across the load is the supply voltage. The voltage across the thyristor is the supply voltage minus the load voltage.

Experiment B6

Design and Testing of a Single Phase Full-wave Mid-point Rectifier

B6.1 OBJECTIVE

To study the performance of a single phase full-wave rectifier with centre-tapped configuration, under the following loads:

1. Purely resistive load
2. RL load

B6.2 THEORY

Full-Wave Rectifier

In a single-phase full-wave rectifier there are two output pulses in a period of the input AC voltage, one each in the positive and negative half cycles. Full-wave controlled rectifiers can be classified as given below.

1. Single-phase rectifiers
   a. Mid-point converter
   b. Bridge converter
      i. Half-controlled converter
      ii. Fully-controlled converter
2. Three-phase rectifiers
   a. Three-pulse converter
   b. Six-pulse converter (bridge converter)
      i. Semi-controlled
      ii. Fully-controlled

Mid-Point Converter

The mid-point configuration of a single-phase full-wave converter employs a single-phase transformer with the secondary having a mid tap. The circuit diagram is shown in figure. Two thyristors are connected in the circuit of which one is triggered during the positive half-cycle and the other during the negative half-cycle. Thyristor T_1 becomes forward biased when the terminal A of the transformer is positive with respect to terminal B. Similarly thyristor T_2 becomes forward biased when the terminal B of the transformer is positive with respect to terminal A.

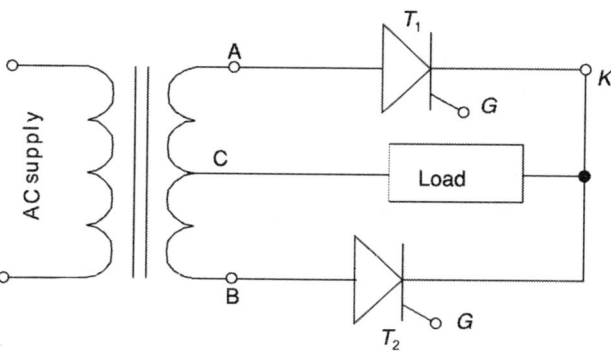

Let the supply voltage be
$$v_s = 2V_m \sin \omega t$$

Consider that during positive half cycle of the supply voltage, the terminal A of the transformer is positive with respect to terminal B.
$$v_{AC} = V_m \sin \omega t \text{ and } v_{BC} = -V_m \sin \omega t$$

Thyristor T_1 is forward biased and when it is fired, it goes into conduction. The load current flows. At the same time thyristor T_2 is reverse biased. The voltage across T_2 is $v_{BA} = v_{BC} - v_{AC} = -2V_m \sin \omega t$. Hence when T_1 conducts, the peak inverse voltage across T_2 is twice the maximum value of the supply voltage.

When v_{BC} is positive, T_2 is forward biased and T_1 is reverse biased. When T_2 is triggered, it conducts and the load current flows through it. Hence when T_2 is triggered, the load current is transferred from T_1 to T_2 and T_1 is turned off. Now when T_1 becomes once again forward biased during the next half cycle, it is triggered and brought into conduction and T_2 is turned off. This process of turning off a conducting thyristor is by natural or line commutation.

B6.3 PURELY RESISTIVE LOAD

1. During positive half cycle of the supply voltage, T_1 is forward biased and T_2 is reverse biased. At $\omega t = \alpha$, T_1 is triggered and is brought into conduction. The load current flows. At $\omega t = \pi$, the load current becomes zero and the thyristor T_1 is commutated. Since the load is purely resistive, the load current is in phase with the voltage across the load so that the wave form of load current is similar to that of the load voltage.
2. During negative half cycle of the supply voltage, T_2 is forward biased and T_1 is reverse biased. At $\omega t = \pi + \alpha$, T_2 is triggered and is brought into conduction. The load current flows. At $\omega t = 2\pi$, the load current becomes zero and the thyristor T_2 is commutated.
3. At $\omega t = 2\pi + \alpha$, T_1 is triggered again and the cycle is thus repeated.

The wave form of load current, during the negative half cycle is exactly is similar that of the positive half cycle.

The average value of the load voltage is

$$V_{av} = \frac{1}{\pi} \int_{\alpha}^{\beta} V_m \sin \theta \, d\theta = \frac{V_m}{\pi}(1 + \cos \alpha)$$

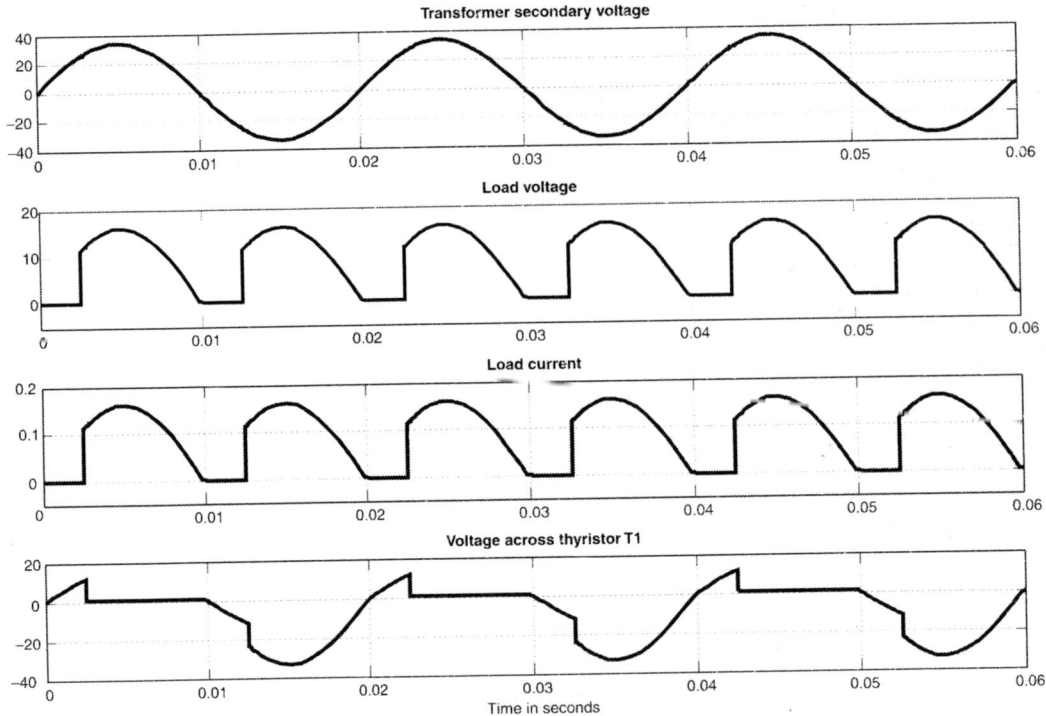

Single phase full-wave rectifier (Mid-point)–R load

B6.4 RL LOAD—DISCONTINUOUS CURRENT MODE

1. Consider that during positive half cycle of the supply voltage, the terminal A of the transformer is positive with respect to terminal B. Thyristor T_1 is forward biased and when it is fired at $\omega t = \alpha$, it goes into conduction. The load current gradually increases from zero value and reaches a maximum value. It then decreases because by this time the supply voltage should have started decreasing from its maximum value.

2. At $\omega t = \pi$, the supply voltage is zero but the load current continues to flow beyond $\omega t = \pi$ because of the inductance effect. If the thyristor triggering angle is large and the load inductance is small, the load current decays to zero value at some angle $\omega t = \beta$; $\pi < \beta < \pi + \alpha$. Thyristor T_1 is commutated. The current remains zero from $\omega t = \beta$ to $\omega t = \pi + \alpha$.

3. For $\omega t > \pi$, thyristor T_2 is forward biased and it is triggered at $\omega t = \pi + \alpha$. The load current now flows through T_2. The load current flows beyond $\omega t = 2\pi$ because of the load inductance. But it decreases to zero at $\omega t = \pi + \beta$ and T_2 is commutated. The load current remains zero from $\omega t = \pi + \beta$ to $\omega t = 2\pi + \alpha$.

4. For $\omega t > 2\pi$, T_1 is forward biased. At $\omega t = 2\pi + \alpha$, T_1 is triggered again. When it conducts, the load current flows through T_1 and T_2 is commutated. The cycle is thus repeated.

The average value of the load voltage is

$$V_{av} = \frac{1}{\pi} \int_{\alpha}^{\beta} V_m \sin\theta \, d\theta = \frac{V_m}{\pi}(\cos\alpha - \cos\beta)$$

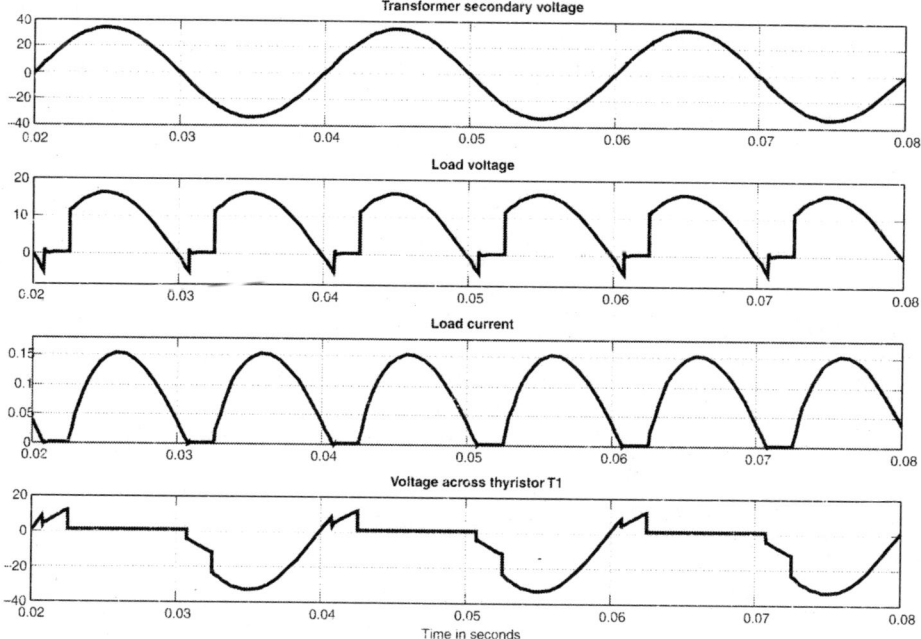

Single phase full-wave rectifier (Mid-point) – RL load (DCM)

B6.5 RL LOAD—CONTINUOUS CURRENT MODE

1. Consider that during positive half cycle of the supply voltage, the terminal A of the transformer is positive with respect to terminal B. Thyristor T_1 is forward biased and when it is fired at $\omega t = \alpha$, it goes into conduction. The load current gradually increases from zero value and reaches a maximum value. It then decreases because by this time the supply voltage should have started decreasing from its maximum value.

2. At $\omega t = \pi$, the supply voltage is zero but the load current continues to flow because of the inductance effect. If the thyristor triggering angle is small and the load inductance is sufficiently large, the load current flows and may not become zero even at the instant when thyristor T_2 is triggered during the negative half cycle. The load current is thus continuous.

3. For $\omega t > \pi$, thyristor T_2 is forward biased and it is triggered at $\omega t = \pi + \alpha$. The thyristor T_1 is commutated and the current is transferred from it to T_2. The load current now flows through T_2.

4. For, T_1 is forward biased and T_2 is reverse biased. At $\omega t = 2\pi + \alpha$, T_1 is triggered again. When it conducts, the load current flows through T_1 and T_2 is now commutated.

5. At $\omega t = 3\pi + \alpha$, thyristor T_2 is triggered again and the cycle of alternate triggering of T_1 and T_2 is repeated and the load current never becomes zero.

Note:

1. When T_1 is turned on, T_2 is commutated and vice versa.

2. From $\omega t = \alpha$ to $\omega t = \pi$, both the load voltage and the load current are positive and so the power flows from the source to the load. From $\omega t = \pi$ to $\omega t = \pi + \alpha$, the load voltage is negative whereas the load current is positive and so the power flows

from the load to the source. This is the flow of reactive power and this reduces the power factor of the circuit.

3. When T_1 is forward biased, T_2 is reverse biased. If $v_{AC} = V_m \sin \omega t$, this reverse voltage is given by, $v_{T2} = v_{BC} - v_{AC} = -2V_m \sin \omega t$. Thus the peak inverse voltage is $2V_m$. The thyristor voltage rating must be twice the maximum value of the supply voltage.

4. The thyristor T_2 remains reverse biased from $\omega t = \alpha$ to $\omega t = \pi$. Hence the circuit turn off time for T_2 is, $t_{c2} = \dfrac{\pi - \alpha}{\omega}$. The thyristor T_1 remains reverse biased from $\omega t - \pi + \alpha$ to $\omega t = 2\pi$. Hence the circuit turn off time for T_1 is, $t_{c1} = \dfrac{2\pi - (\pi + \alpha)}{\omega} = \dfrac{\pi - \alpha}{\omega}$.

Thus the turn off time provided for each of the thyristors is the same. The circuit turn off time must be greater than the turn off time of the thyristor; otherwise commutation failure will occur. Both the thyristors will be conducting together and the secondary winding will be short-circuited.

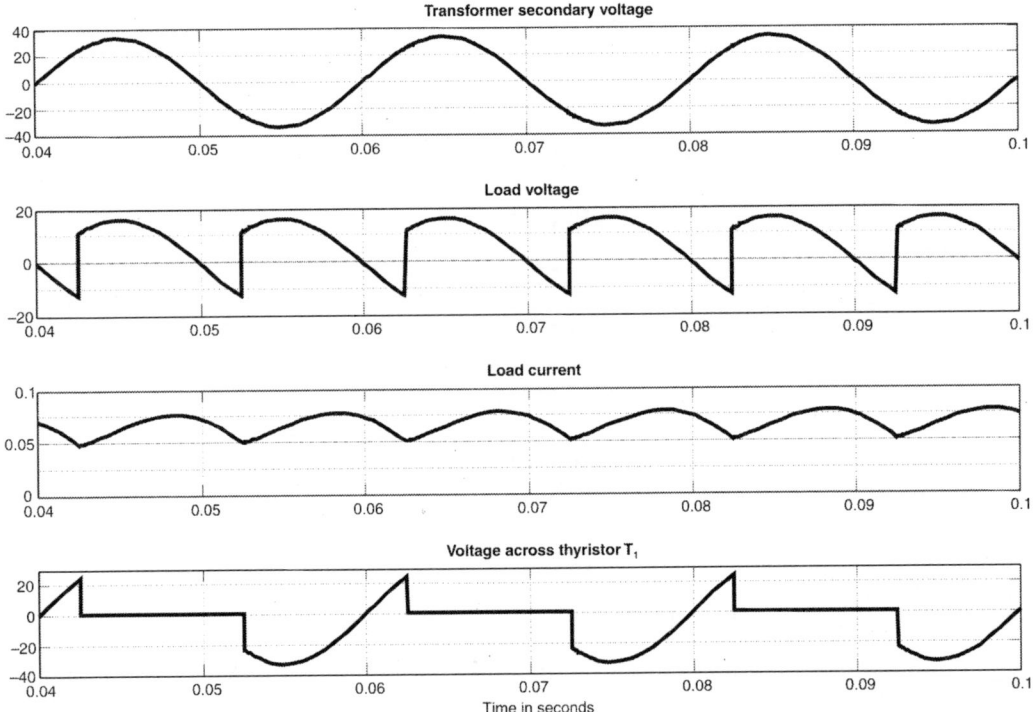

Single phase full-wave rectifier (Mid-point) – RL load (CCM)

The condition for continuous current is that $\alpha < \phi$ and the inductance L is large where $\phi = \tan^{-1}\left(\dfrac{\omega L}{R}\right)$, the load impedance angle.

The average value of the load voltage is

$$V_{av} = \frac{1}{\pi} \int_{\alpha}^{\pi+\alpha} V_m \sin\theta \, d\theta = \frac{V_m}{\pi}(\cos\alpha - \cos(\pi + \alpha)) = \frac{2V_m}{\pi}\cos\alpha$$

B6.6 CIRCUIT DIAGRAM

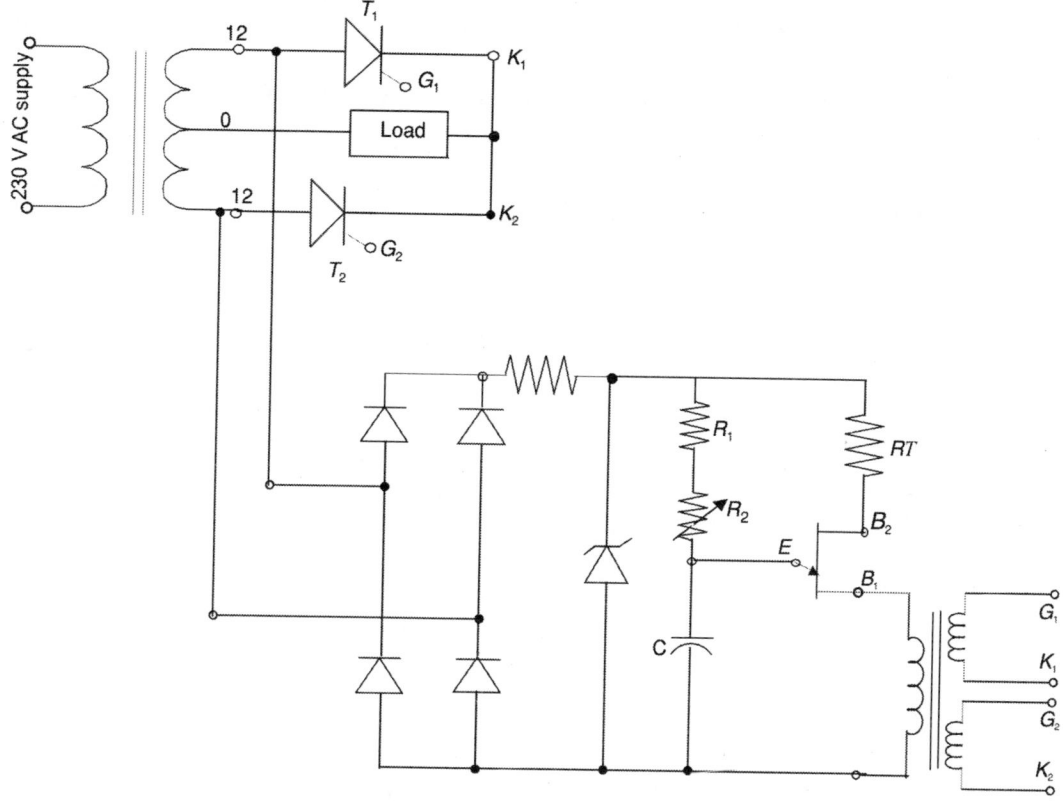

B6.7 DEVICE SPECIFICATION

Thyristor C106D

B6.8 EQUIPMENT AND COMPONENTS

Sl No	Component	Specification	Quantity
1.	Single phase transformer	230 V/12-0-12 V, 50 Hz	1
2.	Resistors (Load)	100 Ω, 20 W	1
3.	Thyristor	C106 D	2
4.	DSO/CRO		
5.	Breadboard		
6.	Thyristor triggering circuit using UJT		

B6.9 PROCEDURE

1. Make the connections as per the circuit diagram.
2. Switch ON the power supply.
3. Keep the variable pot of the triggering circuit at a position corresponding to maximum value.

4. Switch ON the trigger circuit and check whether the circuit produces the required triggering pulses. If necessary, adjust the variable pot of the triggering circuit. (Triggering circuit is the UJT triggering circuit used for a full-wave rectifier. For the triggering circuit design, refer to section B4)

5. Adjust the triggering so as to select a particular delay angle for the thyristor.

6. Observe the waveforms of load voltage, load current and thyristor voltage.

B6.10 OBSERVATION

Draw the following waveforms:

1. Load voltage
2. Load current
3. Thyristor voltage.

Experiment B7

Design and Testing of a Single Phase Full-wave Half Controlled Bridge Rectifier

B7.1 OBJECTIVE

To study the operation and performance of a single phase full-wave half controlled bridge rectifier.

B7.2 THEORY

A single phase full-wave half controlled bridge rectifier is one in which there are two thyristors and two diodes in the bridge circuit. Since there are only two controlled devices, the bridge circuit is called a half controlled rectifier or a semi converter. There are two types of connections for a bridge type single phase semi converter. They are symmetrical connection and asymmetrical connection and are shown below.

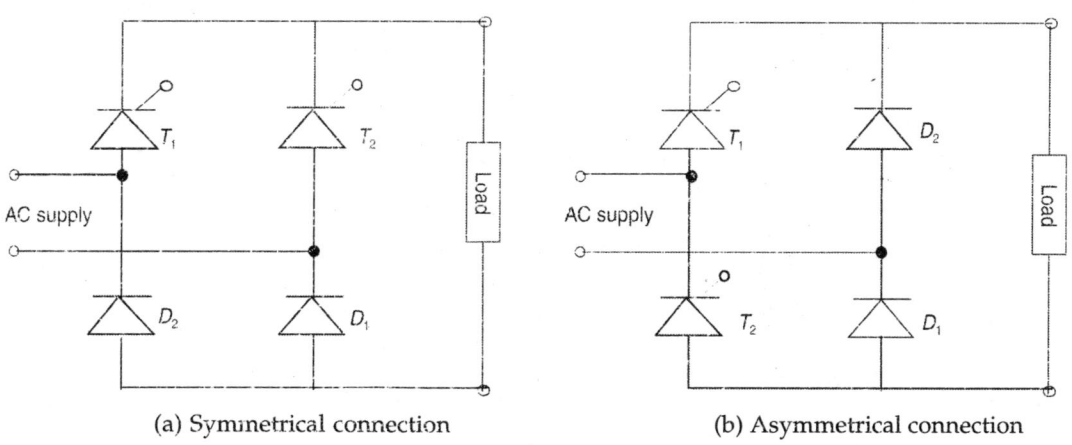

| (a) Symmetrical connection | (b) Asymmetrical connection |

Working: Symmetrical connection

A. R Load

1. During the positive half cycle of the supply voltage, thyristor T_1 is forward biased. When it is triggered at $\omega t = \alpha$, it conducts and the load current flows through T_1 and the diode D_1.
2. At $\omega t = \pi$ the load voltage is zero and the load current is zero. T_1 is commutated.

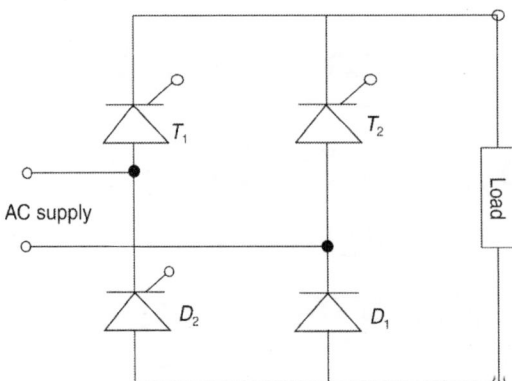

3. For $\omega t = \pi$, T_2 is forward biased and it is triggered at $\omega t = \pi + \alpha$. T_2 is thus turned ON. The load current now flows through T_2 and D_2.
4. At $\omega t = 2\pi$, the load voltage is zero and the load current is zero. T_2 is commutated.
5. At $\omega t = 2\pi + \alpha$, T_1 is triggered again and the cycle is thus repeated.

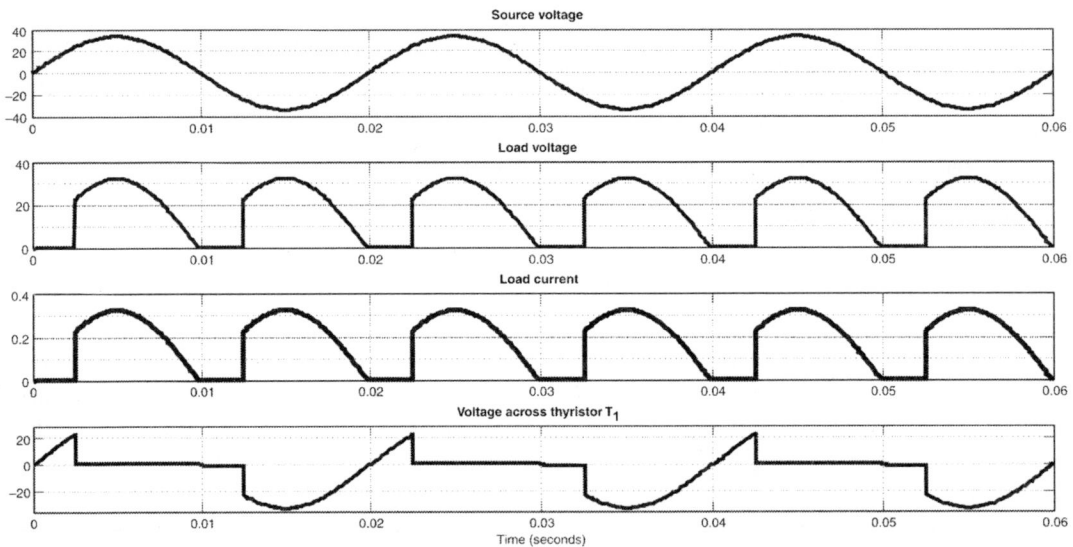

Single phase full-wave half controlled bridge converter with R load

The average value of the load voltage is

$$V_{av} = \frac{1}{\pi} \int_{\alpha}^{\pi} V_m \sin \omega t \, d\omega t = \frac{V_m}{\pi}(1 + \cos\alpha)$$

B. RL load (Discontinuous Current Mode)

1. During the positive half cycle of the supply voltage, thyristor T_1 is forward biased. When it is triggered at $\omega t = \alpha$, it conducts and the load current flows through T_1 and the diode D_1.

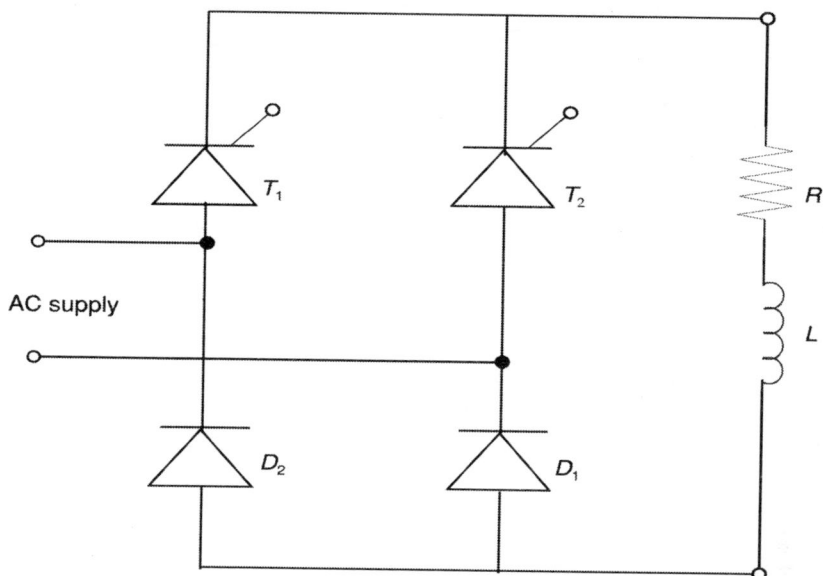

2. At $\omega t = \pi$ the load voltage is zero. But the current cannot be zero because the current through an inductor cannot become zero at the same instant at which the voltage is zero. When the inductor current starts decreasing, the inductor polarity reverses and it maintains the current in the load. The current will continue to flow beyond $\omega t = \pi$. It takes a little more time for the current to fall to zero value.

3. For $\omega t > \pi$, D_1 is reverse biased and D_2 is forward biased. The energy stored in the inductor drives the load current through T_1 and D_2. This is freewheeling action. If the inductance is low, the load current falls to zero value at an angle $\omega t = \beta$; $\pi < \beta < (\pi + \alpha)$. Thus from $\omega t = \alpha$ to $\omega t = \pi$, T_1 and D_1 conduct and from $\omega t = \pi$ to $\omega t = \beta$ T_1 and D_2 conduct.

4. For $\omega t > \pi$, T_2 is forward biased and it is triggered at $\omega t = \pi + \alpha$. T_2 is thus turned ON and T_1 is commutated because the current is transferred from T_1 to T_2. The load current now flows through T_2 and D_2 during the period from $\omega t = \pi + \alpha$ to $\omega t = 2\pi$.

5. For $\omega t > 2\pi$, the diode D_2 is blocked and D_1 conducts. The load current now flows through T_2 and D_1. This is freewheeling current that flows due to the energy stored in the inductor. Thus from $\omega t = \pi + \alpha$ to $\omega t = 2\pi$, T_2 and D_2 conduct and from $\omega t = 2\pi$ to $\omega t = \pi + \beta$ T_2 and D_1 conduct.

6. For $\omega t > 2\pi$, T_1 is forward biased and at $\omega t = 2\pi + \alpha$ it is triggered and brought into conduction. The current now flows through T_1 and D_1 and the cycle is thus repeated.

The current is discontinuous in this case and the mode of operation is called the discontinuous current mode. For discontinuous current mode, the condition is that the firing angle is large and is greater than the load impedance angle $\phi = \tan^{-1}(wL/R)$.

The average value of the load voltage is

$$V_{av} = \frac{1}{\pi} \int_{\alpha}^{\pi} V_m \sin \omega t \, d\omega t = \frac{V_m}{\pi}(1 + \cos \alpha)$$

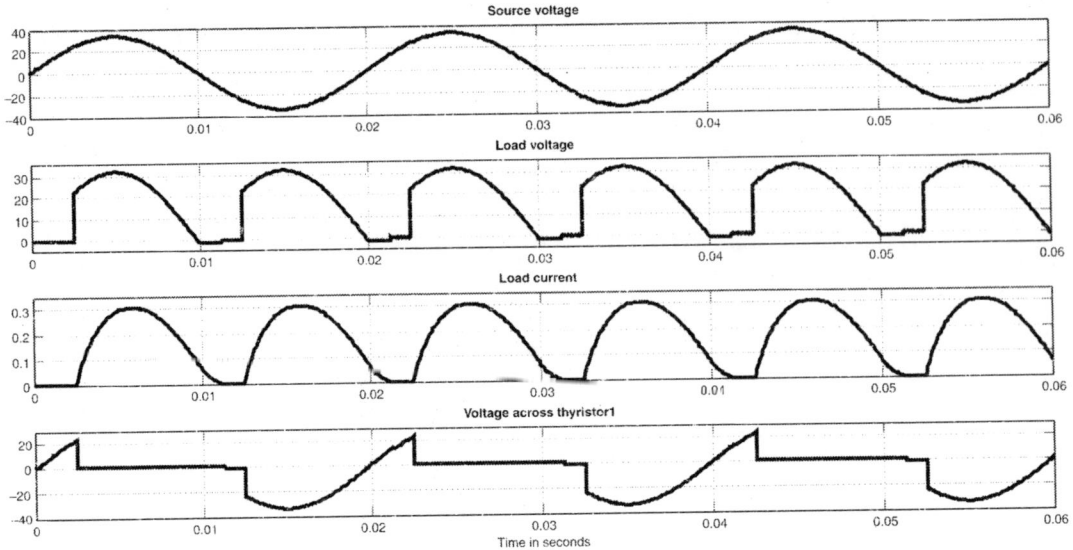

Single phase semi converter with RL load (Discontinuous Current Mode)

C. RL load (Continuous Current Mode)

The load current tends to be continuous for large inductance (or high L/R ratio) and low firing angle ($\alpha < \phi$).

1. During the positive half cycle of the supply voltage, thyristor T_1 is forward biased. When it is triggered at $\omega t = \alpha$, it conducts and the load current flows through T_1 and the diode D_1.

2. At $\omega t = \pi$, the load voltage is zero. But the current cannot be zero because the current through an inductor can not become zero at the same instant at which the voltage is zero. When the inductor current starts decreasing, the polarity of the inductor voltage reverses and it maintains the current in the load. The current will continue to flow beyond $\omega t = \pi$.

3. For $\omega t > \pi$, D_1 is reverse biased and D_2 is forward biased. The energy stored in the inductor drives the load current through T_1 and D_2. This is freewheeling action. Since the inductance is high, the energy stored in the inductor is large enough to drive the current through the load and to maintain the current till $\omega t = \pi + \alpha$. Thus from $\omega t = \pi$ to $\omega t = \pi$, T_1 and D_1 conduct and from $\omega t = \pi$ to $\omega t = \pi + \alpha$ T_1 and D_2 conduct. The current never falls to zero and it is continuous.

4. For $\omega t > \pi$, T_2 is forward biased and it is triggered at $\omega t = \pi + \alpha$. T_2 is thus turned ON and T_1 is commutated because the current is transferred from T_1 to T_2. The load current now flows through T_2 and D_2.

5. For $\omega t > 2\pi$, the diode D_2 is blocked and D_1 conducts. The load current now flows through T_2 and D_1. This is freewheeling current that flows due to the energy stored in the inductor. Thus from $\omega t = \pi + \alpha$ to $\omega t = 2\pi$ T_2 and D_2 conduct and from $\omega t = 2\pi$ to $\omega t = 2\pi + \alpha$ T_2 and D_1 conduct.

6. For $\omega t > 2\pi$, T_1 is forward biased and at $\omega t = 2\pi + \alpha$ it is triggered and brought into conduction. The current now flows through T_1 and D_1 and the cycle is thus repeated.

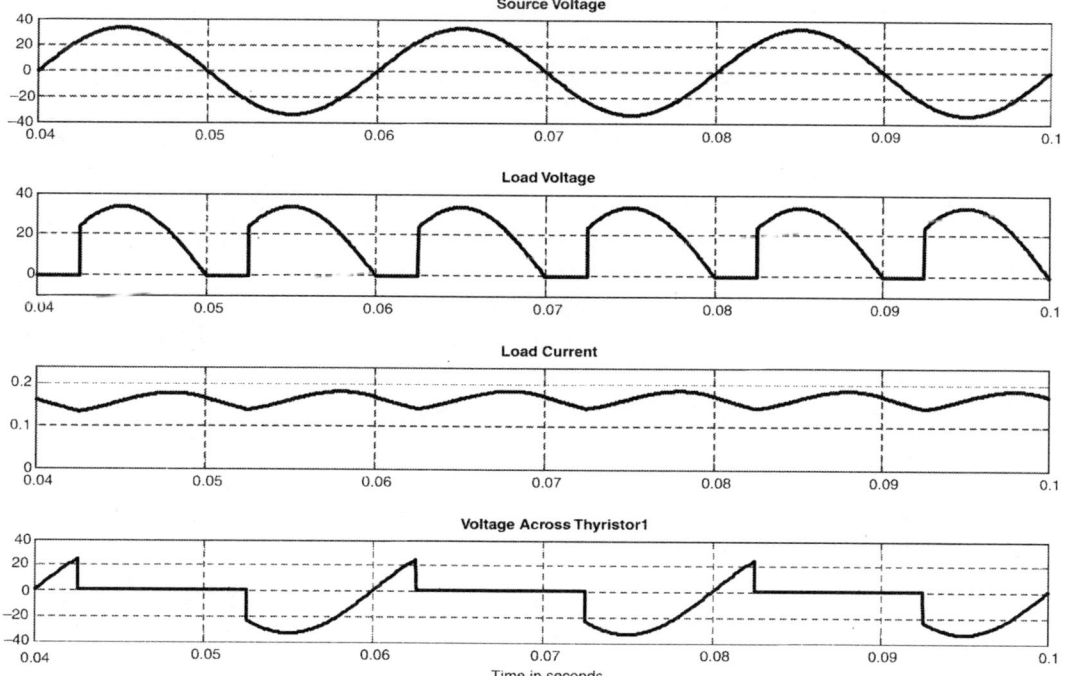

Waveform: Single phase semi converter with RL load (continuous current mode)

The average value of the load voltage is

$$V_{av} = \frac{1}{\pi} \int_{\alpha}^{\pi} V_m \sin \omega t \, d\omega t = \frac{V_m}{\pi}(1 + \cos \alpha)$$

D. RL Load with Freewheeling Diode (Discontinuous Current Mode)

1. During the positive half cycle of the supply voltage, thyristor T_1 is forward biased. When it is triggered at $\omega t = \alpha$, it conducts and the load current flows through T_1 and the diode D_1. The devices T_1 and D_1 conduct from $\omega t = \alpha$ to $\omega t = \pi$.

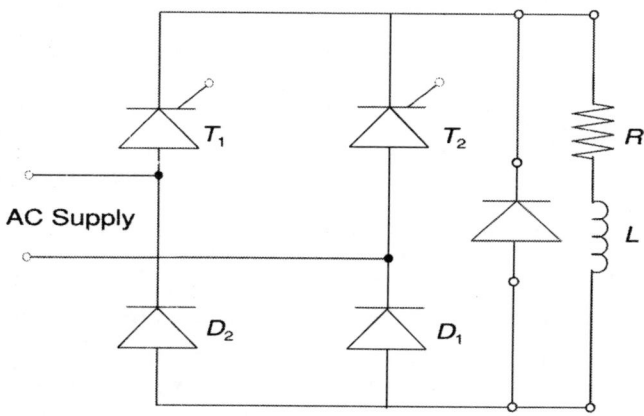

2. For $\omega t > \pi$, the freewheeling diode is forward biased and it conducts. Thyristor T_1 is commutated. The load voltage becomes zero, but the load current is not zero. The energy stored in the inductor drives the load current through the freewheeling diode. This is the freewheeling current and it does not flow through any thyristor.

3. If the inductance is low (or low L/R ratio), the freewheeling current falls to zero at the current extinction angle $\omega t = \beta$; $\pi < \beta + \alpha$. The load voltage is zero from $\omega t = \pi$ to $\omega t = \pi + \alpha$ and load current is zero from $\omega t = \beta$ to $\omega t = \pi + \alpha$.

4. For $\omega t > \pi$, T_2 is forward biased and is triggered at $\omega t = \pi + \alpha$. From $\omega t = \pi + \alpha$ to $\omega t = 2\pi$, T_2 and D_2 conduct and the load current flows.

5. From $\omega t = 2\pi$ to $\omega t = \pi + \beta$ freewheeling occurs. The freewheeling current falls to zero at $\omega t = \pi + \beta$. The load voltage is zero from $\omega t = 2\pi$ to $\omega t = 2\pi + \alpha$ and load current is zero from $\omega t = \pi + \beta$ to $\omega t = 2\pi + \alpha$.

6. At $\omega t = 2\pi + \alpha$, T_1 is triggered and the cycle is repeated.

The average value of the load voltage is

$$V_{av} = \frac{1}{\pi} \int_{\alpha}^{\pi} V_m \sin \omega t \, d\omega t = \frac{V_m}{\pi}(1 + \cos \alpha)$$

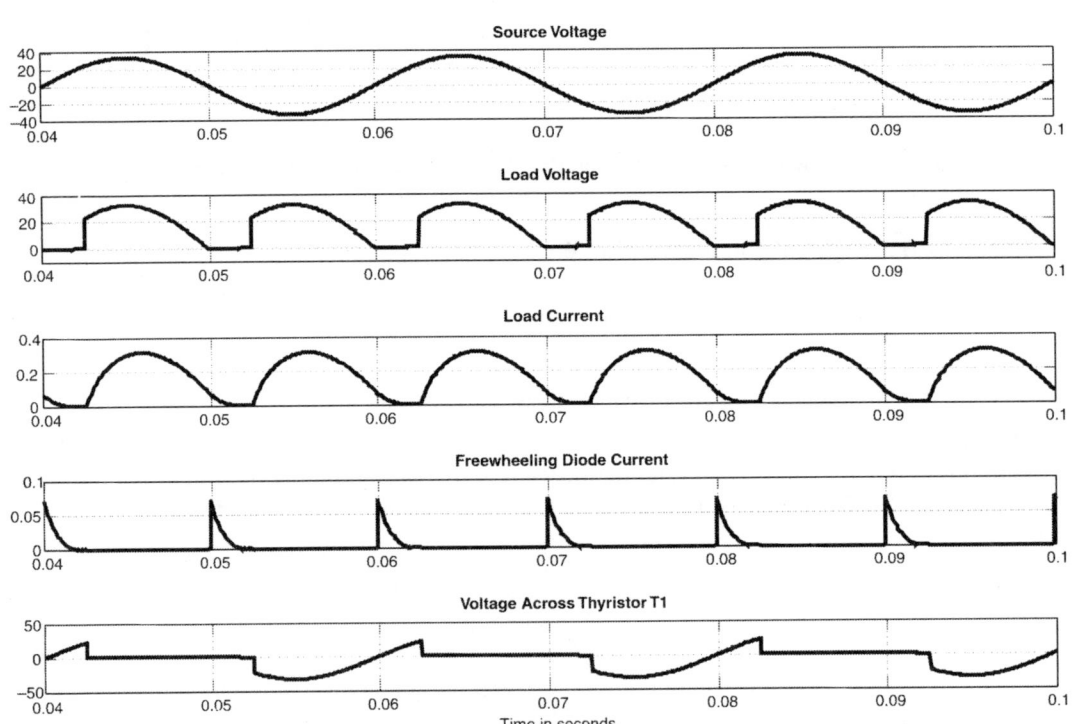

Waveform: Single phase semi converter with RL load and freewheeling diode
(Discontinuous Current Mode)

E. RL Load with Freewheeling Diode (Continuous Current Mode)

1. During the positive half cycle of the supply voltage, thyristor T_1 is forward biased. When it is triggered at $\omega t = \alpha$, it conducts and the load current flows through T_1 and the diode D_1. The devices T_1 and D_1 conduct from $\omega t = \alpha$ to $\omega t = \pi$.

2. For $\omega t > \pi$, the supply voltage is negative, the freewheeling diode is forward biased and it conducts. Thyristor T_1 is commutated. The load voltage becomes zero, but the load current is not zero. The energy stored in the inductor drives the load current through the freewheeling diode. This is the freewheeling current and it does not flow through any thyristor.

3. When the load inductance is large (or high L/R ratio), the freewheeling current continues to flow through the load until $\omega t = \pi + \alpha$ and does not become zero.

4. At $\omega t = \pi + \alpha$, T_2 is triggered and it is turned ON. T_2 and D_2 conduct from $\omega t = \pi + \alpha$ to $\omega t = 2\pi$

5. For $\omega t > 2\pi$ the freewheeling diode is forward biased and it conducts. Thyristor T_2 is commutated. The load voltage becomes zero. The freewheeling current flows from to $\omega t = 2\pi$ to $\omega t = 2\pi + \alpha$.

6. At $\omega t = 2\pi + \alpha$, T_1 is fired again and is brought into conduction and the cycle is repeated. It can be seen that the current never falls to zero value and the operation of the rectifier is said to be in the continuous mode.

7. By selecting a proper value of L, the current ripples can be reduced.

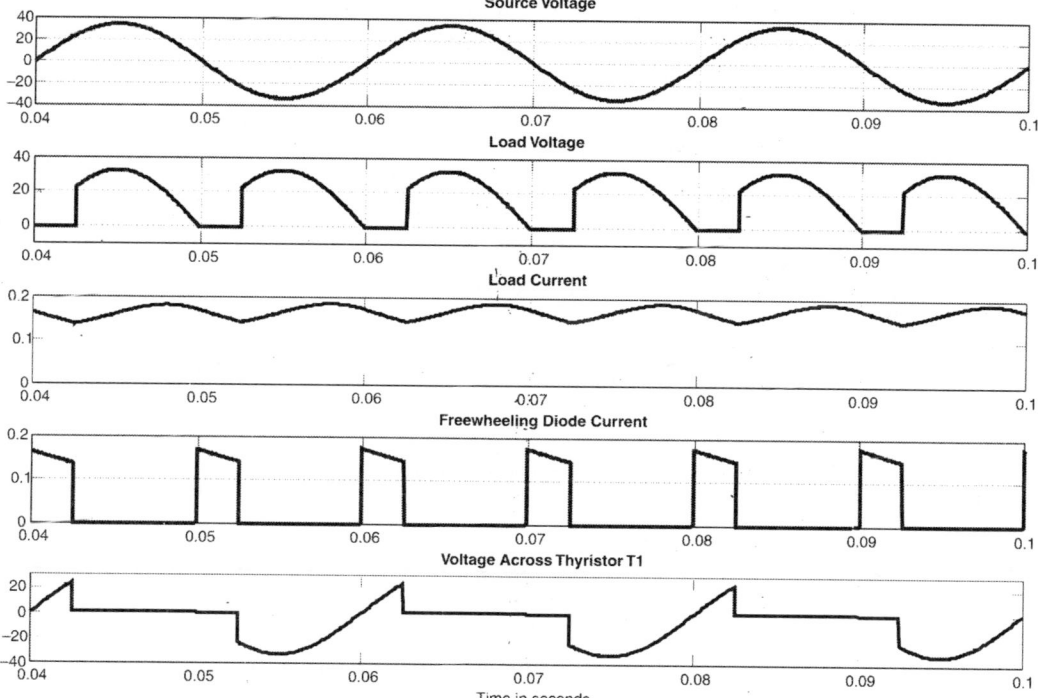

Waveform: Single phase semi converter with RL load and freewheeling diode
(Continuous Current Mode)

B7.3 HALF-WAVING

In a semi converter of symmetrical configuration, with no freewheeling diode across the load, a conducting thyristor carries a freewheeling current during the next half cycle of the supply voltage. If the L/R ratio of the load is high, this freewheeling current may not cease until the next positive half cycle and the same thyristor may continue to conduct. The other thyristor is never turned ON. For example, thyristor T_1

may conduct from $\omega t = \alpha$ to $\omega t = 2\pi + \alpha$ and the thyristor T_2 is not turned ON. The conducting thyristor will continue to conduct beyond $\omega t = 2\pi + \alpha$. This is shown in the figure. In this way a conducting thyristyor never gets commutated and it becomes uncontrollable. This phenomenon is called *half-waving*.

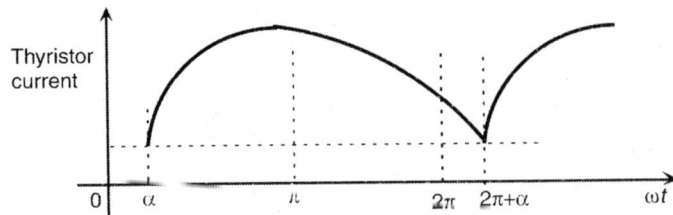

Waveform of thyristor current during half-waving

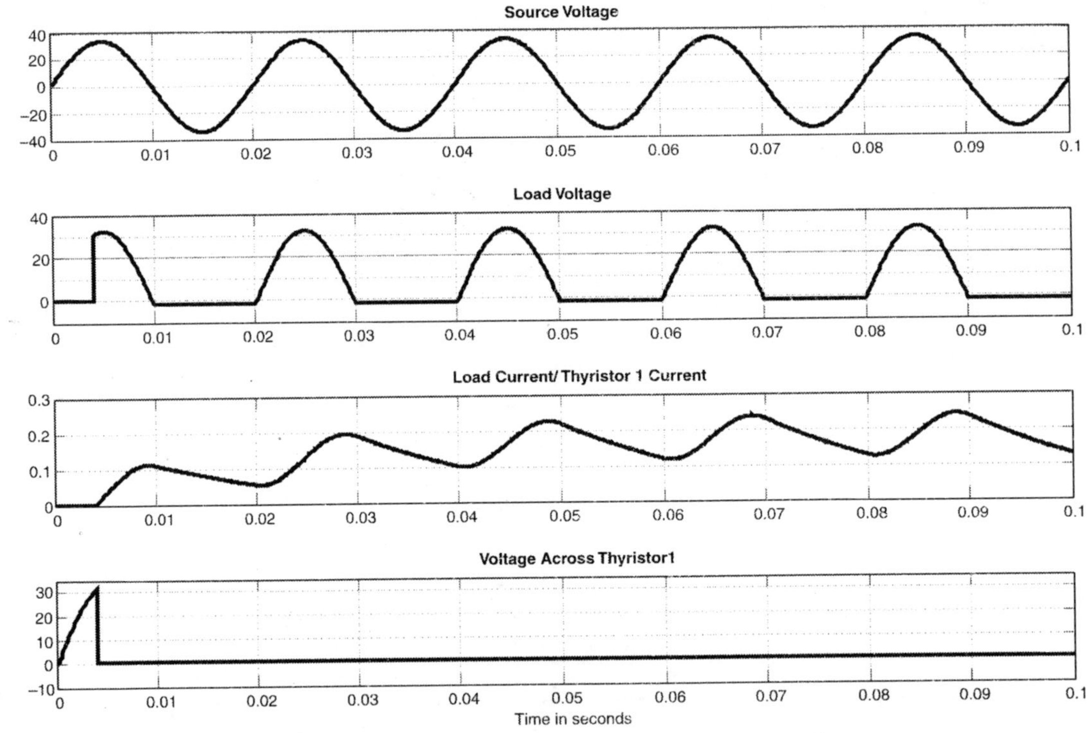

Waveforms: Single phase semi converter with RL load–half-waving phenomenon

Hence a controlled bridge rectifier with symmetrical connection suffers from the problem of *half-waving* when the load inductance is large. Thus this configuration is not suitable for use with high inductive loads. This problem can be avoided when a freewheeling diode is connected across the load. At the end of every half cycle of the supply voltage, the diode conducts. The current through the conducting thyristor is brought to zero and it is commutated. The actual wave forms during half-waving are shown in figure given above. It can be observed from the waveform of voltage across the thyristor that once it is turned ON it continues to conduct and never commutated. Wave forms of thyristor current and thyristor voltage are shown in figure.

B7.4 CONNECTION DIAGRAM

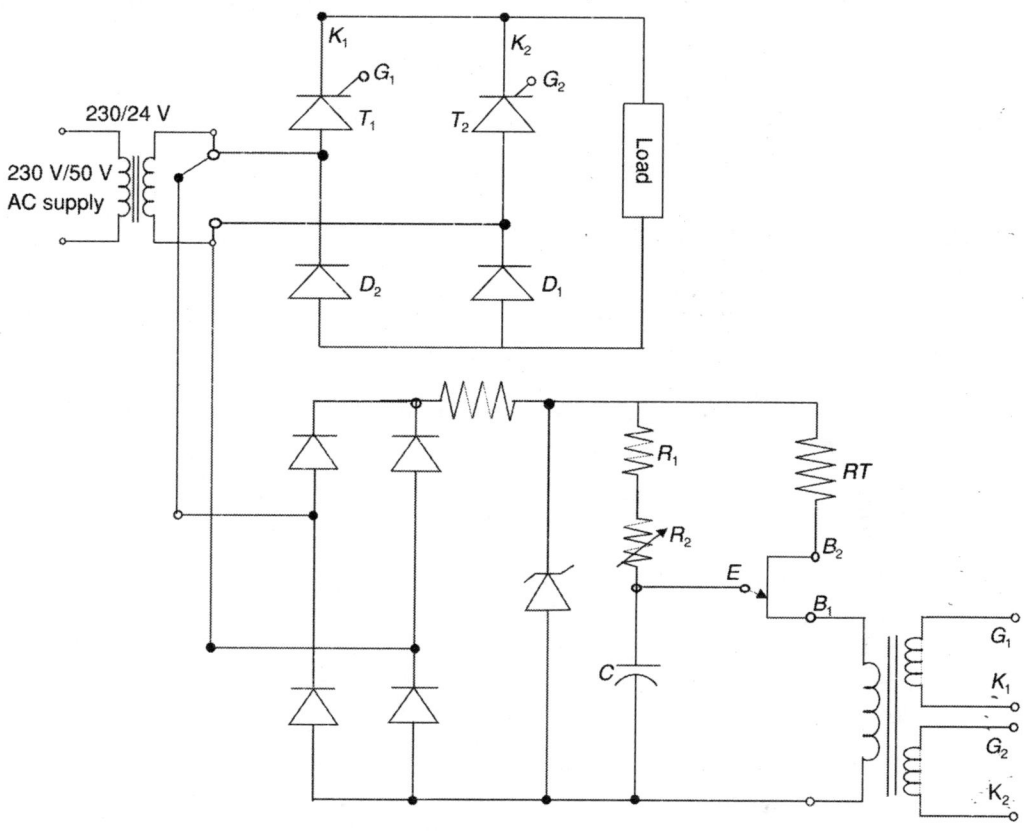

B7.5 EQUIPMENT AND COMPONENTS

Sl No	Component	Specification	Quantity
1.	Single phase transformer	230/24 V, 50 Hz	1
2.	Thyristor	C106D	2
3.	Resistors (Load)	100 Ω, 20 W	1
4.	Inductor (Variable)		1
5.	Diode	1N4007	2
6.	Resistor	1 Ω, 10 W	1
7.	DSO/CRO		
8.	Breadboard		
9.	Thyristor triggering circuit using UJT		

B7.6 PROCEDURE

1. Make the connections as per the circuit diagram.
2. Switch ON the triggering circuit and check for the triggering pulses. (Triggering circuit is the UJT triggering circuit used for a full-wave rectifier. For the triggering circuit design, refer to section B4)
3. Adjust the firing angle to a convenient value.

4. Observe the waveforms of source voltage, load voltage, load current and the thyristor voltage for different cases of loads, namely R load, RL load with and without freewheeling diode.
5. Switch off the supply.

B7.7 OBSERVATION

Plot the observed waveforms on a graph sheet.

B7.8 ASYMMETRICAL CONNECTION

It was shown that there can be a problem of half-waving in the case of a semi converter with symmetrical connection. This happens when a conducting thyristor is allowed to carry a freewheeling current. This is avoided in the asymmetrical connection of the bridge converter because the diodes in the other arm of the bridge carry the freewheeling current and the thyristors are bypassed. Thus a conducting thyristor is freed from carrying the freewheeling current.

The working of the asymmetrical connection of the bridge is similar to that of symmetrical connection with a freewheeling diode connected across the load discussed earlier.

Consider a continuous current conduction mode. During the positive half cycle T_1 is triggered at and it is turned ON. The load current flows through T_1 and D_1. At $\omega t = \pi$ the two diodes starts conducting and freewheeling occurs. T_1 is commutated. The diodes conduct from $\omega t = \pi$ to $\omega t = \pi + \alpha$. At T_2 is turned ON. The load current flows through T_2 and D_2. At freewheeling starts and T_2 is commutated. Freewheeling takes place from $\omega t = 2\pi$ to $2\pi + \alpha$.

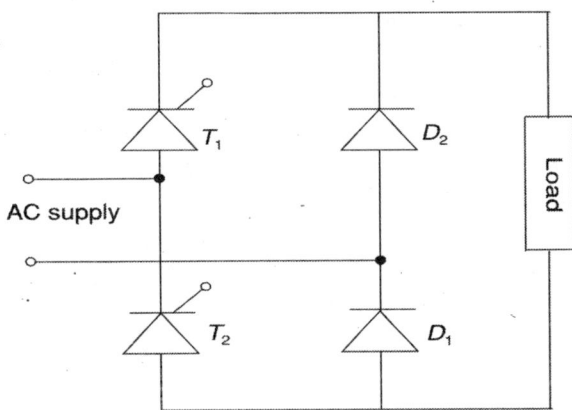

B7.9 PROCEDURE

The procedure for conducting the experiment is similar to that given in the case of symmetrical connection.

B7.10 OBSERVATION

Observe the waveforms of source voltage, load voltage, load current and the voltage across thyristor T_1. The waveforms are identical to those of a symmetrical bridge with a freewheeling diode across the load.

Experiment B8

Design and Testing of a Single Phase Full-wave Fully Controlled Bridge Rectifier

B8.1 OBJECTIVE

To study the operation and performance of a single phase full-wave fully controlled bridge rectifier.

B8.2 THEORY

A single phase fully controlled bridge rectifier is shown in figure given below. It consists of four thyristors. All the four devices are controllable and hence it is called a fully controlled rectifier. During the positive half cycle of the AC voltage, Thyristors T_1 and T_2 are forward biased and they are fired and turned ON. During the negative half cycle T_3 and T4 are turned ON and T_1 and T_2 are commutated.

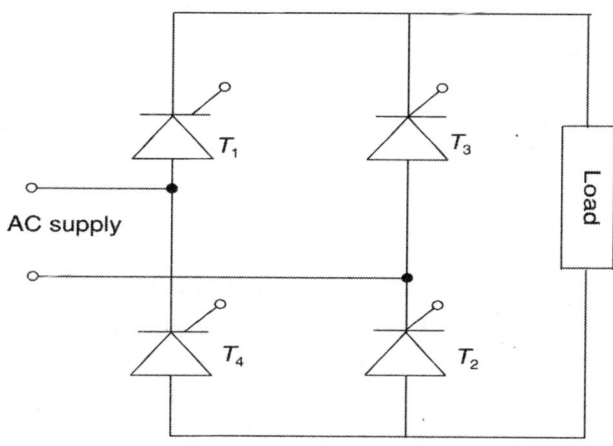

Single phase full converter

A. R Load

1. During the positive half cycle of the supply voltage, thyristor T_1 and T_2 are forward biased. They are triggered at $\omega t = \alpha$ and they conduct and the load current flows through T_1 and T_2.
2. At $\omega t = \pi$ the load voltage is zero and the load current is zero. T_1 and T_2 are commutated.

3. For $\omega t > \pi$, T_3 and T_4 are forward biased and they are triggered at $\omega t = \pi + \alpha$. T_3 and T_4 are turned ON. The load current now flows through T_3 and T_4.
4. At $\omega t = 2\pi$, the load voltage is zero and the load current is zero. T_3 and T_4 are commutated.
5. At $\omega t = 2\pi + \alpha$, T_1 and T_2 are triggered again and the cycle is thus repeated.

The average value of the load voltage is

$$Vav = \frac{1}{\pi} \int_\alpha^\pi V_m \sin \omega t \, d\omega t = \frac{V_m}{\pi}(1 + \cos \alpha)$$

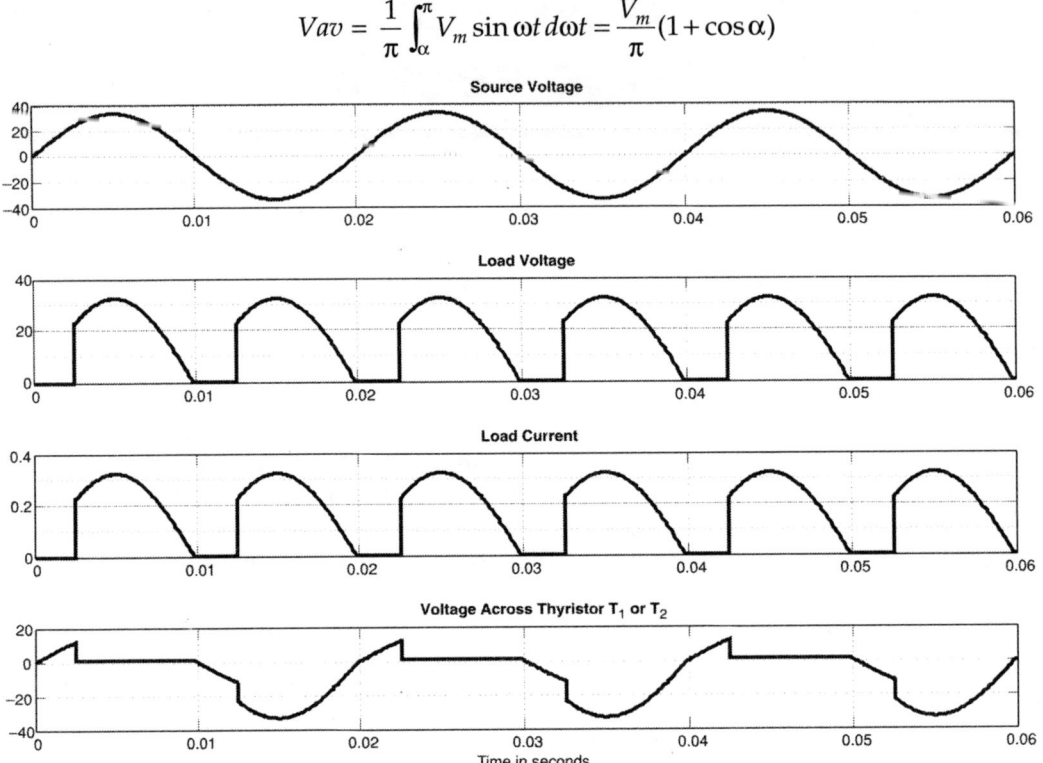

Waveforms: Single phase full-wave fully controlled bridge converter with R load

B. RL Load—Discontinuous Current–Low L/R Ratio $\alpha > \phi$

1. Thyristors T_1 and T_2 are triggered at instant $\omega t = \alpha$, when the supply voltage is positive. The load current flows through these thyristors.
2. At $\omega t = \pi$, the load voltage is zero but the load current is not zero because of the inductance effect. If the load inductance is low, the current will drop down to zero value, a little later at an instant $\omega t = \beta$; $\pi < \beta \leq \pi + \alpha$. From $\omega t = \pi$ to $\omega t = \beta$ the load voltage is negative. β is the current extinction angle.
3. From $\omega t = \beta$ to $\omega t = \pi + \alpha$, the load current is zero and the load voltage is zero. Since the load current is zero, thyristors T_1 and T_2 are commutated.
4. At $\omega t = \pi + \alpha$, T_3 and T_4 are triggered and they are turned ON. Load current flows through T_3 and T_4.
5. At $\omega t = 2\pi$, the supply voltage is zero but the load current doesn't become zero because of inductance effect. It takes a little more time for the load current to fall to zero

value. The current extinction angle is $\omega t = \beta - \pi + 2\pi = \pi + \beta$; $2\pi < \pi + \beta \leq 2\pi + \alpha$. At $\omega t = \pi + \beta$, thyristors T_3 and T_4 are commutated.

6. From $\omega t = \pi + \beta$ to $\omega t = 2\pi + \alpha$ the load current remains zero and the load voltage is zero.

7. At $\omega t = 2\pi + \alpha$, T_1 and T_2 are triggered again and they are turned ON and the cycle is repeated.

The average value of the load voltage is

$$V_{av} = \frac{1}{\pi} \int_{\alpha}^{\pi} V_m \sin \omega t \, d\omega t = \frac{V_m}{\pi}(1 + \cos \alpha)$$

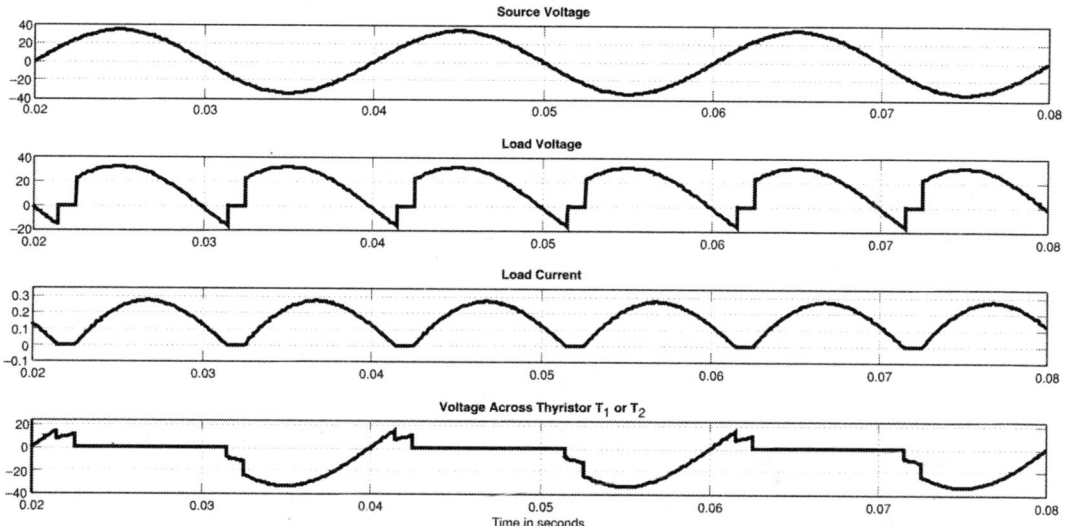

Waveforms: Single phase full-wave fully controlled bridge rectifier with RL load–discontinuous conduction mode

C. RL Load—Continuous Current–High L/R Ratio $\alpha < \phi$

1. Thyristors T_1 and T_2 are triggered at instant $\omega t = \alpha$ when the supply voltage is positive. The load current flows through these thyristors.

2. At $\omega t = \pi$, the load voltage is zero but the load current is not zero because of load inductance. The current decreases but never falls to zero value before $\omega t = \pi + \alpha$.

3. At $\omega t = \pi + \alpha$, T_3 and T_4 are triggered and they are turned ON. The Load current now flows through T_3 and T_4. A reverse voltage appears across each of the thryistors T_1 and T_2 and they are commutated.

4. T_3 and T_4 carry the load current until and the load current never becomes zero.

5. At $\omega t = 2\pi + \alpha$, T_1 and T_2 are triggered and the cycle is repeated.

The average value of the load voltage is

$$V_{av} = \frac{1}{\pi} \int_{\alpha}^{\pi+\alpha} V_m \sin \omega t \, d\omega t = \frac{2V_m}{\pi} \cos \alpha$$

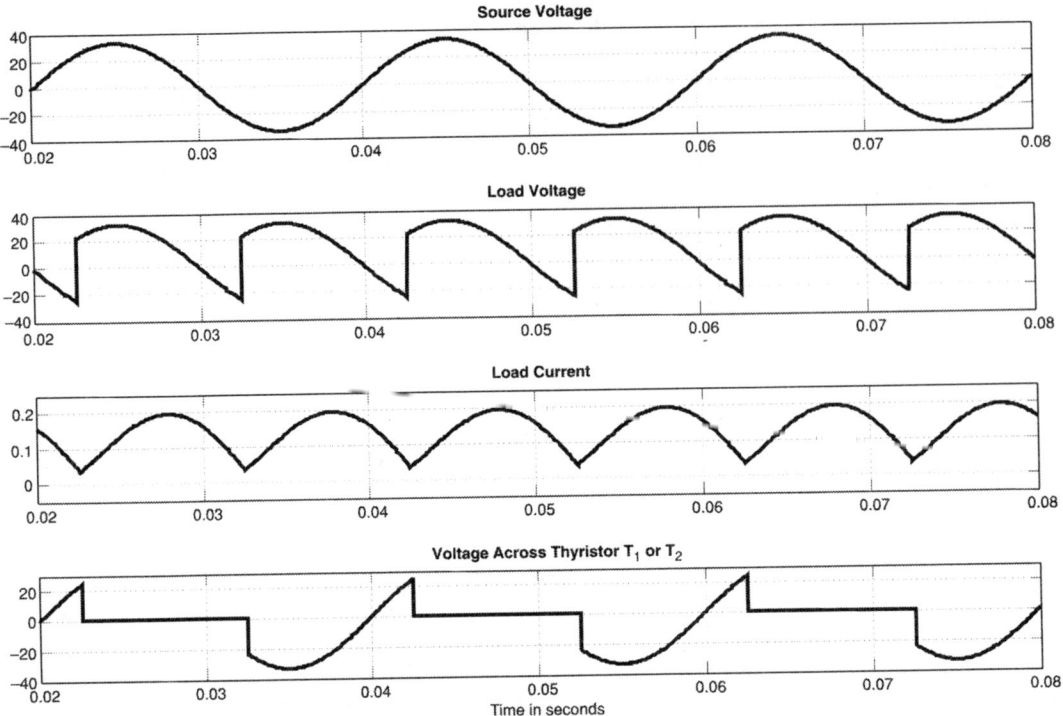

Waveforms: Single phase full-wave fully controlled bridge rectifier with RL load (CCM)

D. RL Load with FW Diode

1. Thyristors T_1 and T_2 are triggered at instant $\omega t = \alpha$ when the supply voltage is positive. The load current flows through these thyristors.

2. At $\omega t = \pi$, the load voltage becomes zero. For $\omega t \geq \pi$, the freewheeling diode conducts and the load is short circuited. The current through the thyristors T_1 and T_2 are zero and they are commutated. The energy stored in the inductor drives the current through the load via the freewheeling diode. This is freewheeling action. Hence from $\omega t = \alpha$ to $\omega t = \pi$ thyristors T_1 and T_2 conduct.

   If the inductance is low (low L/R ratio), the freewheeling current can drop to zero value at an instant $\omega t = \beta$; $\pi < \beta \leq \pi + \alpha$. The diode conducts from $\omega t = \pi$ to $\omega t = \beta$ and during this period the load voltage is zero. This is discontinuous current mode and the waveforms are shown in the figure.

3. From $\omega t = \beta$ to $\omega t = \pi + \alpha$ both the load voltage and load current are zeros.

4. At $\omega t = \pi + \alpha$, T_3 and T_4 are triggered and turned ON. This will commutate the thyristors T_1 and T_2. From $\omega t = \pi + \alpha$ to $\omega t = 2\pi$, T_3 and T_4 conduct and they carry the load current. The diode conducts from $\omega t = 2\pi$ to to $\omega t = 2\pi + \alpha$.

5. At $\omega t = 2\pi + \alpha$, T_1 and T_2 are triggered again and the cycle is repeated.

   If the load inductance value is large (or the load is with high L /R ratio), the current may not become zero at any instant and the freewheeling occurs from to, which makes the current continuous. The waveforms are shown in the figure.

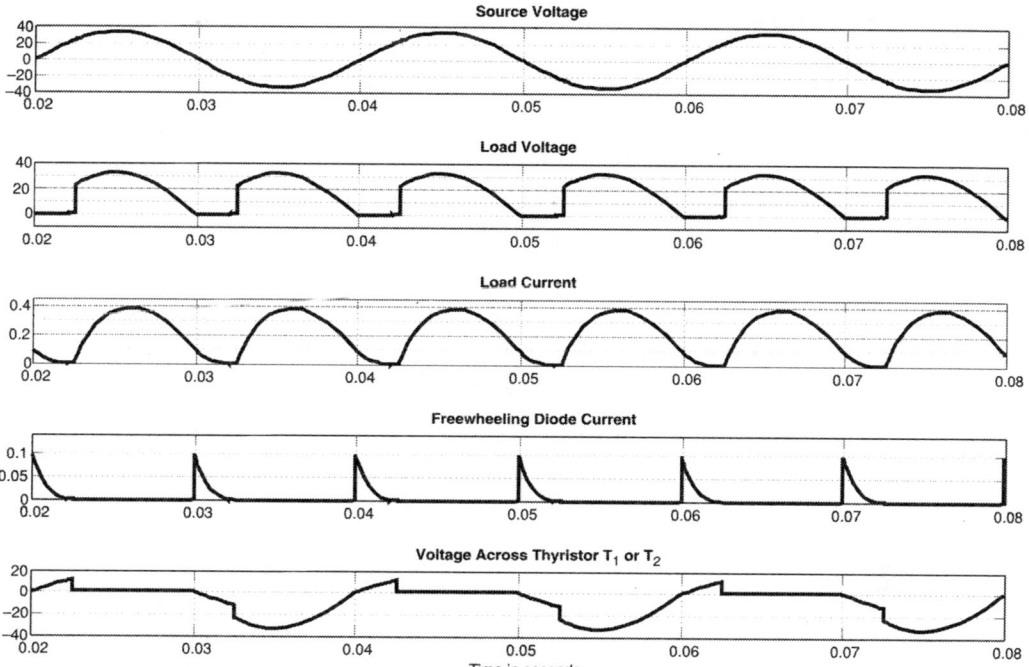

Waveforms: Single phase full-wave fully controlled rectifier with RL load and FWD
(Discontinuous Current Mode)

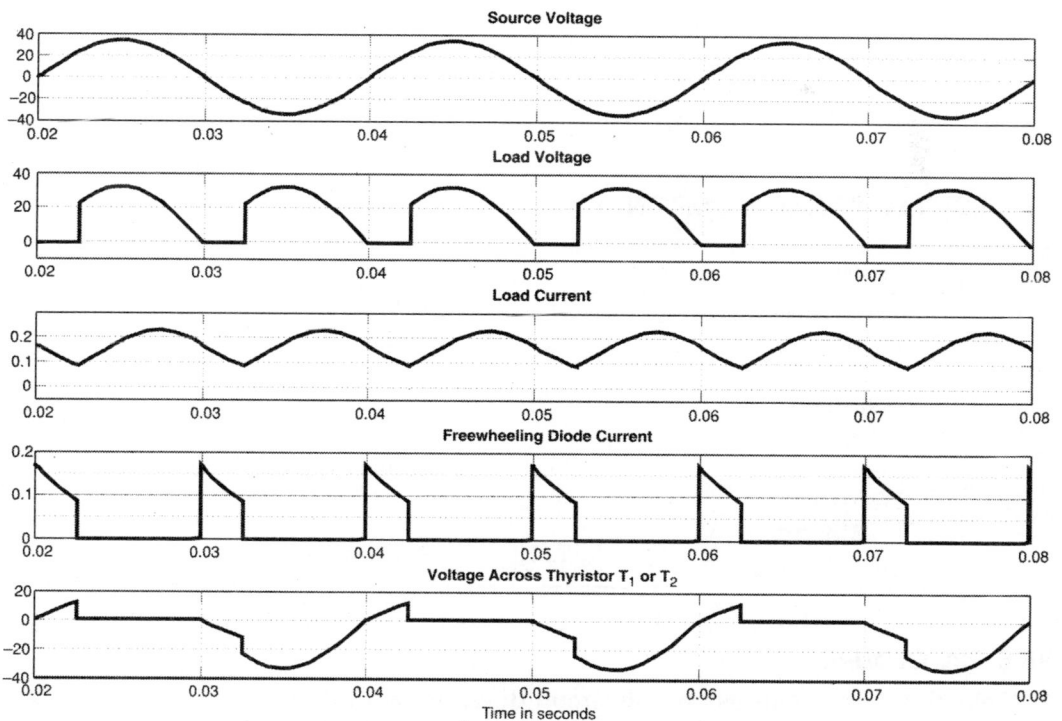

Waveforms: Single phase full-wave fully controlled rectifier with RL load
and FWD (Continuous Current Mode)

B8.3 CIRCUIT DIAGRAM

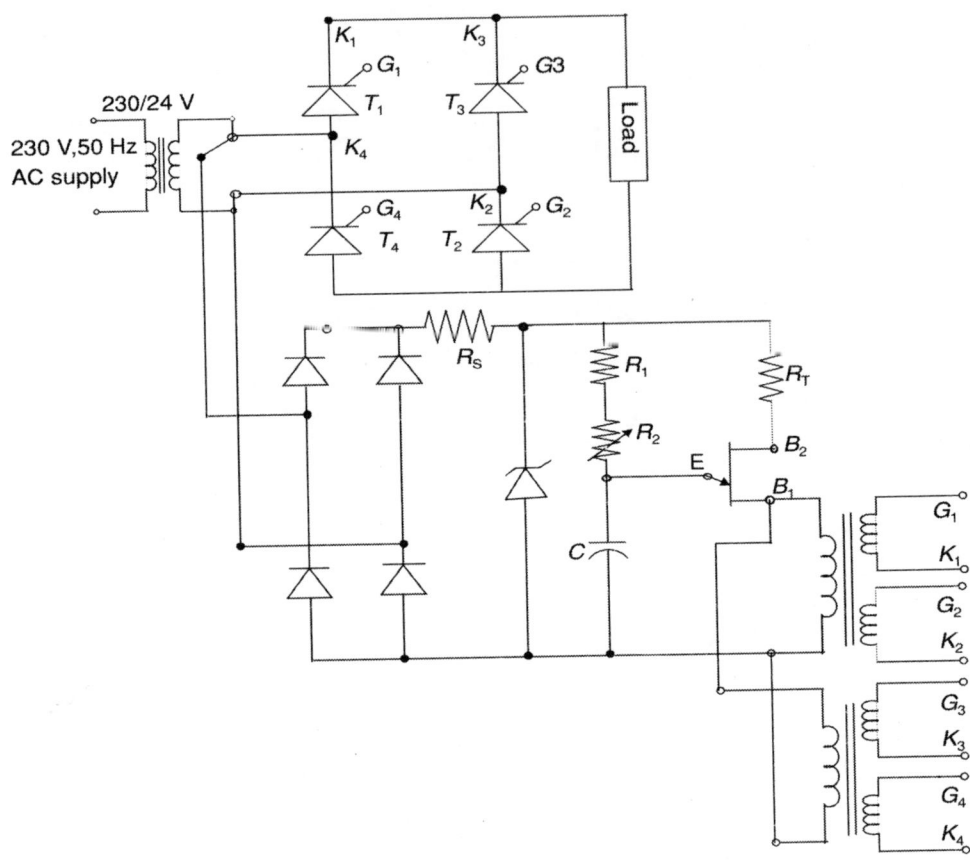

B8.4 EQUIPMENT AND COMPONENTS

Sl No	Component	Specification	Quantity
1.	Transformer	230 V/24 V	1
2.	Thyristor	C106D	4
3.	Load Resistor	100 Ω, 20 W	1
4.	Inductor		1
5.	Diode	1N4007	2
6.	Resistor	1 Ω, 10 W	1
7.	DSO/CRO		
8.	Breadboard		
9.	Thyristor triggering circuit using UJT		

B8.5 PROCEDURE

1. Make the connections as per the circuit diagram, with R load.
2. Switch on the supply to the triggering circuit and check for the triggering pulses. (Triggering circuit is the UJT triggering circuit used for a full-wave rectifier. For the triggering circuit design, refer to section B4)

3. Adjust the firing angle to a convenient value.
4. Observe the waveforms of source voltage, load voltage, load current and the thyristor voltage
5. Switch off the supply.
6. Repeat steps 2 to 5 for all other types of loads, namely, without freewheeling diode across *RL* load and with freewheeling diode across *RL* load.

Note:

1. For making the load current continuous, one or both of the following methods can be adopted:
   (a) Reduce the firing angle of the thyristors.
   (b) Increase the L/R ratio of the load impedance by either increasing L or reducing R.

2. For triggering the four thyristors, the primaries of two pulse transformers are connected in parallel. The secondary terminals (G_1, K_1) and (G_2, K_2) are used to trigger thyristors T_1 and T_2 respectively and the secondary terminals (G_3, K_3) and (G_4, K_4) are used to trigger T_3 and T_4 respectively. Even though gate pulses are available for all the thyristors, only those thyristors which are forward biased are triggered into conduction.

B8.6 OBSERVATION

Plot the observed waveforms on a graph sheet.

Experiment B9

Speed Control of a DC Motor Using a Controlled Rectifier

B9.1 OBJECTIVE

To study the speed control of a DC motor, using a single phase full-wave semiconverter.

B9.2 THEORY

The back emf of a DC motor is given by the expression,

$$E_b = P\phi\frac{Z}{A}n = k\phi n,$$

where n is the speed in *rps*.

and

$$n = \frac{E_b}{k\phi} = \frac{V - I_a R_a}{k\phi}$$

Hence the speed of a DC motor can be controlled in two ways:
1. By varying the armature voltage, keeping the field flux constant (armature voltage control) and
2. By varying the field flux keeping the armature voltage constant (field current control).

In armature voltage control, the field is separately excited and the field current is kept constant at a permissible value. When the field current is kept constant, the speed of the motor is directly proportional to the back emf. The armature is connected across the output terminals of a single phase full-wave half controlled bridge rectifier. The output voltage of the rectifier can be varied by adjusting the firing angle of the thyristor and hence the voltage applied across the armature of the DC motor.

In the experiment to be performed, the field of the DC motor can be a permanent magnet and hence field flux can be assumed to be constant. A semiconverter circuit feeding a DC motor is shown in figure.

The secondary voltage of the transformer is assumed to be, $v = V_m \sin \omega t$. At $\omega t = \alpha$, T_1 is triggered. T_1 and D_1 conduct from $\omega t = \alpha$ to $\omega t = \pi$ and the load voltage is the supply voltage. From $\omega t = \pi$, freewheeling takes place and the devices T_1 and D_2 conduct up to $\omega t = \pi + \alpha$. During the negative half cycle of the supply voltage T_2 becomes forward biased and it is triggered at $\omega t = \pi + \alpha$. Now T_2 and D_2 conduct and the load voltage is the supply voltage. The current through the load is unidirectional and assumed to be continuous.

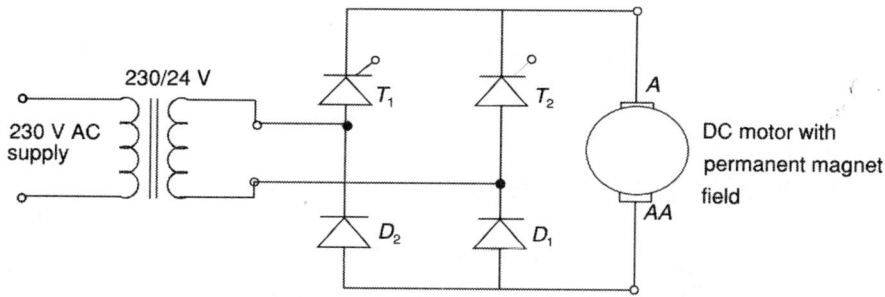

Single phase rectifier feeding a DC motor

At $\omega t = 2\pi$ freewheeling starts and continues up to $\omega t = 2\pi + \alpha$. During freewheeling T_2 and D_1 conduct. At $\omega t = \pi + \alpha$, T_1 is once again fired and the cycle is repeated.

The back emf of the motor can be taken to be the armature voltage, because the armature current is very small. A semiconverter with asymmetrical connection can also be used for the experiment.

B9.3 CIRCUIT DIAGRAM

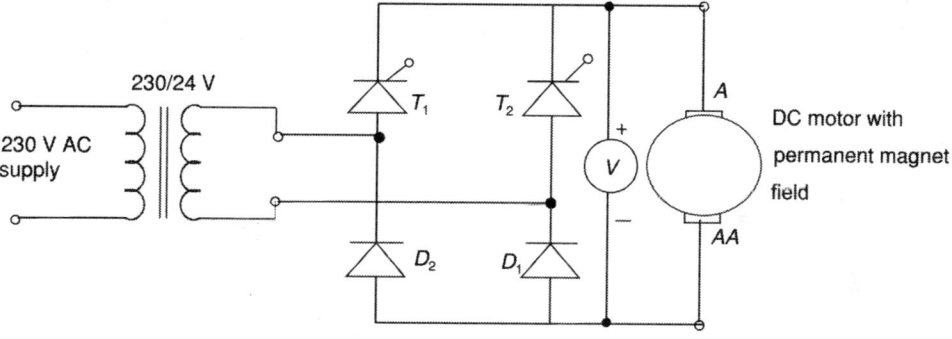

B9.4 DEVICE SPECIFICATION

Thyristors– C106D

Diodes– 1N4007

Dc Motor– 24 V, 1500 rpm   Field – Permanent Magnet   1 A   1.4 kg cm

B9.5 EQUIPMENT AND COMPONENTS

Sl No	Component	Specification	Quantity
1.	Transformer	230 V/24 V	1
2.	Thyristor	C106D	4
3.	Load Resistor	100 Ω, 20 W	1
4.	Inductor		1
5.	Diode	1N4007	2
6.	Resistor	1 Ω, 10 W	1
7.	DSO/CRO		
8.	Breadboard		
9.	Thyristor triggering circuit using UJT		

B9.6 PROCEDURE

1. Make the connections as per the circuit diagram.
2. Adjust the pot of the triggering circuit so as to set the firing angle at the maximum value. This will ensure minimum voltage at the rectifier output and the speed of the motor will be a minimum at starting.
3. Switch ON the AC supply to the rectifier.
4. Switch ON the triggering circuit and check for the pulses using the DSO. (Triggering circuit is the UJT triggering circuit used for a full-wave rectifier, as shown in B7.4. For the triggering circuit design, refer to section B4)
5. Measure the armature voltage and the speed.
6. Reduce the firing angle to a lower value so that the rectifier output voltage is more and hence the speed of the motor.
7. Repeat steps 5 and 6 and make many observations.
8. Switch off the supply.

B9.7 OBSERVATION

Sl.No	1	2	3	4	5	6	7	8
Speed, Rpm								
Armature voltage,V								

B9.8 GRAPH

Plot a graph of speed versus armature voltage.

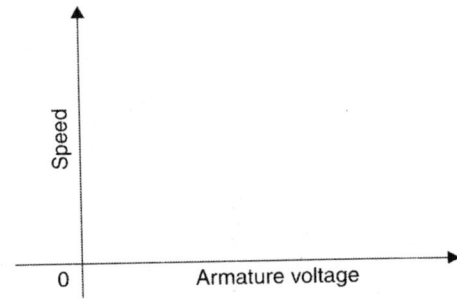

Experiment B10

Design and Testing of an AC Voltage Regulator Using Thyristors

B10.1 OBJECTIVE

To study the performance and operation of a single phase ac voltage regulator using two thyristors connected in anti-parallel.

B10.2 THEORY

A basic phase controlled ac voltage regulator is shown in figure. The thyristors are connected in anti parallel. During positive half cycle T_1 is forward biased and it is triggered and turned ON. During negative half cycle T_2 becomes forward biased and it is gated and turned ON.

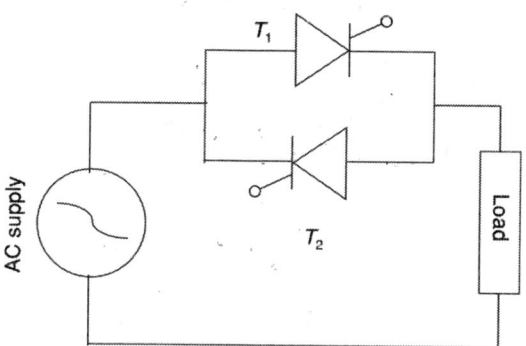

Woking—Voltage Regulator Feeding R Load

1. When T_1 is forward biased during the positive half cycle of the supply voltage it is triggered at $\omega t = \alpha$ and brought into conduction and the load voltage is $v_m \sin \omega t$. T_1 conducts from $\omega t = \alpha$ to $\omega t = \pi$.

2. At $\omega t = \pi$, both the load voltage and load current falls to zero value. T_1 is turned off. The load voltage and load current remains zero from $\omega t = \pi$ to $\omega t = \pi + \alpha$.

3. For $\omega t > \pi$, T_2 is forward biased and it is triggered at $\omega t = \pi + \alpha$. It conducts from $\omega t = \pi + \alpha$ to $\omega t = 2\pi$.

4. At $\omega t = 2\pi$, both the load current and load voltage become zero and T_2 is turned off. The load voltage and the load current are zeros from to

5. At $\omega t = 2\pi$, T_1 is turned ON and the cycle is repeated.

The waveforms are shown in the figure.

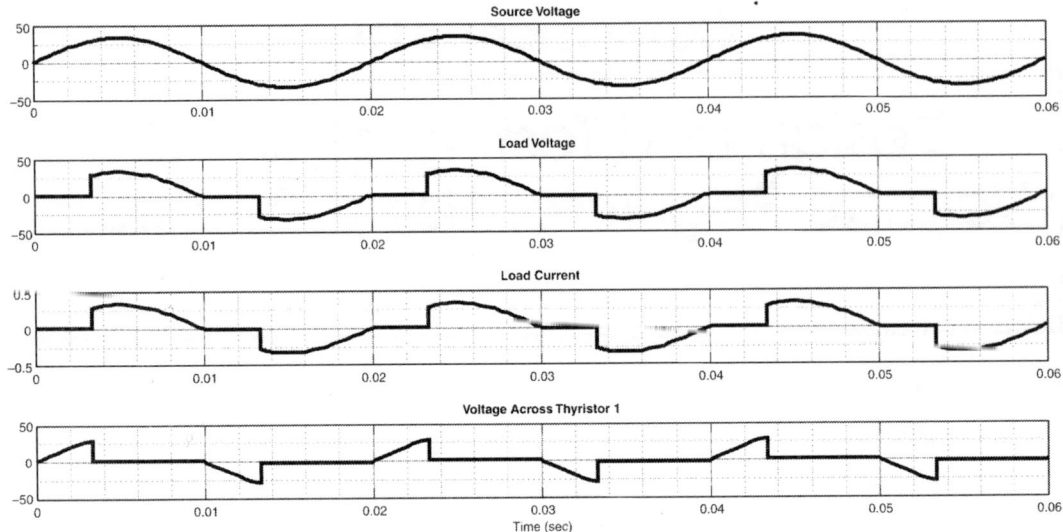

Waveform: Single phase AC voltage controller With R load

Working–Voltage Regulator Feeding RL Load

1. At $\omega t = \alpha$, T_1 is triggered and it starts conducting. At $\omega t = \pi$ the voltage is zero but the current cannot be zero because of the effect of load inductance.

2. At $\omega t = \beta$ the load current becomes zero, if the L/R ratio of the load is low. T_1 conducts from $\omega t = \alpha$ to $\omega t = \beta$. From $\omega t = \beta$ to $\omega t = \pi + \alpha$ both the load voltage and load current are zero. In this case the load current is discontinuous.

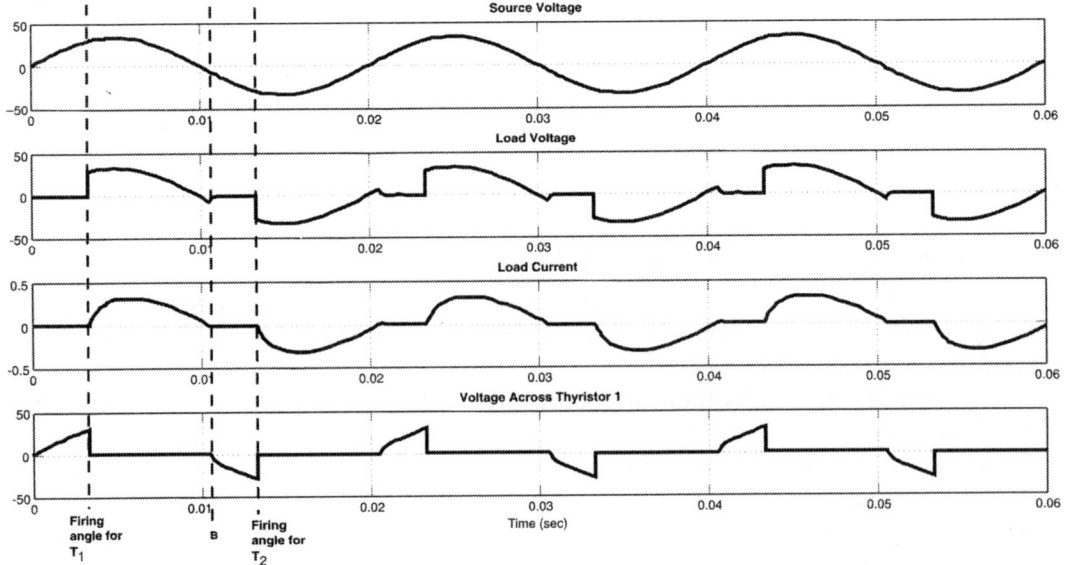

Waveform: Single phase AC voltage controller with RL load

3. At $\omega t = \pi + \alpha$, T_2 is turned ON because it is forward biased. The load voltage is negative and the current flows in the reverse direction.

4. At $\omega t = \pi + \beta$, the load current falls to zero. T_2 conducts from $\omega t = \pi + \alpha$ to $\omega t = \pi + \beta$. From $\omega t = \pi + \beta$ to $\omega t = 2\pi + \alpha$ both the load voltage and load current are zero.

5. At $\omega t = 2\pi + \alpha$, T_1 becomes forward biased and it is triggered and brought to conduction. The cycle is thus repeated.

Note:

1. When T_1 conducts, the voltage drop across it appears as a reverse bias for T_2. Similarly when T_2 conducts, the voltage drop across it appears as reverse bias for T_1.
2. Voltage waveform of T_2 is the inverted form of voltage waveform of T_1.
3. The firing circuits for the two thyristors must be isolated from each other.

B10.3 CIRCUIT DIAGRAM

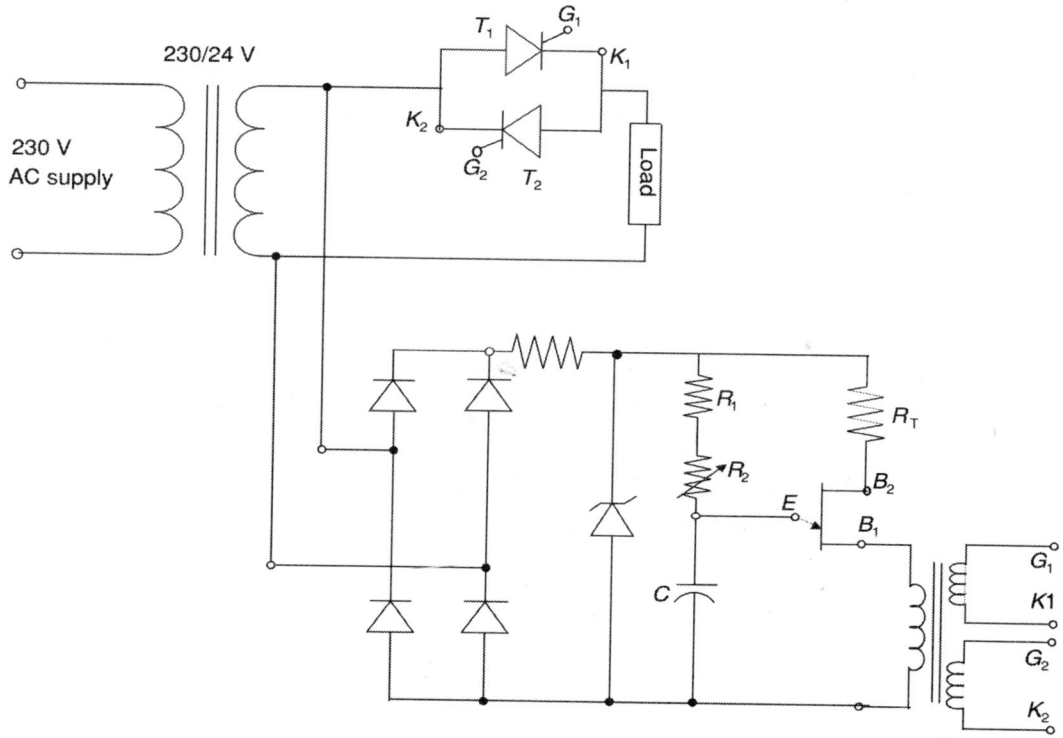

B10.4 EQUIPMENT AND COMPONENTS REQUIRED

Sl No	Component	Specification	Quantity
1.	Thyristors	C106D	2
2.	Load resistor	100 Ω, 20 W	1
3.	Load inductors		
4.	DSO/CRO		
5.	Breadboard		
6.	Thyristor triggering circuit using UJT		

B10.5 PROCEDURE

1. Make the connections as shown in the circuit diagram. The load is pure resistor.
2. Connect a 1 Ohm resistor in series with the load for observing the waveform of load current.
3. Switch ON the supply to the triggering circuit and check for the triggering pulses. (Triggering circuit is the UJT triggering circuit used for a full-wave rectifier. For the triggering circuit design, refer to section B4)
4. Adjust the pot of the triggering circuit for maximum firing angle.
5. Adjust the pot of the triggering circuit to keep the firing angle of the thyristor at a convenient value.
6. Observe the waveforms of the source voltage, load voltage, load current and thyristor voltage on the DSO, for different cases of loads.
7. Connect an RL load, with a low inductance and repeat steps 3 to 6.
8. After making observations, switch off the supply.

B10.6 OBSERVATION

Observe the waveforms of source voltage, load voltage, load current and thyristor voltage for the following cases of load and plot them on a graph sheet.

1. R load.
2. RL load with a very low inductance value.

Experiment B11

Design and Testing of an AC Voltage Regulator Using Triac

B11.1 OBJECTIVE

To study the performance and operation of a single phase ac voltage regulator using a triac.

B11.2 THEORY

A triac is a bidirectional switching semiconductor device which can be turned ON by giving a gate pulse. It can conduct both in the positive and negative half cycle of the supply voltage. The configuration is suitable for low power applications where the load is predominantly resistive and has only a small inductance. The firing circuit of the triac need not be isolated from the power circuit.

A triac can be commutated by reducing the current it carries to a level less than the holding current. The turn-off is usually achieved by reversing the anode voltage. An operating triac must tun-off during the brief instant while the load current passes through zero value. Hence with resistive loads, the process of turning off a triac is simple and easy to accomplish while it is rather difficult with inductive loads.

B11.3 WORKING

Consider that the load is a pure resistor. During positive half cycle of input AC voltage MT_2 is positive with respect to MT_1. The triac is gated with a positive signal at $\omega t = \alpha$ and it conducts and the load voltage is the positive part of the supply voltage. The operation of the triac is in the first quadrant.

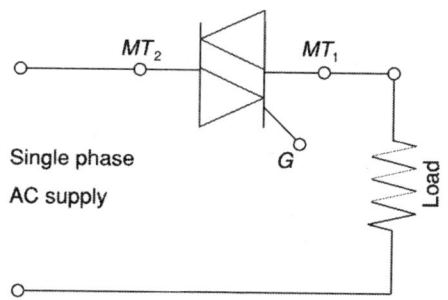

At $\omega t = \pi$, the supply voltage is zero, the load current is zero, and the triac ceases to conduct. During negative half cycle of the supply voltage, MT_2 becomes negative with respect to MT_1. The gate of the triac is now triggered at $\omega t = \pi + \alpha$ with a negative signal and it conducts in the reverse direction. The load voltage is negative because it is the part of the supply voltage from $\omega t = \pi + \alpha$ to $\omega t = 2\pi$. The load current is in the reverse direction. The operation of the thyristor is in the third quadrant. The triac is turned off at $\omega t = 2\pi$. It is turned ON at $\omega t = 2\pi + \alpha$ with a positive gate signal when the supply voltage is positive and the cycle is thus repeated.

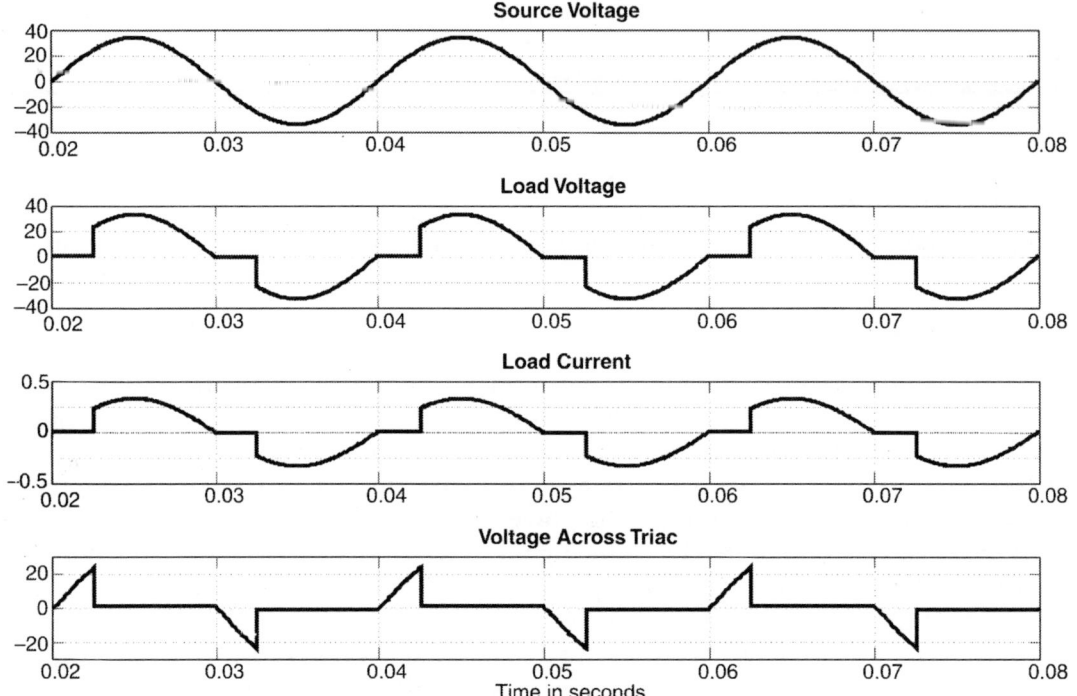

Waveform: AC voltage regulator using triac with R load

B11.4 CIRCUIT DIAGRAM

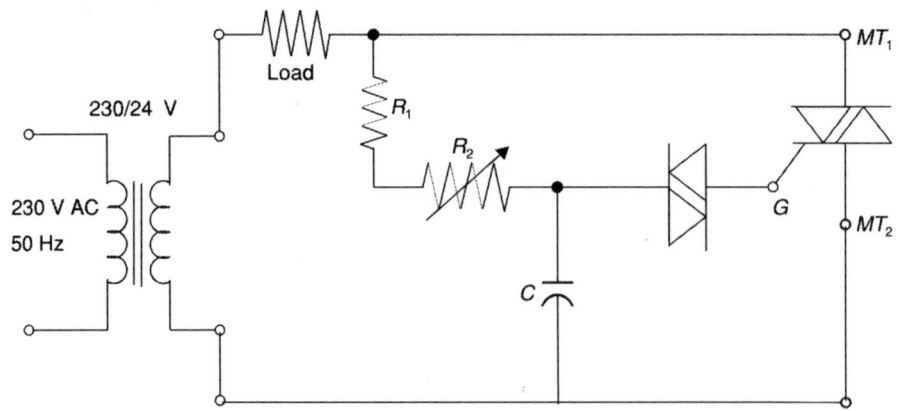

B11.5 DESIGN OF RC TRIGGERING CIRCUIT

The design for the RC triggering circuit is discussed in detail in section B2.

Select $R_1 = 220\ \Omega$

To get the maximum firing angle very close to 180°, a high value of R_2 is to be selected.

Choose $R_2 = 100\ \mathrm{k\Omega}$ pot

$$RC \geq 0.65T, \text{ where } T = 20\ \mathrm{m\ sec} \text{ for 50 Hz supply.}$$

$$C = \frac{0.65T}{(R_1 + R_2)} = \frac{0.65 \times 20 \times 10^{-3}}{100.22 \times 10^3} = 0.1295\ \mu F$$

Select $C = 0.22\ \mu F$

B11.6 EQUIPMENT AND COMPONENTS

1. Single phase transformer   230/24V 50Hz
2. Triac   BT 136
3. Diac   DB 3
4. Capacitor   0.22 ΩF
5. Load Resistor   100 Ω, 20 W
6. Resitor   220 Ω
7. Pot   100 KΩ
8. DSO/CRO
9. Breadboard

B11.7 PROCEDURE

1. Connections are made as per the connection diagram
2. Adjust the pot of the triggering circuit at maximum resistance position.
3. Switch ON the power supply.
4. Switch ON the supply to the triggering circuit and check for triggering pulses.
5. Adjust the pot of the triggering circuit to keep the firing angle to a convenient value.
6. Observe the waveform of load voltage and thyristor voltage on the DSO.

B11.8 OBSERVATION

Observe and plot the waveforms of source voltage, load voltage, load current and voltage across the triac on a graph sheet.

Experiment B12

Design and Testing of a Complementary Commutation Circuit

B12.1 OBJECTIVE

To study the complementary commutation process, to turn OFF of a conducting thyristor.

B12.2 THEORY OF OPERATION

The complementary commutation is a method of impulse commutation in which a reverse voltage is suddenly applied across a conducting thyristor to turn it OFF. The complementary commutation circuit is shown in figure. The main thyristor is connected in series with the load resistor and a complementary thyristor is connected in series with another resistor. A capacitor is connected between the cathodes of the two thyristors as shown in figure.

When the main thyristor T_1 is turned ON, the load R_1 is connected across the supply voltage and the load current flows. At the same time the capacitor gets charged to the supply voltage, through the resistor R_2, with v_{ab} positive. When T_2 is triggered, the resistor R_2 is connected across the supply. The capacitor voltage now appears across T_1. This is a reverse voltage across T_1 and it is turned OFF. With T_2 conducting, the capacitor now gets charged through R_1 with v_{ba} positive. If T_1 is fired again, this capacitor voltage appears across T_2, which is a reverse voltage across T_2 and it is turned OFF. When one thyristor is turned ON, the other is commutated and hence the name complementary commutation.

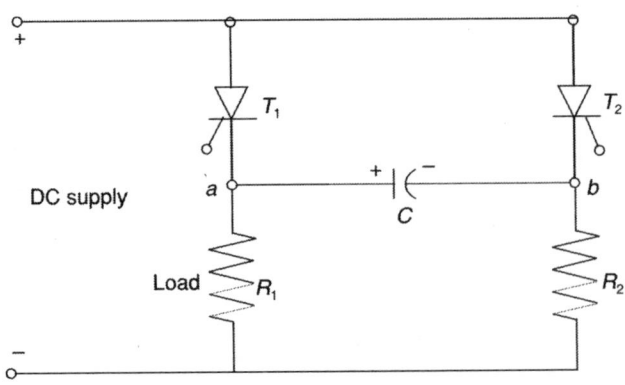

B12.3 ANALYSIS

When T_1 is turned ON with T_2 OFF

The circuit is shown in the figure. The circuit equation is

$$V_S = \frac{1}{C}\int_0^t i_C(t)\,dt + i_C(t)R_2$$

Taking laplace transforms,

$$\frac{I_C(s)}{Cs} + I_C(s)R_2 = \frac{V_S}{s}$$

$$IC(s) = \frac{V_S}{R_2}\frac{1}{s + \dfrac{1}{R_2C}}$$

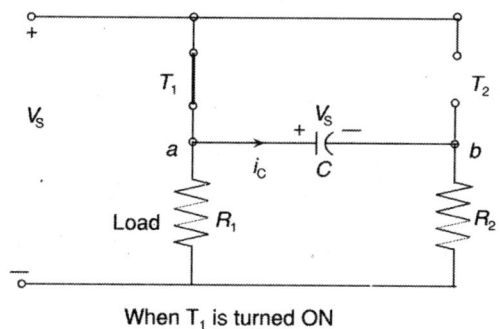

When T_1 is turned ON

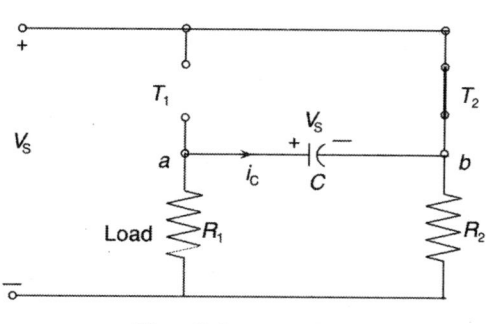

When T_2 is turned ON

Taking inverse laplace transforms

$$i_C(t) = \frac{V_S}{R_2}e^{-t/R_2C}$$

The capacitor voltage is

$$v_c = V_S - i_C(t)R_2 = V_S - V_S e^{-t/R_2C}V_S e^{-t/R_2C} = V_S(1 - e^{-t/R_2C})$$

When T_2 is turned ON

When T_2 is turned ON, T_1 is commutated.

$$V_S = -V_S - \frac{1}{C}\int_0^t i_C(t)\,dt - i_C(t)R_1$$

Taking Laplace transforms,

$$\frac{I_C(s)}{Cs} + I_C(s)R_1 = -2V_S$$

$$I_C(s) = -\frac{2V_S}{R_1}\frac{1}{s + \dfrac{1}{R_1C}}$$

Taking inverse Laplace transforms,

$$i_C(t) = -\frac{2V_S}{R_1}e^{-t/R_1C}$$

Voltage across thyristor T_1 is

$$V_{T_1} = v_{ba} = V_S + i_C(t)R_1 = V_S - 2V_S e^{-t/R_1 C} = V_S(1 - 2e^{-t/R_1 C})$$

The circuit turn OFF time of T_1 can be determined by equating $V_{T_1} = 0$.

$$v_{T_1}\big|_{t=t_{OFF1}} = V_S(1 - 2e^{-t\,OFF1/R_1 C}) = 0$$

$$t_{OFF1} = R_1 C \ln 2$$

When T_1 is turned ON again

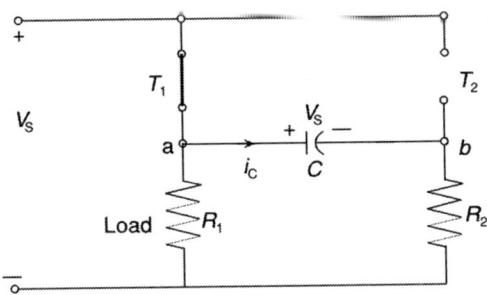

When T_1 is turned ON again

The circuit equation is

$$V_S = -V_S + \frac{1}{C}\int_0^t i_C(t)\,dt + i_C(t)R_2$$

Taking laplace transforms,

$$\frac{I_C(s)}{Cs} + I_C(s)R_2 = 2V_S$$

$$I_C(s) = \frac{2V_S}{R_2}\frac{1}{s + \dfrac{1}{R_2 C}}$$

Taking inverse Laplace transforms,

$$i_C(t) = \frac{2V_S}{R_2}e^{-t/R_2 C}$$

Voltage across thyristor T_2 is

$$V_{T_1} = v_{ba} = V_S + i_C(t)R_2 = V_S - 2V_S e^{-t/R_2 C} = V_S(1 - 2e^{-t/R_2 C})$$

The circuit turn OFF time of T_2 can be determined by equating $V_{T_2} = 0$.

$$v_{T_2}\big|_{t=t_{OFF2}} = V_S(1 - 2e^{-t\,OFF2/R_2 C}) = 0$$

$$t_{OFF2} = R_2 C \ln 2$$

Thyristor current is

$$i_{T_1} = \frac{V_S}{R_1} + i_C(t) = \frac{V_S}{R_1} + \frac{2V_S}{R_2}e^{-t/R_2 C}$$

Maximum steady state thyristor current

$$I_{T_1} = \frac{V_S}{R_1} + i_C(t) = \frac{V_S}{R_1} + \frac{2V_S}{R_2}$$

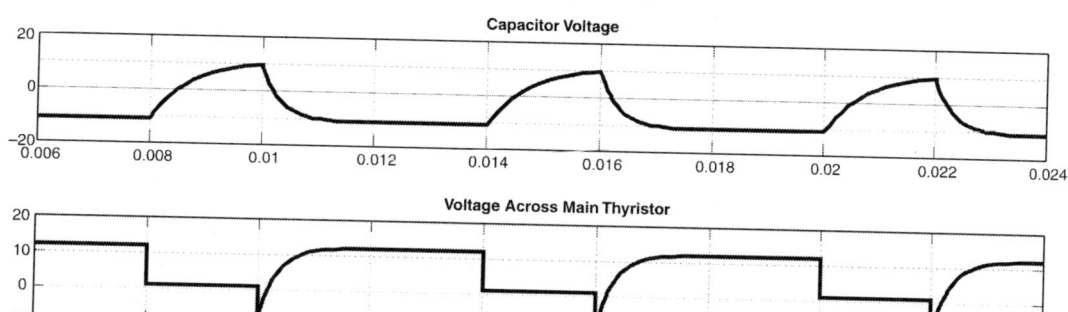

Capacitor Voltage

Voltage Across Main Thyristor

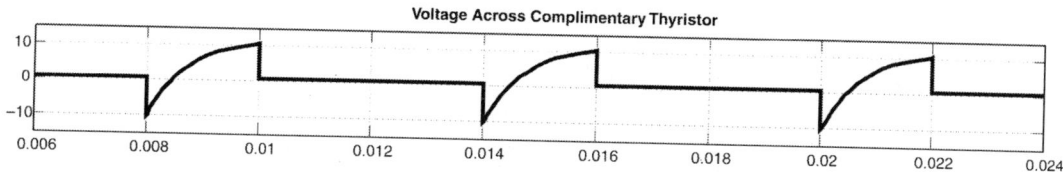

Voltage Across Complimentary Thyristor

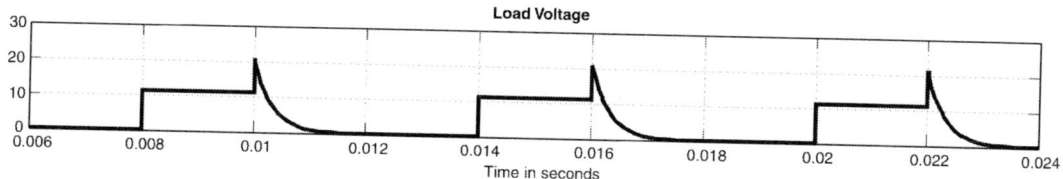

Load Voltage

Time in seconds

B12.4 DESIGN EQUATIONS

Maximum thyristor current

$$t_{OFF1} = R_1 C \ln 2$$
$$t_{OFF2} = R_2 C \ln 2$$

R_2 is selected such that (V_S/R_2) is greater than the holding current of the complementary thyristor.

B12.5 CIRCUIT DESIGN

B12.5.1 Design of C and R_2

Supply voltage = 12 V DC

Select load resistance $R_1 = 100\ \Omega$, 10 W

Thyristor current = $\dfrac{12}{100}$ = 0.12 A

Holding current of thyristor = 30 mA

The turn OFF time of thyristor $T_1 = 100\ \mu s$. Selecting a factor of safety of 2, the circuit turn OFF time is $t_{OFF} = 200\ \mu s$.

$$t_{OFF1} = R_1 C \ln 2$$

$$C = \frac{t_{OFF1}}{R_1 \ln 2} = \frac{200 \times 10^{-6}}{100 \times \ln 2} = 2.885 \, \mu F$$

Select $C = 3.3 \, \mu F$

R_2 is selected such that the current through the complementary thyristor is greater than its holding current.

$$R_2 < \frac{V_S}{I_h} = \frac{12}{30 \times 10^{-3}} = 400 \, \Omega$$

Select $R_2 = 220 \, \Omega$

B12.5.2 Triggering Circuit Design (Refer to section B3.3)

Select UJT 2N2646

$$\eta = 0.65, \; I_p = 100 \, \mu A, \; I_V = 4 \, \mu A,$$
$$V_P = \eta V_{BB} + V_D = 0.65 \times 15 + 0.7 = 9.05$$

$$R_{max} = \frac{V_{BB} - V_p}{I_p} = \frac{15 - 9.05}{100 \times 10^{-6}} = 59.5 \, k\Omega$$

Select $R_1 = 220 \, \Omega$ and $R_2 = 100 \, k\Omega$ pot

$$C = \frac{-T}{R \ln(1 - \eta)} = -\frac{10 \times 10^{-3}}{100 \times 10^3 \ln(1 - 0.65)} = 0.1 \mu F$$

$$R_T = R_T = \frac{10^4}{\eta V_{BB}} = \frac{10^4}{0.65 \times 15} = 1025.6$$

Select $R_T = 1.2 \, k\Omega$

Note: For the main thyristor T_1, the period may be taken as 6 ms. For triggering the complementary thyristor T_2, a similar circuit, with the same design parameters, is used. If T_1 is triggered at, say, $t = 1 ms$, then T_2 is triggered at, say at $t = 7$ ms.

B12.6 CIRCUIT DIAGRAM

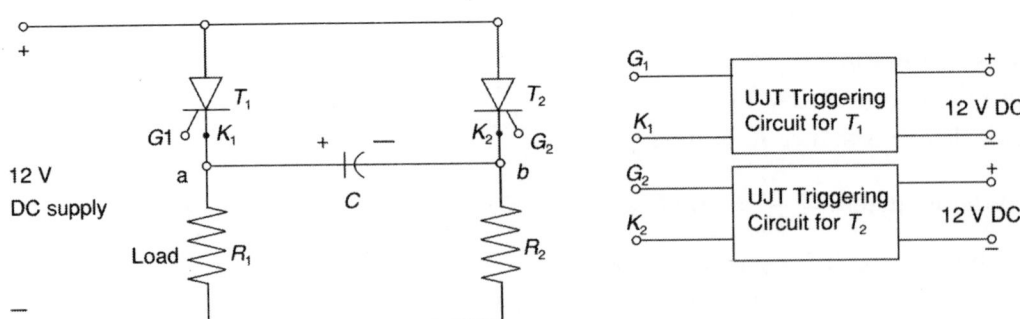

B12.7 EQUIPMENT AND COMPONENTS

Sl No	Component	Specification	Quantity
1.	Thyristor firing circuit using UJT		2
2.	Thyristor	C106D	2
3.	Load Resistor	100 Ω, 20 W	1
4.	Resistor	220 Ω	1
5.	Capacitor	0.1 μF	1
6.	DC supply	12 V	1
7.	Breadboard		

B12.8 PROCEDURE

1. Make the connections as per the circuit diagram and using the design values of R_1, R_2 and C.
2. Switch ON the DC supply.
3. First trigger thyristor T_1, by adjusting the variable pot R_2 of the triggering circuit for T_1 to allow the capacitor to charge fully.
4. Trigger the complementary thyristor T_2 next, by adjusting the variable port R_2 of the triggering circuit for T_2.
5. Ensure that the circuit operates in the proper way. Observe the waveforms and see that the thyristor T_1 is commutated just after T_2 is turned ON and vice versa.
6. Observe the waveforms of capacitor voltage, voltage across main thyristor, voltage across complementary thyristor and the load voltage.

B12.9 OBSERVATION

Plot the observed waveforms on a graph sheet.

Experiment B13

Design and Testing of a Resonance Commutation Circuit

B13.1 OBJECTIVE

To study the parallel resonance commutation process of a thyristor.

B13.2 THEORY

In *resonance commutation*, the current through a thyristor is brought down to a value less than the holding current by sending a current through it in the reverse direction using an LC resonant circuit. The circuit diagram of a parallel resonance commutation circuit is shown in the figure. The LC resonating circuit is connected in parallel to the thyristor. It is a forced commutation circuit in which the thyristor is forced to turn OFF by sending a reverse current through it. When the net current is less than the holding current, the thyristor is turned OFF. The process is also called *current commutation*.

When the DC supply is turned ON, the capacitor gets charged to the level of the supply voltage through R and L with v_{ab} positive. The thyristor is forward biased but remains in the forward blocking mode.

When the gate is triggered, the thyristor is turned ON and the supply voltage is applied across the load resistor. The load current flows. At the same time the capacitor

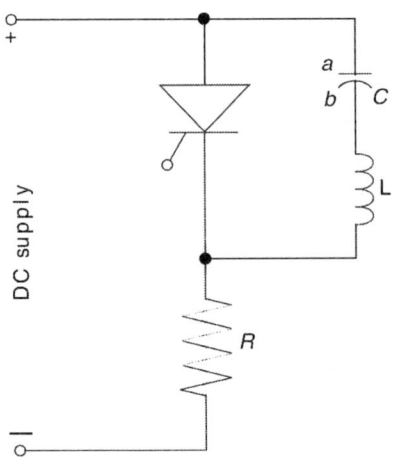

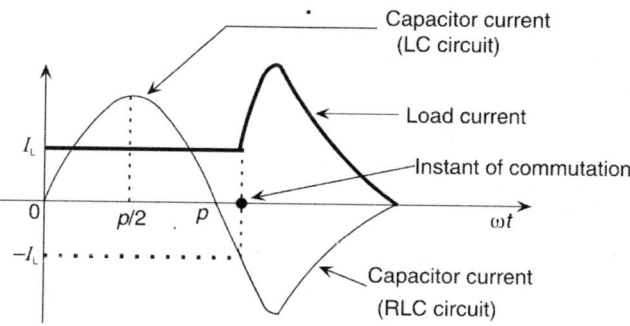

discharges through the thyristor and the energy is transferred to the inductor. The thyristor current is now the sum of the load current and the capacitor discharge current. During this time, the capacitor starts charging in the reverse direction with v_{ba} positive.

After the capacitor is charged to a maximum reverse voltage, it starts discharging through the thyristor in the opposite direction. As this discharge current increases, the net current through the thyristor decreases. When the resultant current through the thyristor is less than the holding current, it is turned off. The point at which the thyristor is turned off is shown in the figure. The remaining charge of the capacitor is discharged through the resistor and the current decays down to zero. The capacitor once again charges to the level of the supply voltage through L and R and the cycle is repeated.

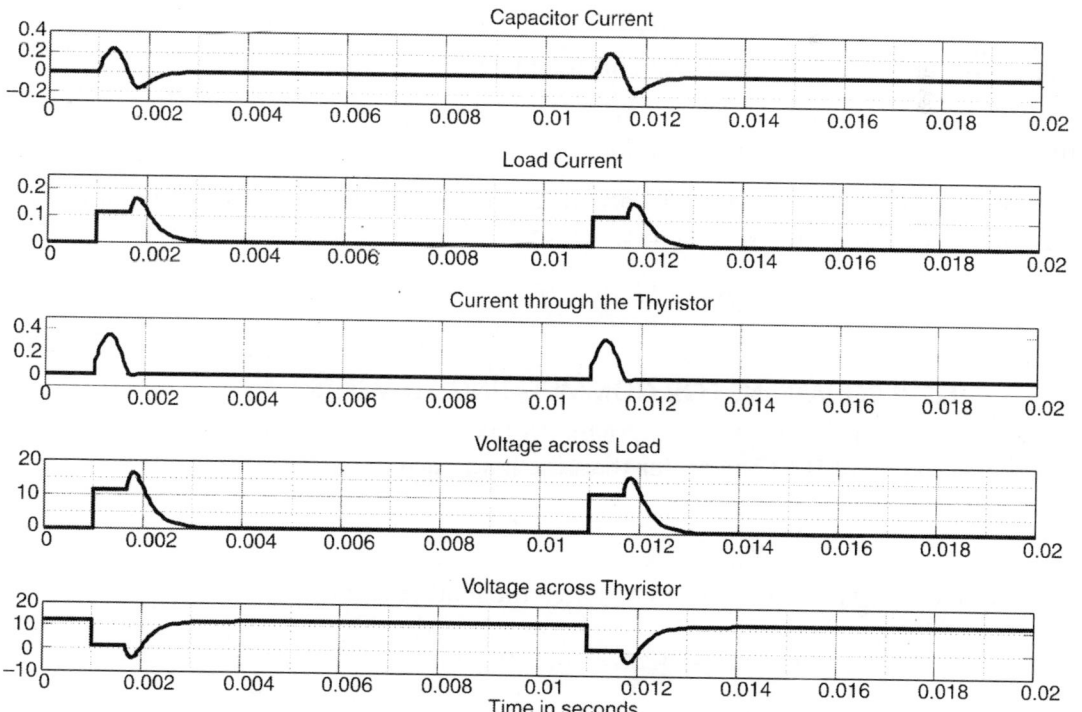

Waveforms: Resonance commutation

B13.3 ANALYSIS

Let the capacitor be charged to the supply voltage V_{dc}. When the thyristor is turned ON,

$$V_{dc} = L\frac{di_c(t)}{dt} + \frac{1}{C}\int_0^t i_c(t)dt$$

Taking Laplace transforms,

$$\frac{V_{dc}}{s} = LsI_c(s) + \frac{1}{Cs}I_c(s) = \left(Ls + \frac{1}{Cs}\right)I_c(s)$$

$$I_c(s) = \frac{V_{dc}C}{LCs^2 + 1} = \frac{V_{dc}}{L} \cdot \frac{1}{s^2 + \frac{1}{LC}}$$

$$= \frac{V_{dc}}{L} \cdot \sqrt{Lc} \cdot \frac{\left(\frac{1}{\sqrt{LC}}\right)}{s^2 + \frac{1}{Lc}}$$

$$= V_{dc}\sqrt{\frac{C}{L}} \cdot \frac{\omega}{s^2 + \omega^2}$$

Taking inverse Laplace transform,

$$i_c(t) = V_{dc}\sqrt{\frac{C}{L}} . \sin \omega t$$

$$\omega = \frac{1}{\sqrt{LC}} \text{ rad/sec}$$

The capacitor current is sinusoidal and the maximum value is

$$I_{c\,max} = V_{dc}\sqrt{\frac{C}{L}}$$

When the capacitor current is equal and opposite to the thyristor current, the device is turned off. The capacitor voltage is given by,

$$v_c(t) = L\frac{di}{dt}v_{dc}\cos\omega t$$

B13.4 DESIGN EQUATIONS FOR COMMUTATING ELEMENTS

1. For the commutation of the thyristor to take place, the maximum value of the capacitor current is to be greater than the load current.

$$V_{dc}\sqrt{\frac{C}{L}} > \frac{V_{dc}}{R}$$

Choose the capacitor current to be twice the load current.

$$V_{dc}\sqrt{\frac{C}{L}} = 2\frac{V_{dc}}{R} \quad \text{or} \quad R = 2\sqrt{\frac{L}{C}} \quad \text{or} \quad \left(\frac{R}{2L}\right)^2 = \frac{1}{LC}$$

This makes the RLC circuit critically damped. In practice the circuit is made to be over-damped by suitably selecting the values of L and C.

2. At the instant when commutation starts, $i_c + I_L = 0$ or $i_c = -I_L$

$$V_{dc}\sqrt{\frac{C}{L}} = -2I_L \quad \text{or} \quad V_{dc}\sqrt{\frac{C}{L}}\left(-\frac{1}{2}\right) = I_L$$

This can be written as

$$V_{dc}\sqrt{\frac{C}{L}}\sin\left(\pi + \frac{\pi}{6}\right) = I_L$$

This shows that the thyristor gets commutated at $\omega t = 210°$
The capacitor voltage at this instant is

$$V_c = V_{dc}\cos 210° = -\frac{\sqrt{3}}{2}V_{dc}$$

3. During turn-off time of the thyristor, the capacitor voltage changes from

$$V_c = -\frac{\sqrt{3}}{2}V_{dc} \text{ to } V_c = 0$$

Hence at this time

$$C\frac{dv_c}{dt} = I_L \quad \text{or} \quad \frac{C\frac{\sqrt{3}}{2}V_{dc}}{t_{off}} = I_L = \frac{V_{dc}}{R}$$

Thus

$$C = \frac{2t_{off}}{\sqrt{3}R}$$

For an over-damped RLC circuit $L < CR^2/4$

B13.5 CIRCUIT DESIGN

B13.5.1 Design of Commutating Elements

The turn OFF time of the thyristor is $t_q = 100 \text{ μF}$.
Assume a factor of safety of 2 for the thyristor turn OFF time.
The circuit turn OFF time is $t_{OFF} = 200 \text{ μs}$.
Supply voltage = 12 V
Select the load resistor $R = 100 \text{ Ω}$, 10 W

$$C = \frac{2t_{off}}{\sqrt{3}R} = \frac{2 \times 200 \times 10^{-6}}{\sqrt{3} \times 100} = 2.3 \text{ mF} \qquad \text{Select } C = 2.7 \text{ μF}$$

$$L < \frac{CR^2}{4} = \frac{2.7 \times 10^{-6} \times (100)^2}{4} = 6.75 \text{ mH} \quad \text{Select } L = 6 \text{ mH}$$

B13.5.2 Triggering Circuit Design (Refer to B3)

Select UJT 2N2646
$n = 0.65$, $I_p = 100 \text{ μA}$, $I_V = 4 \text{ mA}$
Base to base voltage $V_{BB} = 12 \text{ V}$
$V_P = \eta V_{BB} + V_D = 0.65 \times 12 + 0.7 = 8.5 \text{ V}$

$$R_{max} = \frac{V_{BB} - V_p}{I_p} = \frac{12 - 8.5}{100 \times 10^{-6}} = 35 \text{ kΩ}$$

Select $R_1 = 220 \text{ Ω}$ and $R_2 = 50 \text{ kΩ}$ pot

The period for charging and discharging the capacitor is taken to be 10 ms.

$$C = \frac{-T}{R\ln(1-\eta)} = -\frac{10\times10^{-3}}{50\times10^3\ln(1-0.65)} = 0.2\ \mu F \qquad \text{Select } C = 0.22\ \mu F$$

$$R_T = \frac{10^4}{\eta V_{BB}} = \frac{10^4}{0.65\times12} = 1282\ W \qquad \text{Select } R_T = 1.2\ k\Omega$$

B13.6 CIRCUIT DIAGRAM

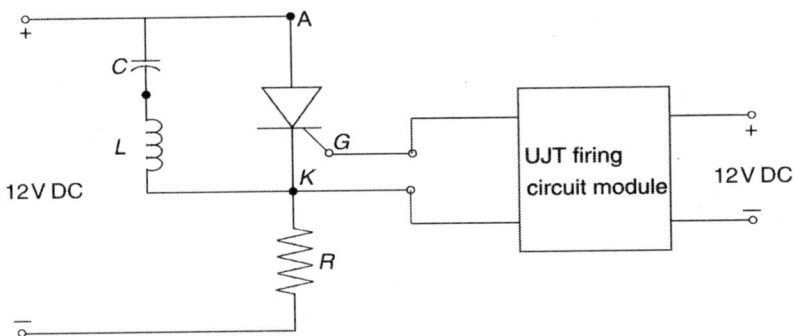

B13.7 EQUIPMENT AND COMPONENTS

Sl No	Component	Specification	Quantity
1.	UJT firing circuit		1
2.	Thyristor	C106D	1
3.	Load Resistor	100 Ω, 20 W	1
4.	Inductor	6 mH	1
5.	Capacitor	2.7 μF	1
6.	DC supply	12 V	1
7.	DSO/CRO		
8.	Breadboard		

B13.8 PROCEDURE

1. Select the values of the commutating elements L and C as per the design.
2. Connections are made as shown in the circuit diagram.
3. The DC supply is switched ON
4. Adjust the pot in the UJT triggering circuit and set a proper value of the chopping period T. This can be found out by observing the waveform of load voltage using a DSO.
5. Ensure that the circuit operates in the proper way.
6. Observe the waveforms of capacitor current, load current, thyristor current, load voltage and the voltage across the thyristor.
7. Switch off the DC supply.

B13.9 OBSERVATION

Observe the waveforms of capacitor current, load current, thyristor current, load voltage and the voltage across the thyristor and draw them on a graph sheet.

Experiment B14

Design and Testing of an Impulse Commutation Circuit

B14.1 OBJECTIVE

To study the turning OFF process of a thyristor using impulse commutation technique.

B14.2 THEORY OF OPERATION

The circuit diagram of an impulse commutated chopper is shown in the figure. The supply source, the main thyristor T_1 and the load form the power circuit. Auxiliary thyristor T_2, the diode, the Inductor and the capacitor form the commutation circuit.

   For proper operation of the chopper, pre-charging of the capacitor with polarity as shown in the circuit is essential. Auxiliary thyristor T_2 is triggered first by giving a gate signal. Once capacitor is charged fully, the charging current reduces to zero and hence T_2 will be naturally commutated. Now trigger main thyristor T_1 by giving gate signal. The capacitor discharges through T_1. The current through T_1 will be the sum of load current I_L and the capacitor discharge current. The capacitor discharge current, called the ringing current, is through the diode and the inductor. This ringing current charges the capacitor once again, but in the reverse direction with bottom plate positive and top plate negative. Once charging is complete, the capacitor current becomes zero. The diode prevents the reverse flow of capacitor current through the inductor. Load current I_L continues to flow through the main thyristor T_1. Now the auxiliary thyristor T_2 is triggered by giving a gate signal. When T_2 is turned ON the capacitor

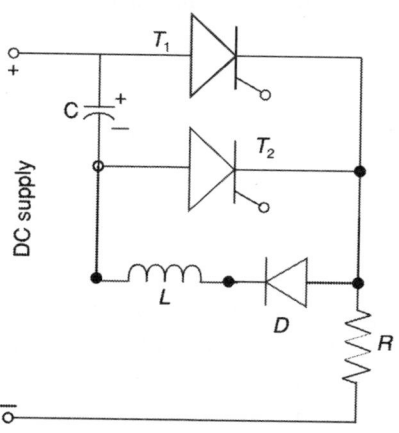

voltage is directly applied across T_1. This reverse voltage, suddenly applied across it, reverse biases it and it is turned OFF. Since a reverse voltage is suddenly applied across a conducting thyristor to turn it OFF, it is called *impulse commutation*. It is also called *voltage commutation* or *auxiliary commutation*.

The discharge current of the capacitor after commutation of the main thyristor T_1 will be diverted through the load. This current reduces to zero after some time and this depends on the capacitance and the load resistance. The capacitor will be once again charged to the supply voltage with the upper plate positive. The capacitor current now becomes zero ant the thyristor T_2 is naturally commutated. The thyristor T_1 is triggered again and this process is repeated.

B14.3 ANALYSIS

Initial State

With the main thyristor T_1 in the OFF state, let the auxiliary thyristor T_2 be triggered first. The capacitor charges through the load resistance R. Let the initial voltage across the capacitor be V_0. When the capacitor is fully charged, the charging current becomes zero and the thyristor T_2 is turned OFF naturally.

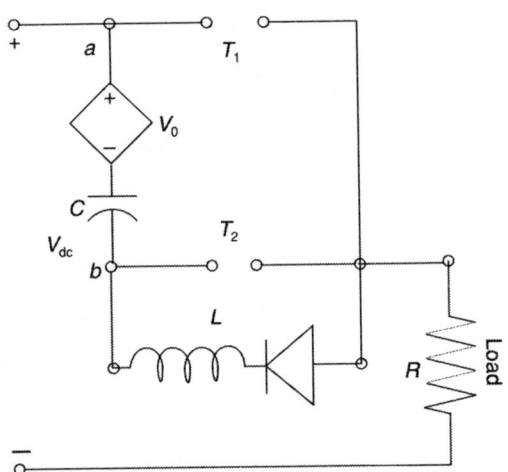

Impulse commutation-initial state

Under steady state conditions,

$$i_c = 0 \qquad v_c = V_0 \qquad I_L = 0 \qquad i_{T_1} = 0$$

$$i_{T_1} = V_{dc} \qquad i_{T_2} = 0 \qquad i_{T_2} = V_{dc} - V_0$$

Mode 1

At time $t = t_1$, the main thyristor T_1 is triggered and it conducts. The load is connected across the DC supply. The capacitor discharges through T_1, the inductor and the diode.

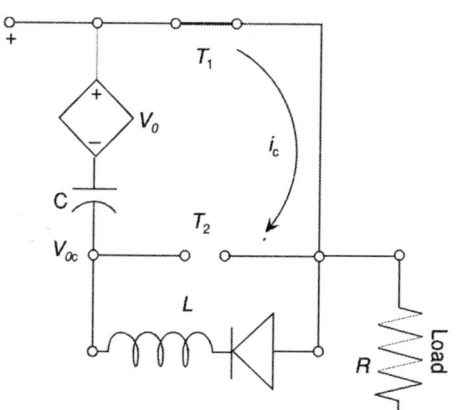

Impulse commutation Mode 1

The capacitor discharge current can be obtained by solving the equation,

$$L\frac{di_c}{dt} + \frac{1}{C}\int_0^t i_c dt - V_0 = 0$$

Taking laplace transforms,

$$L\{sI_c(s) - i(0)\} + \frac{I_c(s)}{Cs} = \frac{V_0}{s}$$

Since the capacitor is initially fully charged, $i(0) = 0$.

Hence

$$I_c(s) = V_0\sqrt{\frac{C}{L}}\frac{(1/\sqrt{LC})}{s^2 + (1/LC)}$$

Taking inverse laplace transforms,

$$i_c(t) = V_0\sqrt{\frac{C}{L}}\sin\left(\frac{1}{\sqrt{LC}}t\right)$$

$$v_c(t) = v_{ab} = L\frac{di_c}{dt} = v_0\cos\left(\frac{1}{\sqrt{LC}}t\right)$$

The capacitor current is sometimes known as ringing current and the frequency $\omega_r = \dfrac{1}{\sqrt{LC}}$ is called the ringing fequency.

The capacitor current is maximum at $t = \dfrac{\pi}{2}\sqrt{LC}$ and at this instant the capacitor voltage is zero. The load current is given by,

$$I_L = \frac{V_{dc}}{R}$$

Main thyristor current, $i_{T_1} = I_L + i_c = \dfrac{V_{dc}}{R} + V_0\sqrt{\dfrac{C}{L}}\sin\left(\dfrac{1}{\sqrt{LC}}t\right)$

Voltage across the main thyristor, $v_{T_1} = 0$

Auxiliary thyristor current, $i_{T_2} = 0$

Voltage across the auxiliary thyristor, $v_{T_2} = -v_{ab} = -V_0 \cos\left(\dfrac{1}{\sqrt{LC}}t\right)$

MODE 2

The capacitance charges to a reverse voltage. The capacitor current decays to zero. The main thyristor continues to conduct and the current through it is now the load current.

$$i_c = 0 \qquad v_c = -V_0 \qquad i_{T_1} = V_{dc}/R \qquad v_{T_1} = 0$$

$$I_L = V_{dc}/R \qquad i_{T_2} = 0 \qquad v_{T_2} = V_0$$

Impulse commutation Mode 2

Mode 3

At time $t = t_2$, the auxiliary thyristor T_2 is triggered and is turned ON. A reverse voltage is applied across the main thyristor and it is turned OFF. The capacitor then discharges through the load and once again charges to a voltage V_0. When the capacitor is fully charged, T_2 is turned OFF by self commutation.

The circuit equation is,

$$-\frac{1}{C}\int_0^t i_c dt - Ri_c = V_{dc} + V_0$$

Taking laplace transforms,

$$\frac{I_c(s)}{Cs} + RI_c(s) = -\frac{(V_{dc} + V_0)}{s}$$

$$I_c(s) = \frac{(V_{dc} + V_0)}{R} \cdot \frac{1}{s + \dfrac{1}{RC}}$$

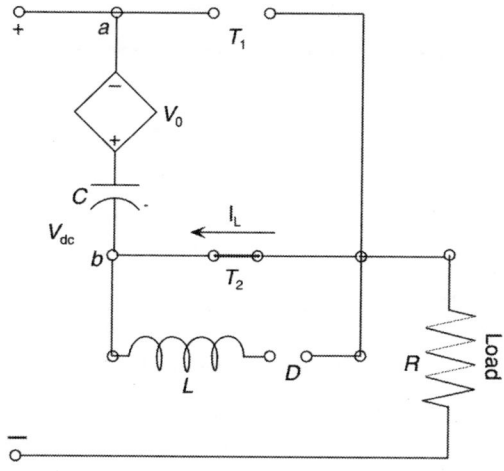

Impulse commutation Mode 3

Taking laplace inverse,

$$i_c(t) = -\frac{(V_{dc} + V_0)}{R} e^{-t/RC}$$

$$v_c(t) = v_{ab} = V_{dc} + Ri_c = V_{dc} - (V_{dc} + V_0)e^{-t/RC}$$

The load current is given by, $I_L = i_c$

$$i_{T_1} = 0 \qquad i_{T_1} = V_{ab} = V_{dc} - (V_{dc} + V_0)\,e^{-t/RC}$$

$$i_{T_2} = i_c \qquad v_{T_2} = 0$$

At the instant when T_2 is triggered, the capacitor current is,

$$i_c = -\frac{(V_{dc} + V_0)}{R} \quad \text{and} \quad v_c = -V_0$$

Now

$$V_{dc} = V_0 + v_{T_2} + V_{load}$$

When T_2 is turned OFF, the load current is zero and the load voltage is zero. At this instant, voltage across T_2 is,

$$v_{T_2} = V_{dc} - V_0$$

This is shown in the waveforms in figure.

1. Ringing frequency, $\omega_r = 1/\sqrt{LC}$
2. Ringing time $t_r = \pi\sqrt{LC}$
3. Minimum time for which the main thyristor T1 should conduct is equal to the time required by the capacitor to charge to full reverse voltage and this is,

$$t_{T_1ON\,min} = \pi\sqrt{LC}$$

4. The circuit turn OFF time for T_1 can be obtained by solving the equation for v_{T_1} in mode 2.

$$v_{T_1} = V_{dc} - (V_{dc} + V_0)e^{-t\,OFF1/RC} = 0$$

$$t_{OFF1} = RC\ln\left(\frac{V_{dc} + V_0}{V_{dc}}\right)$$

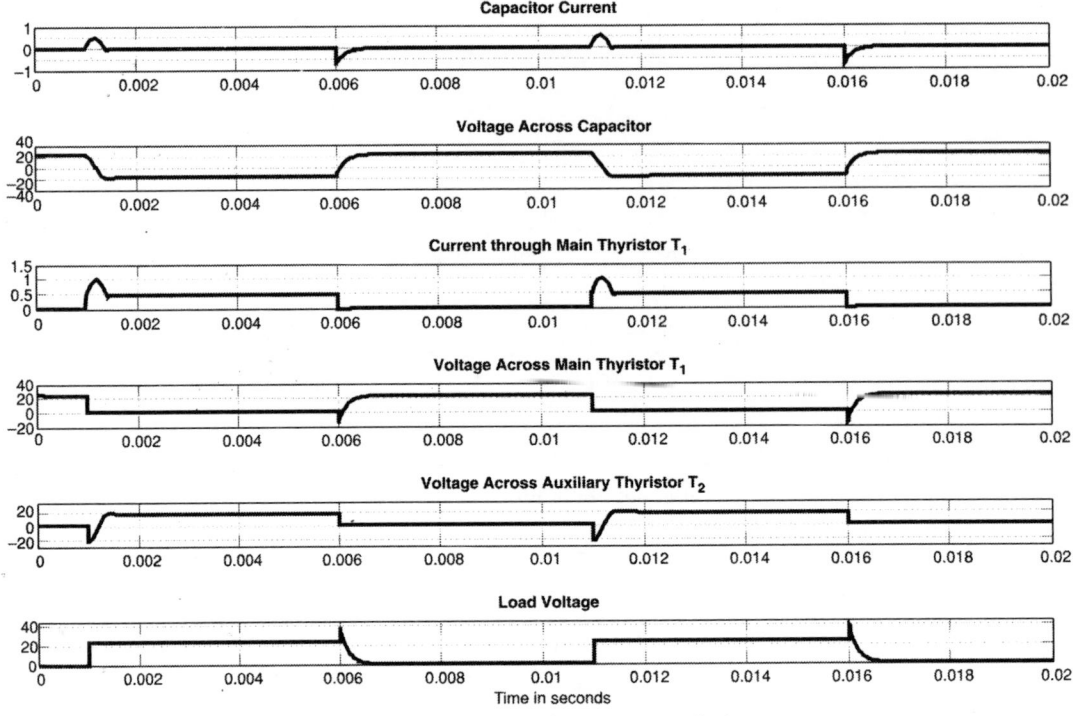

Waveforms: Impulse commutation

The method of estimating the circuit turn off time is shown in figure given below.

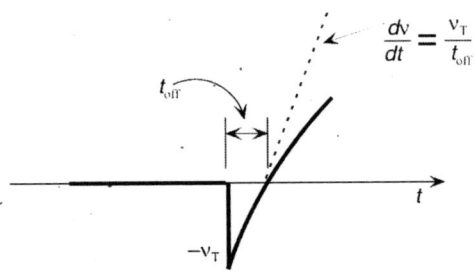

Impulse commutation–Circuit Turn OFF time

5. The circuit turn OFF time for T_2 can be obtained by solving the equation for v_{T_2} in mode 1.

$$v_{T_2} = -V_0 \cos\left(\frac{1}{\sqrt{LC}} t_{\text{OFF2}}\right) = 0 \quad t_{\text{OFF2}} = \frac{\pi}{2}\sqrt{LC}$$

6. The rate of rise of rise of voltage across T_1

$$\frac{dv_{T_1}}{dt}\bigg|_{t=t_{\text{OFF1}}} = \frac{d}{dt}\left[V_{dc} - (V_{dc} + V_0)e^{-t/RC}\right]_{t=t_{\text{OFF1}}} = \frac{V_{dc}}{RC} V/\sec$$

7. The rate of rise of rise of voltage across T_2

$$\frac{dv_{T_2}}{dt}\bigg|_{t=t_{OFF2}} = \frac{d}{dt}\left[\frac{V_0}{\sqrt{LC}}\sin\omega t\right]_{t=t_{OFF2}} = \frac{V_0}{\sqrt{LC}}V/\sec$$

8. Peak current through the main thyristor is,

$$i_{T_1\text{peak}} = I_L + i_c = \frac{V_{dc}}{R} + V_0\sqrt{\frac{C}{L}} = \frac{V_{dc}}{R} + V_{dc}\sqrt{\frac{C}{L}}$$

9. Peak current through the auxiliary thyristor is,

$$i_{T_2\text{peak}} = \frac{(V_{dc} + V_0)}{R} = \frac{2V_{dc}}{R}$$

B14.4 CIRCUIT DIAGRAM

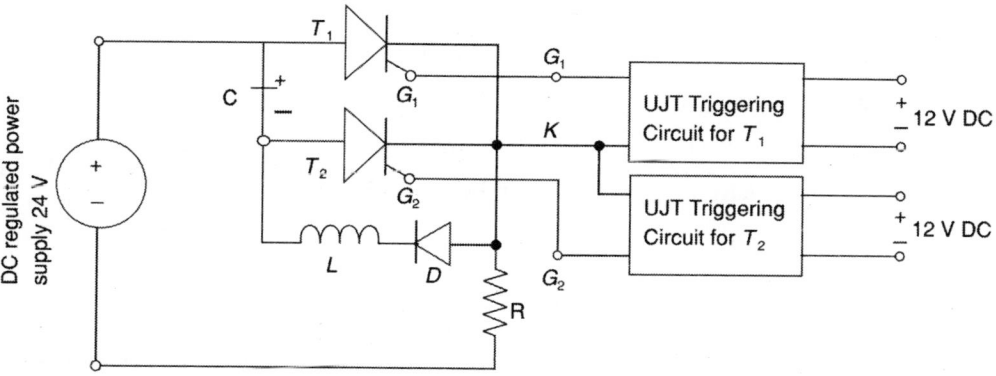

B14.5 DEVICE SPECIFICATIONS

Thyritor C106D
Turn OFF time of thyristor $t_{OFF} = 100$ μs

B14.6 DESIGN EQUATIONS FOR COMMUTATING ELEMENTS

$$t_{OFF1} = RC \ln 2$$

$$t_{OFF2} = \frac{\pi}{2}\sqrt{LC}$$

B14.7 CIRCUIT DESIGN

B14.7.1 Design of Commutating Elements

Turn off time of thyristors = 100 μ sec
Let the supply voltage $V_{dc} = 24$ V
Select load resistance $R_L = 100$ Ω, 10 W

The turn –OFF time of the thyristor is 100 μs. Adopting a factor of safety of 2, the circuit turn-OFF time selected as $t_{OFF} = 200$ μs.

$$t_{OFF1} = RC \ln 2$$

$$C = \frac{t_{OFF1}}{R \ln 2} = \frac{200 \times 10^{-6}}{100 \ln 2} = 2.9 \ \mu F$$

Select $C = 3.3 \ \mu F$

$$t_{OFF2} = \frac{\pi}{2} \sqrt{LC}$$

$$L = \left(\frac{2t_{OFF2}}{\pi}\right)^2 \frac{1}{C} = \left(\frac{2 \times 200 \times 10^{-6}}{\pi}\right) \frac{1}{3.3 \times 10^{-6}} = 5 \ mH$$

B14.7.2 Triggering Circuit Design (Refer to B3.3)

Select UJT 2N2646

$\eta = 0.65$, $I_P = 100 \ \mu A$, $I_V = 4 \ mA$, $V_{BB} = 12 \ V$,
$$V_P = \eta V_{BB} + V_D = 0.65 \times 12 + 0.7 = 8.5 \ V$$

$$R_{max} = \frac{V_{BB} - V_P}{I_p} = \frac{12 - 8.5}{100 \times 10^{-6}} = 35 \ k\Omega$$

Select $R_1 = 220 \ \Omega$ and $R_2 = 50 \ k\Omega$ pot

$$C = \frac{-T}{R \ln(1-n)} = -\frac{10 \times 10^{-3}}{50 \times 10^3 \ln(1 - 0.65)} = 0.2 \ mF \quad \text{Select } C = 0.22 \ \mu F$$

$$R_T = \frac{10^4}{\eta V_{BB}} = \frac{10^4}{0.65 \times 12} = 1282 \ \Omega$$

Select $R_T = 1.2 \ k\Omega$

Note: For the main thyristor T_1, the period may be taken as 5 ms. For triggering the auxiliary thyristor T_2, a similar circuit, with the same design parameters, is used. It must be remembered that T_2 is triggered first followed by T_1. As an example, if T_2 is triggered, say, at $t = 1 ms$ then T_1 is triggered at, say at $t = 6 ms$.

B14.8 EQUIPMENT AND COMPONENTS

Sl No	Component	Specification	Quantity
1.	UJT firing circuit		1
2.	Thyristor	C106D	1
3.	Load Resistor	100 Ω, 20 W	1
4.	Inductor	5 mH	1
5.	Capacitor	3.3 μF	1
6.	DC supply	12 V	1
7.	DSO/CRO		
8.	Breadboard		

B14.9 PROCEDURE

1. Make the connections as per the circuit diagram and using the design values of R, L and C.
2. Switch ON the DC supply.

3. First trigger the auxiliary thyristor T_2, by adjusting R_2 of the triggering circuit for T_2 to allow the capacitor to charge fully.
4. Trigger the main thyristor T_1 next, by adjusting R_2 of the triggering circuit for T_1.
5. Ensure that the circuit operates properly. Observe the waveforms and see that the thyristor T_1 is commutated just after T_2 is turned ON again.
6. Observe the waveforms of capacitor current, capacitor voltage, main thyristor current, voltage across main thyristor, voltage across the auxiliary thyristor and the load voltage.

B14.10 OBSERVATION

Observe the following waveforms and plot them on a graph sheet.
(a) capacitor current
(b) capacitor voltage
(c) main thyristor current
(d) voltage across main thyristor
(e) voltage across the auxiliary thyristor and
(f) the load voltage.

Experiment B15

Design and Testing of A DC–DC Buck Regulator

B15.1 OBJECTIVE

To design and test a DC–DC buck regulator.

B15.2 BASIC PRINCIPLE OF OPERATION

A DC–DC buck regulator is basically a DC–DC chopper with an output voltage that can be maintained at a particular value and which is less than the input voltage. In a DC–DC buck regulator, the semiconductor switch can be a MOSFET or an IGBT or a BJT. If a BJT is selected, it is operated in saturated region (ON) or cutoff region (OFF). A basic DC–DC convereter is shown in Fig. B15.1. When the switch is turned ON, the output is the same as input and the output is zero when the switch is turned OFF.

The output voltage waveform is shown in the figure. The average or DC component of the output voltage can be calculated by taking the area under the waveform divided by the time period.

$$V_0 = \frac{1}{T} \int_0^{t_{ON}} V_s \, dt = V_s \frac{t_{ON}}{T} = \delta V_s$$

where δ is known as the duty ratio or duty cycle which is the ratio of the time for which the switch is ON to the switching period.

$$\delta = \frac{t_{ON}}{t_{ON} + t_{OFF}} = \frac{t_{ON}}{T} = t_{ON} f$$

where f is the switching frequency or the chopping frequency.

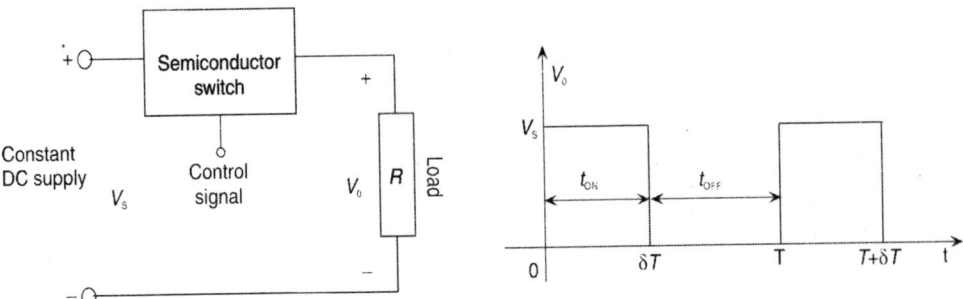

Fig. B15.1: (a) A basic DC–DC converter   (b) The output voltage waveform

The output voltage can be varied by varying the duty ratio. In a buck converter output voltage is less than the input voltage. Hence it is also called step - down DC-DC converter.

B15.3 THEORY OF BUCK REGULATOR

A buck regulator consists of a semiconductor switch, an inductor, a capacitor and a freewheeling diode as shown in Fig. B15.2.

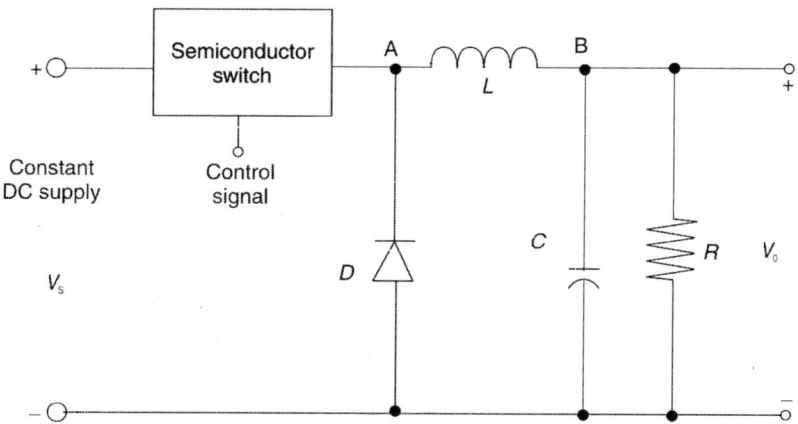

Fig. B15.2: The circuit diagram of buck regulator

When the switch is turned ON, the current flows through the inductor, the capacitor and the load. The diode is reverse biased. The inductor current increases and energy is stored in it with positive polarity at terminal A and negative polarity at terminal B. The load voltage is V_0. When the switch is turned OFF, the inductor voltage quickly reverses its polarity and the diode is forward biased. The energy stored in the inductor is released to the load via the freewheeling diode. The load current continues to flow even after the DC supply is turned OFF. The inductor current gradually decreases and before it reaches zero value, the switch is turned ON again and the cycle is repeated.

B15.4 ANALYSIS

The following assumptions are made for the analysis of the buck regulator.
1. The components are ideal.
2. The circuit is operating in the steady state.
3. The inductor current is continous.
4. The value of output capacitor is very large so that the output voltage is maintained at a constant value.

When the switch is turned ON

The voltage across the inductor is given by

$$V_{L_1} = V_s - V_o; 0 < t < t_{ON}$$

When the switch is turned OFF

The voltage across the inductor is

$$V_{L_2} = V_o; t_{ON} < t < T$$

Since the average voltage across the inductor over a period is zero

$$V_{L_1}\frac{t_{ON}}{T} + V_{L_2}\frac{t_{OFF}}{T} = 0$$

$$(V_S - V_O)\frac{t_{ON}}{T} - V_O\frac{t_{OFF}}{T} = 0$$

$$(V_S - V_O)\,\delta - V_O(1-\delta) = 0$$

Solving this equation $V_O = \delta V_S$

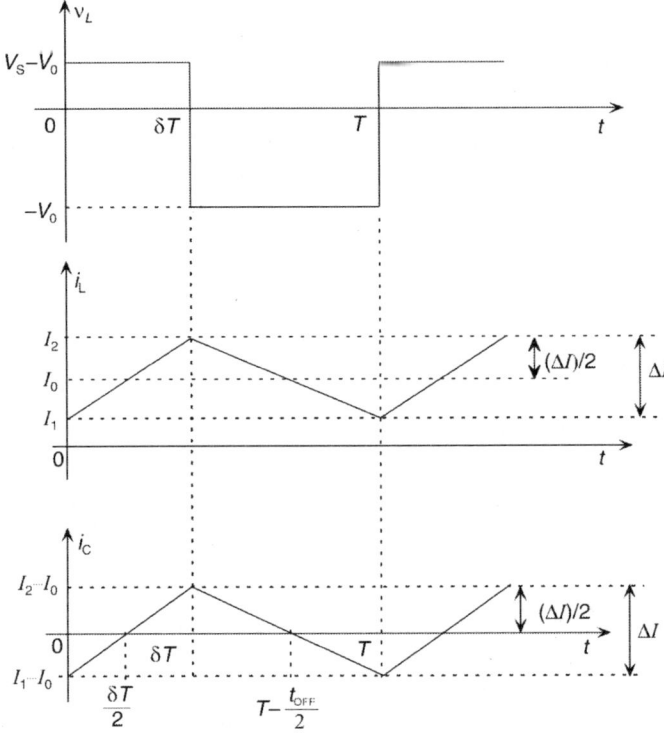

Fig. B15.3: Waveforms (a) Voltage across inductor (b) Inductor current (c) Capacitor current

Since $0 < \delta < 1$, the output voltage is less than the input voltage and hence the name *buck regulator* or *step-down regulator*. The above equation shows that by varying the duty ratio δ the output voltage can be varied.

B15.5 CURRENT RIPPLE ΔI

When the switch is turned ON

$$= V_S - V_O = L\frac{di_{L_1}}{dt}; 0 < t < t_{ON}$$

$$\frac{di_{L_1}}{dt} = \frac{V_S - V_O}{L} \quad \text{or} \quad \frac{\Delta i_{L_1}}{t_{ON}} = \frac{V_S - V_O}{L}$$

Since $\qquad\qquad \delta = \frac{t_{ON}}{T}, \quad \frac{\Delta i_{L_1}}{\delta T} = \frac{V_S - V_O}{L}$

The rate of change of inductor current is constant and it can be assumed to increase from a minimum value I_1 to a maximum value I_2.
Let the current ripple be

$$\Delta i_{L_1} = \Delta I = I_2 - I_1$$

Hence

$$\frac{\Delta i_{L_1}}{dt} = \frac{I_2 - I_1}{\delta T} = \frac{\Delta I}{\delta T} = \frac{V_S - V_O}{L}$$

$$\Delta I = \frac{V_S - V_O}{L}\delta T = \frac{V_S - \delta V_S}{L}\delta T = \frac{V_S\delta(1-\delta)}{Lf} = \frac{V_O(1-\delta)}{Lf}$$

When the switch is turned OFF

$$v_{L_2} = -V_O ; t_{ON} < t < T$$

The current falls from I_2 to I_1 in time $t_{OFF} = T - t_{ON} = (1 - d)T$ so that the voltage across the inductor is

$$v_{L_2} = L\frac{I_1 - I_2}{t_{OFF}} = -L\frac{\Delta I}{(1-\delta)T} = -V_O ; t_{ON} < t < T$$

The waveforms of output voltage, inductor current and capacitor current are shown in the figure. Since the average value of the capacitor current over a period is zero, the average value of the inductor current is equal to the average value of the load current.

$$I_L = I_0 = \frac{I_2 + I_1}{2} \quad \text{and} \quad \Delta I = I_2 - I_1$$

Solving these equations

$$I_2 = I_L + \frac{\Delta I}{2} = I_0 + \frac{\Delta I}{2} \quad I_1 = I_L - \frac{\Delta I}{2} = I_0 - \frac{\Delta I}{2}$$

B15.6 VOLTAGE RIPPLE

In the waveform of the capacitor current, the positive area from $t = t_{ON}/2$ to $t = T - (t_{OFF}/2)$ is equal to the change in charge of the capacitor and is equal to

$$\Delta Q = \frac{1}{2}\frac{\Delta I}{2}\left(T - \frac{t_{OFF}}{2} - \frac{t_{ON}}{2}\right) = \frac{1}{2}\frac{\Delta I}{2}\left(T - \frac{T}{2}\right) = \frac{(\Delta I)T}{8} = \frac{(\Delta I)}{8f}$$

The peak to peak ripple voltage is

$$\Delta v_C = \frac{\Delta Q}{C} = \frac{(\Delta I)}{8fC}$$

$$\Delta I = \frac{V_0(1-\delta)}{Lf}$$

$$C = \frac{(1-\delta)V_0}{8f^2 L(\Delta v_C)}$$

B15.7 DESIGN EQUATIONS

Buck converter is usually designed for continous current operation. To keep the inductor current continous, a minimum value of inductance is required which can be obtained from the expression for the lower limit of inductor current I_1. For a minimum inductance the current falls to zero.

$$\Delta I = \frac{V_0(1-\delta)}{Lf}$$

$$L = \frac{V_0(1-\delta)}{(\Delta I)f}$$

$$I_1 = I_0 - \frac{\Delta I}{2} = 0 \quad \text{or} \quad \Delta I = 2I_0$$

The value of inductance L is normally selected as 25% greater than I_{min} to ensure continuous conduction.

$$C = \frac{(1-\delta)V_0}{8f^2L(\Delta v_C)}$$

It can be observed from the above equations that when the switching frequency increases the size of both inductor and capacitor reduces.

B15.8 DESIGN OF A BUCK REGULATOR

Specifications

Supply voltage, $V_S = 12$ V DC Output voltage, $V_O = 9$ V
Peak to peak ripple voltage = 0.5%
Select a load resistance of $R = 100\ \Omega$, 10 W

$$\Delta V_O = \Delta V_C = 0.5\%$$

$$= \frac{0.5}{100} \times V_O = \frac{0.5}{100} \times 9 = 0.045 \text{ V}$$

Duty ratio, $\delta = \dfrac{V_O}{V_S} = \dfrac{9}{12} = 0.75$

Select switching frequency $f = 1$ kHz

$$L_{min} = \frac{(1-\delta)R}{2f} = \frac{(1-0.75) \times 100}{2 \times 1 \times 10^3} = 12.5 \text{ mH}$$

Select the value of inductance L to be about 25% greater than L_{min}.

$$L = 12.5 \times 1.25 = 15.625 \text{ mH}$$

Choose $L = 20$ mH.
Select the switching frequency $f = 1$ kHz

$$C = \frac{V_0(1-\delta)}{8Lf^2\Delta v_C} = \frac{9 \times (1-0.75)}{8 \times 20 \times 10^{-3} \times (1 \times 10^3)^2 \times 0.045} = 312.5 \ \mu\text{F}$$

Choose the value of capacitance (from the standard capacitor values table), $C = 330\ \mu$F.
Select most commonly used diode 1N4007 which works fairly well for the frequencies ranging from 0Hz to 10 KHz. If the value of switching frequency is high, a diode with low reverse recovery time has to be selected. Examples of fast recovery diodes are 1N4933 to 1N4937.

B15.9 CIRCUIT DIAGRAM

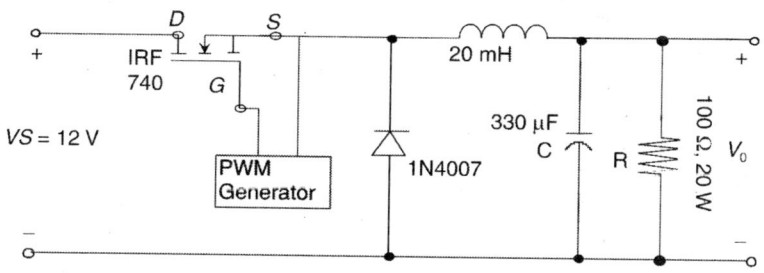

Fig. B15.4: DC-DC buck regulator

B15.10 EQUIPMENT AND COMPONENTS

Sl No	Equipments	Specification	Quantity
1.	DC supply	0–30 V	1
2.	Bread Board		1
3.	MOSFET	IRF 740	1
4.	Inductor		1
5.	Capacitor	330 µF	1
6.	Resistance	100 Ω, 20 W	1
7.	PWM generator		1
8.	Diode	IN4007	1
9.	DSO/CRO		

B15.11 PROCEDURE

1. Make the connections on the bread board as per the circuit diagram, shown in the Fig. B15.4.
2. Apply switching pulses at the gate of MOSFET. Adjust the frequency of the PWM pulse to 1 kHz. The pulse magnitude should be in the range of 12–15 V. Ensure the same by viewing the gate pulse on the DSO.
3. Switch ON the DC supply. Adjust the input voltage to 12 V.
4. Observe the output voltage across the load resistor R on the DSO and note down the values.
5. Also observe the voltage across the inductor and the inductor current.
    Note 1: When DSO/CRO probe is connected between inductor terminals or the switch, all other channels of DSO should be disconnected since most of the DSOs are non-isolated types, i.e. it has a common ground.
    Note 2: If PWM generator is not available, a function generator (FG) can be used to generate the gating pulses for the MOSFET. One may choose square wave as function in the FG and select frequency as 1 kHz. Here duty ratio is always 0.5 since waveform is a square wave.

An isolation circuit is needed to provide isolation between power circuit and FG. The magnitude of the pulse required to turn ON the MOSFET is 12 to 15 V. The figure given in Fig. B15.5 shows an optoisolator or optocoupler based isolation circuit using the IC MCT2E. Other optocoupler ICs are TLP250, 4N25 to 4N28 and 6N137.
Note 3: PWM signals can also be obtained using 555IC

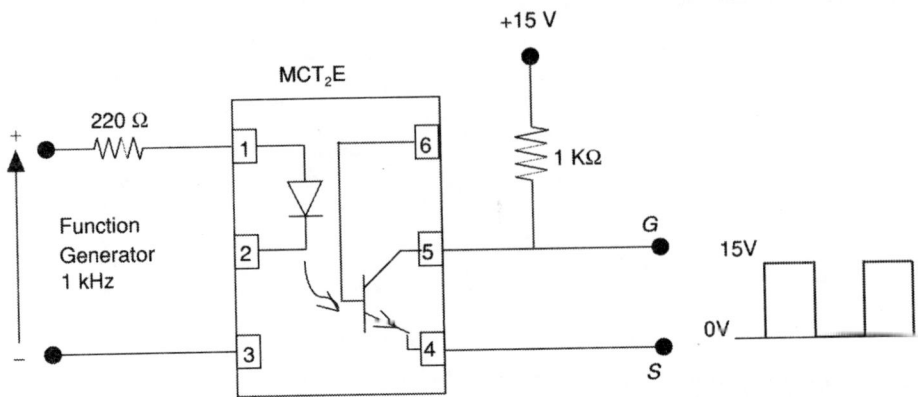

Fig. B15.5: An optocoupler based isolation circuit

B15.12 OBSERVATION

Observe the waveforms of load voltage, inductor voltage and inductor current and plot them on a graph sheet.

Experiment B16

Design and Testing of a DC–DC Boost Regulator

B16.1 OBJECTIVE

To design and test a DC–DC boost regulator.

B16.2 PRINCIPLE OF OPERATION

A boost regulator is basically a DC–DC chopper in which the output voltage can be maintained at a particular value and which is greater than the input voltage. Hence it is named as step up regulator or boost regulator. A boost regulator consists of a semiconductor switch, an inductor, a capacitor and a diode as shown in Fig. B16.1.

When the semiconductor switch is turned ON, the inductor is connected across the supply and the diode D is reverse biased. The current through the inductor increases and energy is stored in it. When the semiconductor switch is turned OFF, the polarity of inductor voltage quickly reverses. The diode becomes forward biased and provides a path for the inductor current. The energy stored in the inductor is released to the load through the diode and the load current flows. The inductor current decays to a lower value. Now the output voltage is the sum of input voltage and the inductor voltage and is greater than supply voltage. Hence this DC–DC regulator is known as a *boost regulator*.

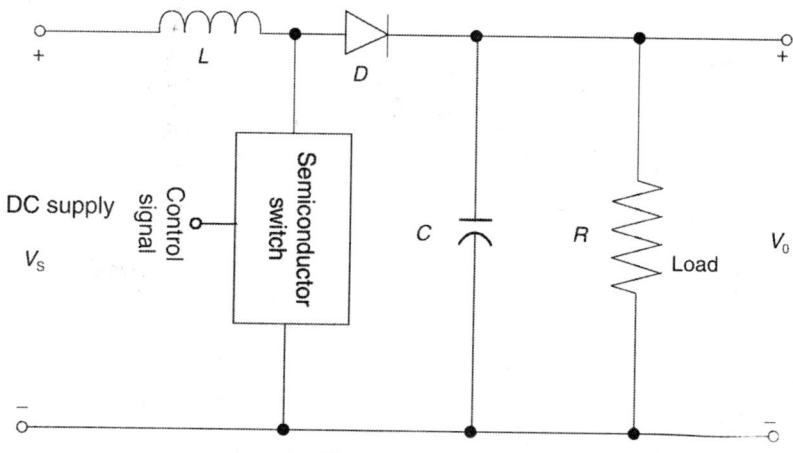

Fig. B16.1: A boost regulator

121

B16.3 ANALYSIS

The following assumptions are made for the analysis of boost regulator.
1. The components are ideal.
2. The circuit is operating in the steady state and so the voltage and current waveforms are periodic.
3. The inductor current is continous.
4. The value of output capacitor is very large so that the output voltage is maintained at a constant value.

When the switch is turned ON

Assume that the semiconductor switch conducts for a time t_{ON}. During this period the inductor is connected across the DC supply.

$$v_{L_1} = L\frac{di_{L_1}}{dt} = V_S$$

When the switch is turned OFF

The switch is turned OFF for a time t_{OFF}. The inductor voltage is

$$v_{L_2} = L\frac{di_{L_2}}{dt} = V_S - V_0$$

Let the period be T so that

$$T = t_{ON} + t_{OFF}$$

The average value of the inductor voltage over a period is zero.

Hence $\qquad v_{L_1}\dfrac{t_{ON}}{T} + v_{L_2}\dfrac{t_{OFF}}{T} = 0$

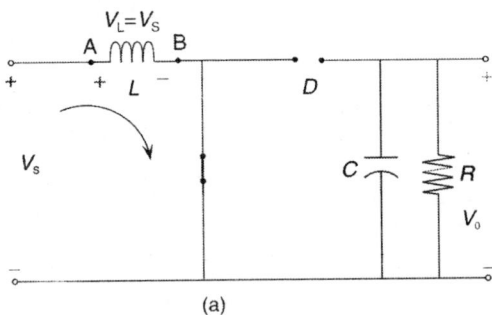

(a)

Fig. B16.2 Boost regulator—When the switch is turned ON

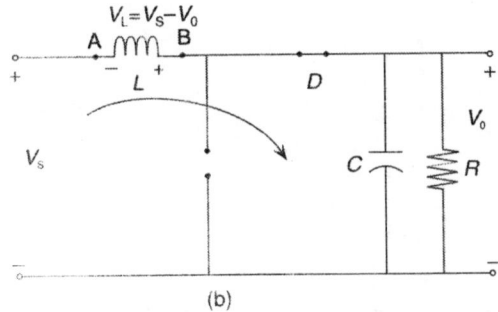

(b)

Fig. B16.3 Boost regulator—When the switch is turned OFF

$$v_{L_1}\frac{t_{ON}}{T} + v_{L_2}\left(1 - \frac{t_{ON}}{T}\right) = 0$$

$$v_{L_1} + v_{L_2}(1 - \delta) = 0$$

$$V_S\delta + (V_S - V_0)(1 - \delta) = 0$$

Solving this equation, the equation for the output voltage is

$$V_0 = \frac{V_S}{1 - \delta}$$

Since $0 < \delta < 1$, the output voltage is always greater than the input voltage and so regulator is known as *boost regulator*.

B16.4 CURRENT RIPPLE ΔI

When the switch is turned ON

$$\frac{di_{L_1}}{dt} = \frac{V_S}{L}$$

The rate of change of inductor current is constant. Since the the inductor current is assumed to be continuous, it can be considered to be increasing linearly from a minimum value I_1 to a maximum value I_2 during a time t_{ON}.

Hence
$$\frac{di_{L_1}}{dt} = \frac{\Delta I_{L_1}}{t_{ON}} = \frac{I_2 - I_1}{\Delta T} = \frac{V_S}{L}$$

Let the peak to peak current ripple be
$$\Delta I = I_2 - I_1$$

Then
$$\frac{\Delta I}{\Delta T} = \frac{V_S}{L}$$

or
$$\Delta I = \frac{V_S\delta T}{L} = \frac{V_S\delta}{Lf} = \frac{V_0(1 - \delta)\delta}{Lf}$$

When the switch is turned OFF

$$VL_2 = L\frac{di_{L_2}}{dt} = V_S - V_O$$

$$\frac{di_{L_2}}{dt} = \frac{V_S - V_0}{L}$$

The rate of change of inductor current is constant and it decreases linearly from I_2 to I_1 during a time t_{OFF}. The waveforms of inductor voltage, inductor current and capacitor current are shown in Fig. B16.4. The capacitor current is equal to the diode current minus the load current.

The maximum and minimum values of inductor current can be obtained by an inspection of the waveforms.

$$I_L = \frac{I_2 + I_1}{2} \qquad \Delta I = I_2 - I_1$$

Solving these equations

$$I_2 = I_L + \frac{\Delta I}{2} \qquad I_1 = I_L - \frac{\Delta I}{2}$$

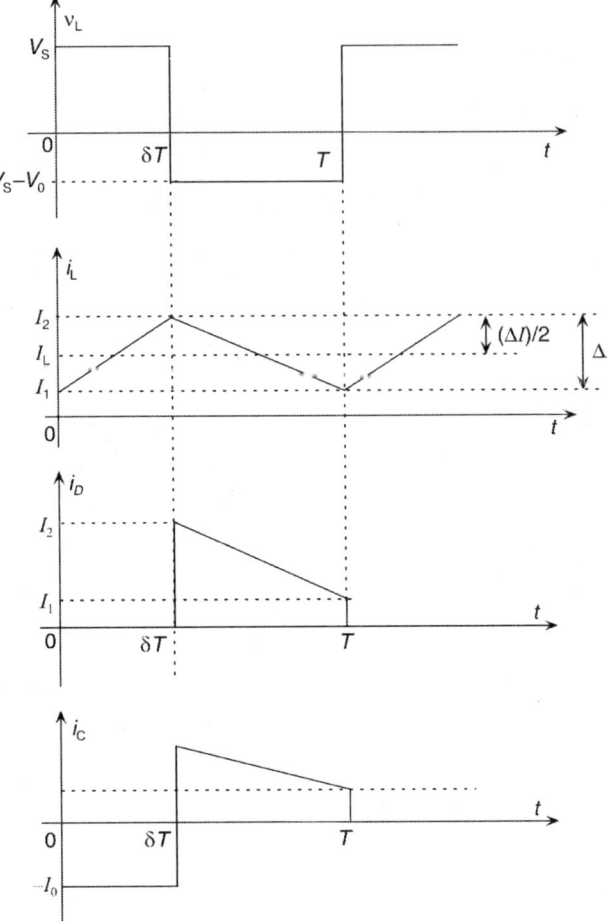

Fig. B16.4: Waves for a boost regulator

Assuming a loss free regulator, the average value of the output power is equal to the average value of the input power.

$$V_0 I_0 = V_S I_L$$

$$I_0 = \frac{V_S I_L}{V_0} = (1 - \delta) I_L$$

B16.5 OUTPUT VOLTAGE RIPPLE ΔV_0

After a few cycles, when the switch is turned ON the capacitor discharges through the load so that the average value of the capacitor current is equal to the average value of the load current.

$$I_C = I_0$$

$$\frac{\Delta Q}{t_{ON}} = I_0$$

$$\Delta Q = I_0 t_{ON} = I_0 \delta T$$

$$C(\Delta v_C) = I_0 \Delta T$$

$$\Delta v_C = \frac{I_0 \delta T}{C} = \frac{I_0 \delta}{Cf}$$

Hence the output voltage ripple is

$$\Delta v_0 = \frac{I_0 \delta}{Cf}$$

B16.6 DESIGN EQUATIONS FOR FILTER ELEMENTS

At minimum value of inductance, the current falls to zero.

Hence
$$I_1 = I_L - \frac{\Delta I}{2} = 0$$

$$I_L = \frac{\Delta I}{2}$$

$$\frac{I_0}{1-\delta} = \frac{V_S \delta}{2L_{min} f}$$

$$L_{min} = \frac{V_S \delta (1-\delta)}{2I_0 f} = \frac{V_0 \delta (1-\delta)^2}{2I_0 f}$$

A higher value of L is selected to make the current continuous.
Output voltage ripple

$$\Delta v_0 = \frac{I_0 \delta}{Cf} \quad \text{or} \quad C = \frac{I_0 \delta}{\Delta v_0 f}$$

If the switching frequency is high, lower values of L and C can be selected.

B16.7 DESIGN OF BOOST REGULATOR

Specifications
Input voltage = 12 V
Output voltage of 20 V
Load resistor = 100 Ω.
Allowable peak to peak ripple voltage = 1%.
The inductor current is to be continuous.

$$V_0 = \frac{V_S}{1-\delta}$$

Duty ratio $\qquad \delta = \frac{V_0 - V_S}{V_0} = \frac{20-12}{20} = 0.4$

Load current $\qquad I_0 = \frac{V_O}{R} = \frac{20}{100} = 0.2 \text{ A}$

Select the switching frequency $f = 1$ kHz.

$$L_{min} = \frac{\delta(1-\delta)^2 R}{2f} = \frac{0.4(1-0.4)^2 \times 100}{2 \times 1 \times 10^3} = 7.2 \text{ mH}$$

The value of inductance L is selected as 25% greater than L_{min} to ensure continuous conduction.

$$L = 7.2 \times 1.25 = 9 \text{ mH} \qquad \text{Choose } L = 12 \text{ mH}$$

Average value of inductor current,

$$I_L = \frac{I_O}{1-\delta} = \frac{0.2}{1-0.4} = 0.333 \, A$$

Peak to peak inductor current ripple

$$\Delta I = \frac{V_s \delta}{Lf} = \frac{12 \times 0.4}{12 \times 10^{-3} \times 1 \times 10^3} = 0.4 \, A$$

Maximum inductor current,

$$I_{L\,max} = I_L + \frac{\Delta i_L}{2} = 0.333 + \frac{0.4}{2} = 0.533 \, A$$

Minimum inductor current,

$$I_{L\,max} = I_L - \frac{\Delta i_L}{2} = 0.333 - \frac{0.4}{2} = 0.133 \, A$$

The value of the capacitance

$$C = \frac{I_O \delta}{\Delta V_O f}$$

Given peak to peak ripple voltage,

$$\Delta V_O = \Delta V_C = 1\%$$

$$\Delta V_O = \frac{V_O \times 1}{100} = \frac{20}{100} = 0.2 \, V$$

$$C = \frac{I_O \delta}{(\Delta V_O) f} = \frac{0.2 \times 0.4}{0.2 \times 1 \times 10^3} = 400 \, \mu F$$

Select $C = 470 \, \mu F$

Choose the value of capacitance (from the standard capacitor values table),

$$C = 470 \, \mu F$$

Select most commonly used diode 1N4007 which works fairly well for the frequencies ranging from 0 Hz to 10 KHz. If the value of switching frequency is high, a diode with low reverse recovery time has to be selected. Examples of fast recovery diodes are 1N4933 to 1N4937.

B16.8 CIRCUIT DIAGRAM

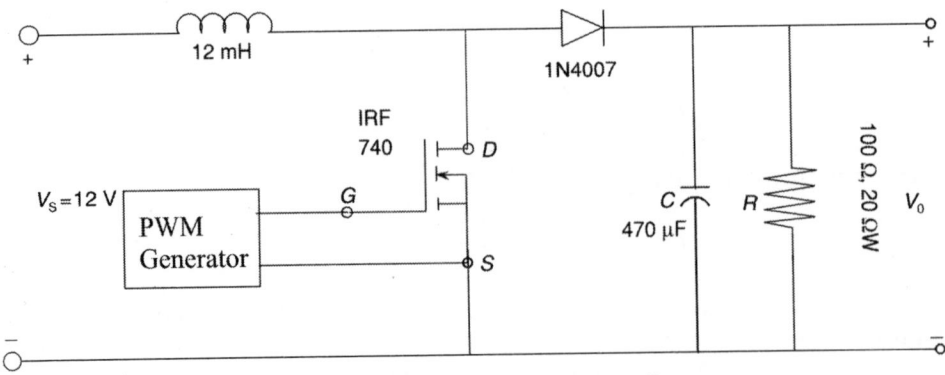

Fig. B16.5: Boost regulator connection diagram

B16.9 INSTRUMENTS/DEVICES REQUIRED

Sl No	Equipment/Components	Specification	Quantity
1.	DC supply	0–15 V	1
2.	Bread board		
3.	MOSFET	IRF740	1
4.	Inductance		1
5.	Capacitor	470 μF	1
6.	Resistor	100 Ω, 20 W	1
7.	PWM generator		1
8.	Diode	1N4007	1
9.	DSO/CRO		

B16.10 PROCEDURE

1. Make the connections as per the circuit diagram, shown in the Fig. B16.5, on the bread board.
2. Give switching pulses at the gate of MOSFET. Adjust the frequency of the PWM pulse to 1 kHz. The pulse magnitude should be in the range of 12–17 V. Ensure the same by viewing the gate pulse on the DSO
3. Switch ON the DC supply. Adjust the input voltage to 12 V.
4. Observe the output voltage across the load resistor R on the DSO and note down the values
5. Also measure the voltage across the inductor and current through the inductor.

Note: For PWM generation refer to B15.11.

B16.11 OBSERVATION

Observe the load voltage, inductor voltage and inductor current and plot them on a graph sheet.

Experiment B17

Design and Testing of a Current Commutated Chopper

B17.1 OBJECTIVE

To study the performance of a current commutated chopper, both under constant frequency control and variable frequency control.

B17.2 THEORY

The parallel resonance commutation circuit, explained in section B13, can be operated as a current commutated chopper. The principle of operation, analysis and design are all explained in the same section. A basic introduction to a DC–DC chopper is given in section B15 (buck regulator) and in section B16 (boost regulator). A buck converter is also called a step-down chopper in which the output voltage is less than the input voltage. A boost converter is also called a step-up chopper in which the output voltage is greater than the input voltage. In both cases, the output voltage is controlled by varying the duty ratio of the chopper. The duty ratio is the ratio of the time for which the semiconductor switch is ON to the switching period. The switching period is also called the chopping period.

Duty ratio $d = \dfrac{t_{ON}}{T} = t_{ON}f$

where T is the chopping period and $f = 1/T$ is the chopping frequency.

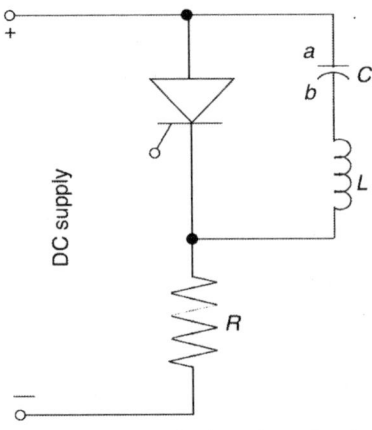

Resonance commutation circuit

128

B17.3 OPERATION AS A CHOPPER

The chopper circuit is shown in the figure given above. It is a step-down chopper, as the duty ratio is less than unity. It was shown earlier that the output voltage can be varied by varying the duty ratio. Duty ratio can be varied either by varying t_{ON}, the time for which the semiconductor switch is ON with the switching frequency constant or by keeping t_{ON} constant and varying the frequency f. The former is called the *constant frequency control* or *pulse width modulation control* and the latter is called the *variable frequency control*. Hence the output voltage of the chopper can be controlled by two different ways and these are called *time ratio controls*.

a. **Constant frequency operation:** In this method, the thyristor ON time (time for which the thyristor is ON) is varied but the chopping period T (or chopping frequency f) is kept constant. This can be achieved by varying either L or C or both in the thyristor commutation circuit. Note that the minimum ON time for the thyristor is $t_{ON} = \pi\sqrt{LC}$ sec and chopping period is $T = -RC \ln (1 - \eta)$, where R and C are the elements of the UJT firing circuit and η is the intrinsic stand-off ratio of the UJT.

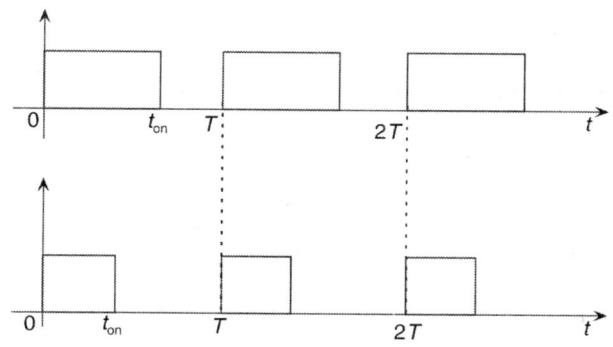

Constant frequency with fixed T

b. **Variable frequency operation:** In this method, the elements values of L and C in the thyristor commutation circuit are kept constant. Hence t_{ON} is constant. The chopping period T is varied by adjusting the charging time of the capacitor in the triggering circuit; this is by adjusting the variable resistor R_2 in the UJT triggering circuit.

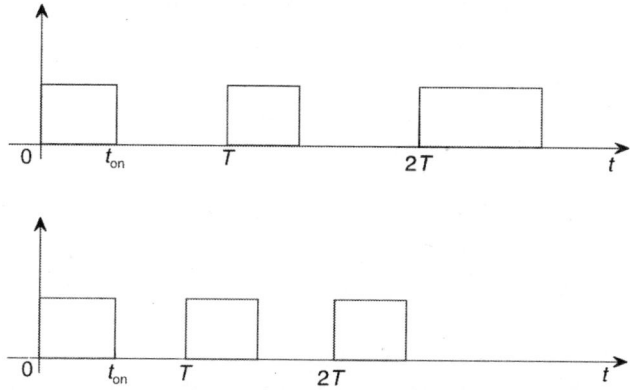

Varying frequency with fixed t_{on}

B17.4 CIRCUIT DIAGRAM

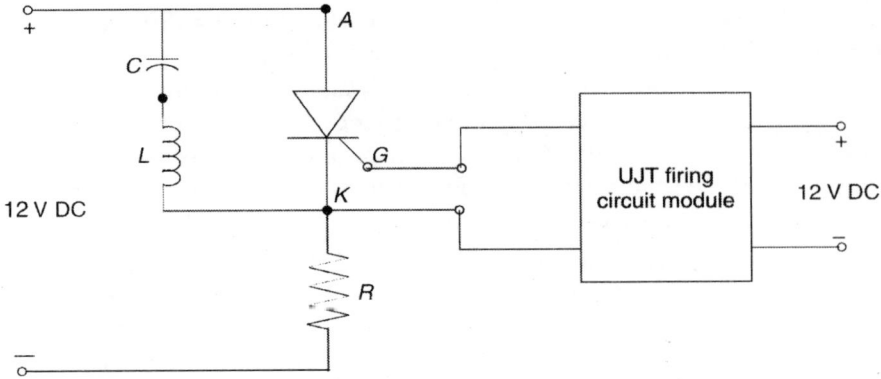

B17.5 DESIGN

The design of commutating elements and the UJT triggering circuit are given in section B13.5 and B3.6.

B17.6 EQUIPMENT AND COMPONENTS

Sl No	Component	Specification	Quantity
1.	Thyristor triggering circuit using UJT		1
2.	Thyristor	C106D	1
3.	Load Resistor	100 Ω, 20 W	1
4.	Inductor	6 mH	1
5.	Capacitor	2.7 µF	1
6.	DC supply	0–15 V	1
7.	Breadboard		

B17.7 PROCEDURE

B17.7.1 Constant Frequency Operation

1. Select the values of the commutating elements L and C as per the design.
2. Connections are made as shown in the circuit diagram.
3. The DC supply is switched ON
4. Adjust the pot in the UJT triggering circuit and set a proper value of the chopping period T. This can be found out by observing the waveform of load voltage using a DSO.
5. Ensure that the circuit operates in the proper way.
6. Observe the waveform of load voltage and plot on a graph sheet.
7. Vary the inductance L to vary the t_{ON} and observe the values of t_{ON} and T. The value of T must be the same as in the previous setting.
8. Observe again the waveform of load voltage and plot on the same graph sheet.
9. Steps 7 and 8 may be repeated to get different waveform for different value of t_{ON}.
10. Switch off the DC supply

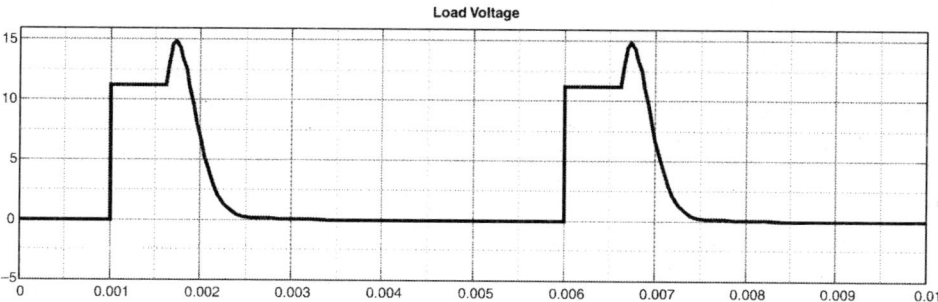

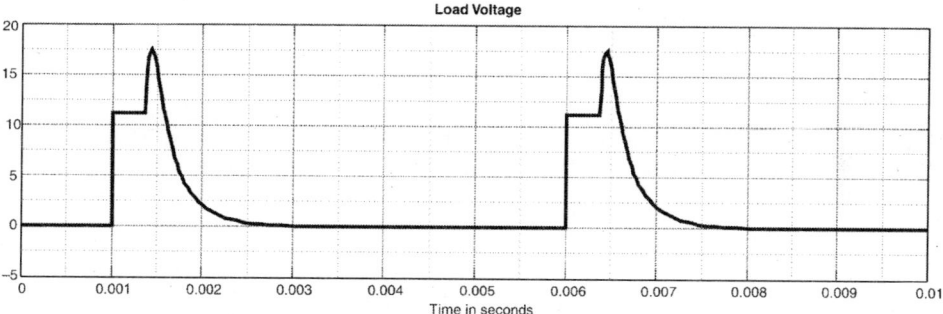

Constant frequency control

B17.7.2 Variable Frequency Operation

1. Select the values of the commutating elements L and C as per the design.
2. Connections are made as shown in the circuit diagram.

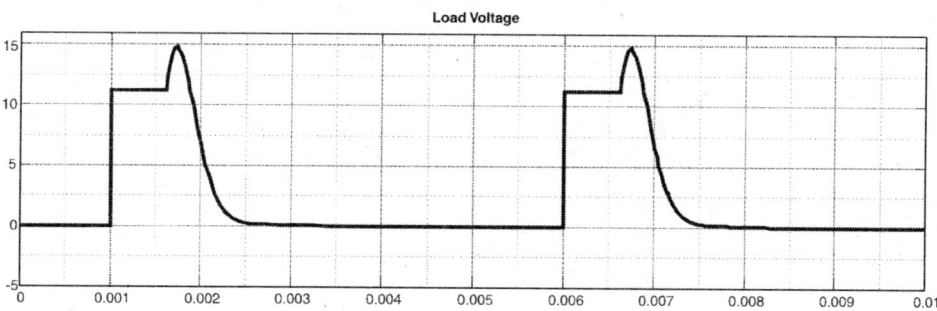

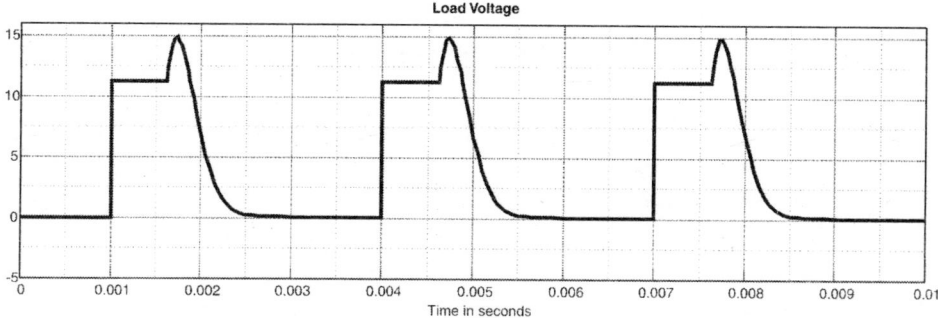

Variable frequency control

3. The DC supply is switched ON.
4. Adjust the pot ON the UJT triggering circuit and set a proper value of the chopping period T. This can be found out by observing the waveform of load voltage using a DSO.
5. Ensure that the circuit operates in the proper way.
6. Observe the waveform of load voltage and plot on a graph sheet.
7. Now keeping the values of L and C the same, vary the pot in the UJT triggering circuit to change the chopping period.
8. Observe again the waveform of load voltage and plot on the same graph sheet.
9. Steps 7 and 8 may be repeated to get different waveform for different value of T.
10. Switch off the DC supply.

B17.8 OBSERVATION

Observe the output voltage waveforms under constant frequency control and variable frequency control under different duty ratios and plot them on graph sheets.

Experiment B18

Design and Testing of a Voltage Commutated Chopper

B18.1 OBJECTIVE

To study the performance of a voltage commutated chopper, both under constant frequency control and variable frequency control.

B18.2 THEORY

The auxiliary commutation circuit, explained in section B14, can be operated as a voltage commutated chopper. The reason for this is that the main thyristor is turned OFF by forced commutation by applying a reverse voltage across it. The principle of operation, analysis and design are all explained in the same section. In a DC-DC chopper, the output voltage can be controlled by varying the duty ratio of the chopper. The duty ratio is the ratio of the time for which the semiconductor switch is ON to the switching period. The switching period is also called the chopping period.

Duty ratio $$d = \frac{t_{ON}}{T} = t_{ON}f$$

Where, T is the chopping period and $f = 1/T$ is the chopping frequency.

A voltage commutated chopper is a buck converter. The output voltage of a buck converter or step down chopper is given by

$$V_0 = \delta V_S$$

where, V_0 is the output voltage and V_S is the input voltage to the converter.

B18.3 OPERATION AS A CHOPPER

A voltage commutated chopper is shown in the figure given below. Duty ratio can be varied either by varying t_{ON}, the time for which the semiconductor switch is ON with the switching frequency constant or by keeping t_{ON} constant and varying the frequency f. The former is called the constant frequency control or pulse width modulation control and the latter is called the variable frequency control. Hence the output voltage of the chopper can be controlled by two different ways and these are called time ratio controls.

B18.3.1 Constant Frequency Control or Pulse Width Modulation Control

The frequency and hence the time period T of the chopper is kept constant and the output pulse width t_{ON} is varied. This method is known as PWM control or Pulse

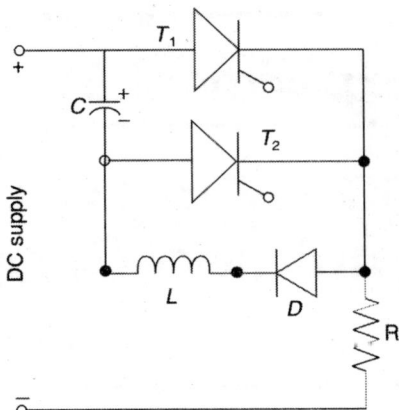

Width modulation control. In this method, the thyristor ON time (time for which the thyristor is ON) is varied but the chopping period T (or chopping frequency f) is kept constant. This can be achieved by varying either L or C or both in the thyristor commutation circuit. Note that the minimum ON time for the thyristor is $t_{ON} = \pi\sqrt{LC}$ sec.

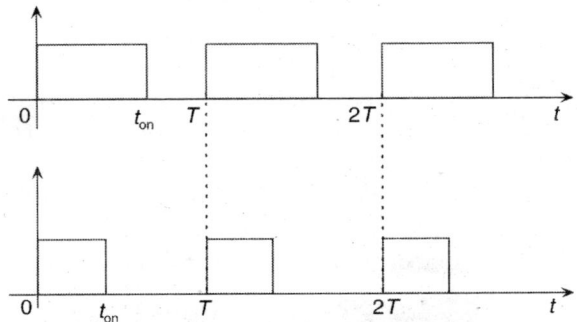

B18.3.2 Variable Frequency Control

In this method, either the pulse width t_{ON} is kept constant varying t_{OFF} or the pulse width is varied keeping t_{off} constant. Thus the frequency can be varied, thereby changing the duty cycle. This is known as frequency control method.

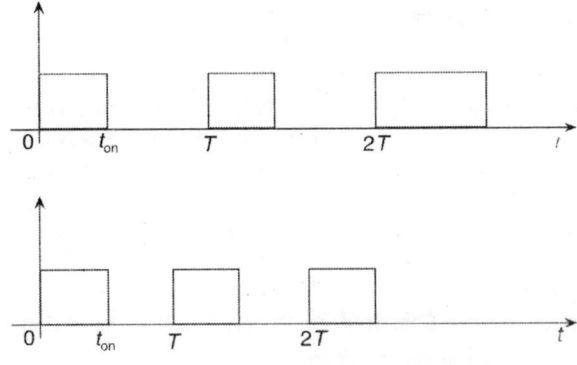

Varying frequency with fixed t_{on}

B18.4 CIRCUIT DIAGRAM

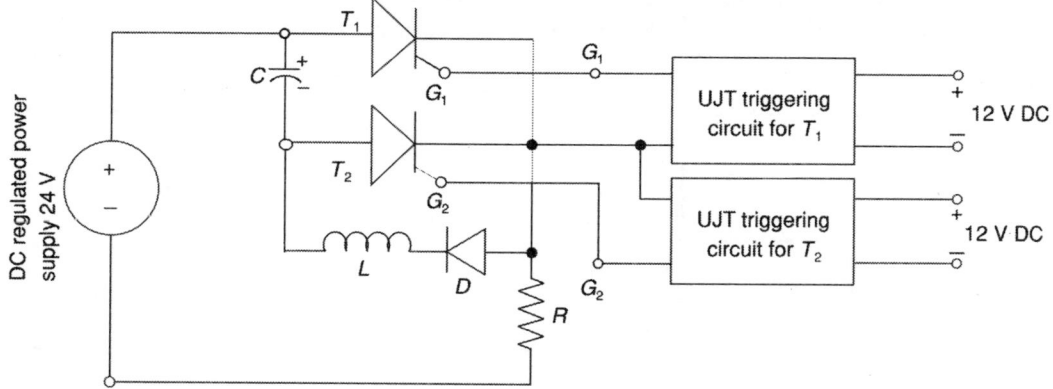

B18.5 DESIGN

The design of commutating elements L and C and the design of UJT triggering circuit are given in Section B14.6 and B3.6.

B18.6 EQUIPMENT AND COMPONENTS

Sl No	Components	Specification	Quantity
1.	Thyritor	C106D	2
2.	Thyristor triggering circuit using UJT		2
3.	Load resistor	100 Ω, 10 W	1
4.	Inductor	5 mH	1
5.	Capacitor	3.3 μF	1
6.	Diode	1N4007	1
7.	DC supply	24 V	1
8.	DSO		1
9.	Breadboard		

B18.7 PROCEDURE

B18.7.1 Constant Frequency Control or PWM Control (t_{ON} is Varied and T is Kept Constant)

1. Make the connections as per the circuit diagram, selecting a set of values for L and C.
2. Switch ON the main supply.
3. Switch ON the triggering circuit and ensure the existence of triggering pulses.
4. Switch ON the auxiliary thyristor.
5. Switch ON the main thyristor.
6. Observe the waveform of load voltage on the DSO to ensure that the circuit operates in a proper way.
7. Looking at the DSO, vary the variable resistor R_2 of the triggering circuit for the main thyristor and keep the chopping period constant T.

8. Change the values of L and/or C. This is to vary the t_{ON}.
9. Observe the load voltage for different values of t_{ON}.
10. Switch off the DC supply and remove circuit connections.

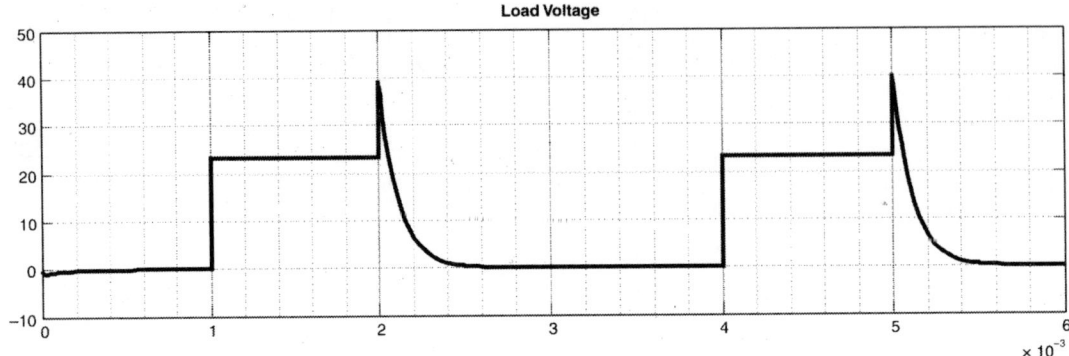

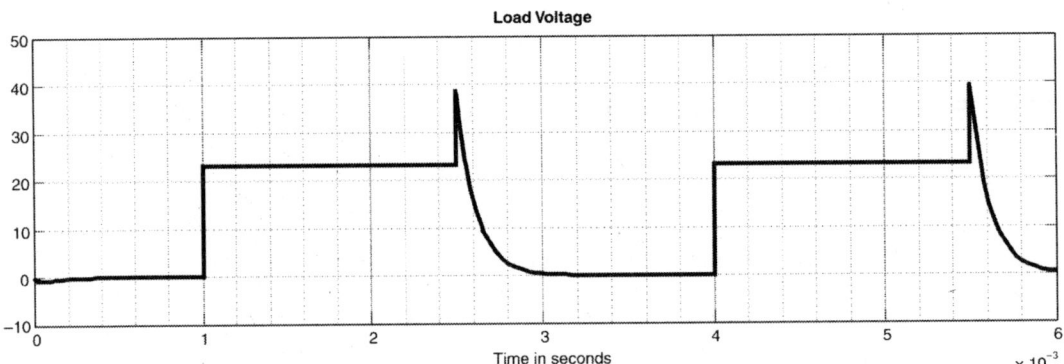

Constant frequency operation of voltage commutated chopper
(t_{ON} is varied and T is kept constant)

B18.7.2 Frequency Control (t_{ON} is Kept Constant and T is Varied)

1. Keep the same circuit connections as in the previous case. Select proper values of L and C and keep these values constant. This will make t_{ON} constant.
2. Switch ON the main supply.
3. Switch ON the triggering circuit and ensure the existence of triggering pulses.
4. Switch ON the auxiliary thyristor.
5. Switch ON the main thyristor.
6. Observe the waveform of load voltage on the DSO to ensure that the circuit operates in a proper way.
7. Looking at the DSO, vary the variable resistor R_2 of the triggering circuit for the main thyristor. This is to vary the period T of the chopper.
8. Observe the load voltage waveform for different values of T.
9. Switch off the supply and remove circuit connections.

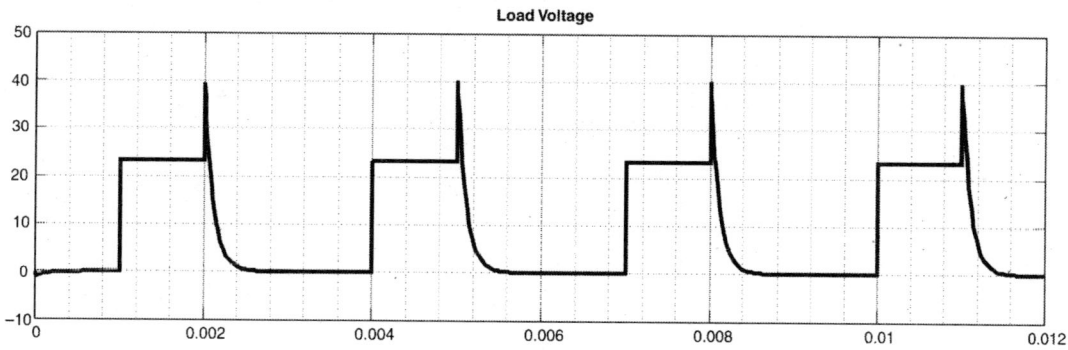

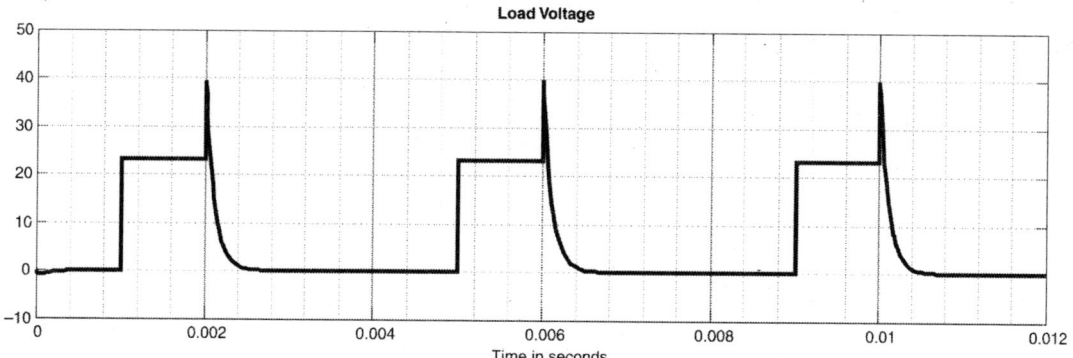

Variable frequency operation of voltage commutated chopper
(t_{ON} is constant and T is varied)

B18.8 OBSERVATION

Observe the output waveforms under constant frequency control and variable frequency control under different duty ratios and plot them on graph sheets.

Section

C

Circuit Simulation Using MATLAB/Simulink

Experiment C1

Simulation of an R-Triggering Circuit

C1.1 OBJECTIVE

To simulate the resistance triggering circuit, for a thyristor, using matlab/simulink.

C1.2 R-TRIGGERING CIRCUIT USING MATLAB/SIMULINK

Theory and design of R-triggering circuit is discussed in B1, in section B.

Figure C1.1 shows the basic MATLAB/Simulink model of a resistance triggering circuit.

C1.3 MODELLING AND SIMULATION OF A R-TRIGGERING CIRCUIT USING MATLAB/SIMULINK

Following steps are to be followed for analyzing and modeling the R-triggering circuit.

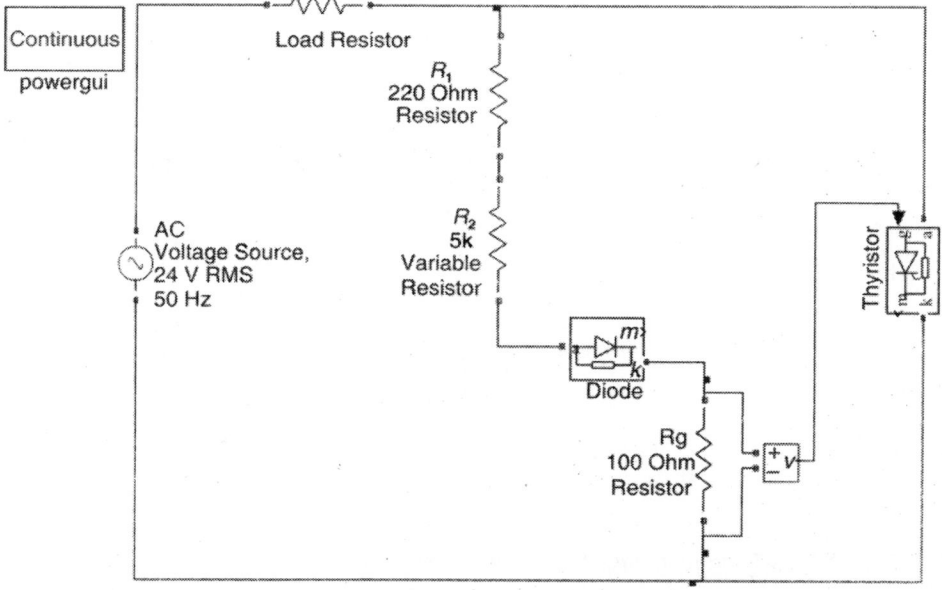

Fig. C1.1: Basic model of R-triggering circuit using MATLAB/Simulink

Step 1: Enter "simulink" in the MATLAB command window to open the simulink library browser or click the simulink button in the MATLAB toolbar.

Step 2: Select File > New > Model in the simulink library browser to create a new model or press Ctrl + N. An empty model window is opened and it can be saved by selecting File > Save in the model window.

Step 3: Drag and place the following building blocks from *simulink library* and *Sim Power Systems Library* to the simulink model file.

    (a) Powergui block [Sim Power Systems]

    (b) AC voltage source [Sim Power Systems> Electrical Sources]

    (c) Series RLC branch (configure it to *R*) [Sim Power Systems > Elements]

    (d) Diode [Sim Power Systems > Power Electronics]

    (e) Thyristor [Sim Power Systems > Power Electronics]

    (f) Voltage Measurement [Sim Power Systems > Measurements]

    (g) Scope [Simulink > Sinks]

    (h) From [Simulink > Signal Routing]

    (i) Goto [Simulink > Signal Routing]

Step 4: Connect the blocks according to the circuit diagram. Also connect required number of voltage measurement blocks to the *scope* through *Goto* and *from* blocks, as shown in Fig. C1.2, to observe the waveforms. The *Goto* and *from* blocks need not be necessarily used, but it will reduce the number of connections in the circuit.

Step 5: Set the parameters of the individual blocks. Components and specifications used in the R-triggering circuit are:

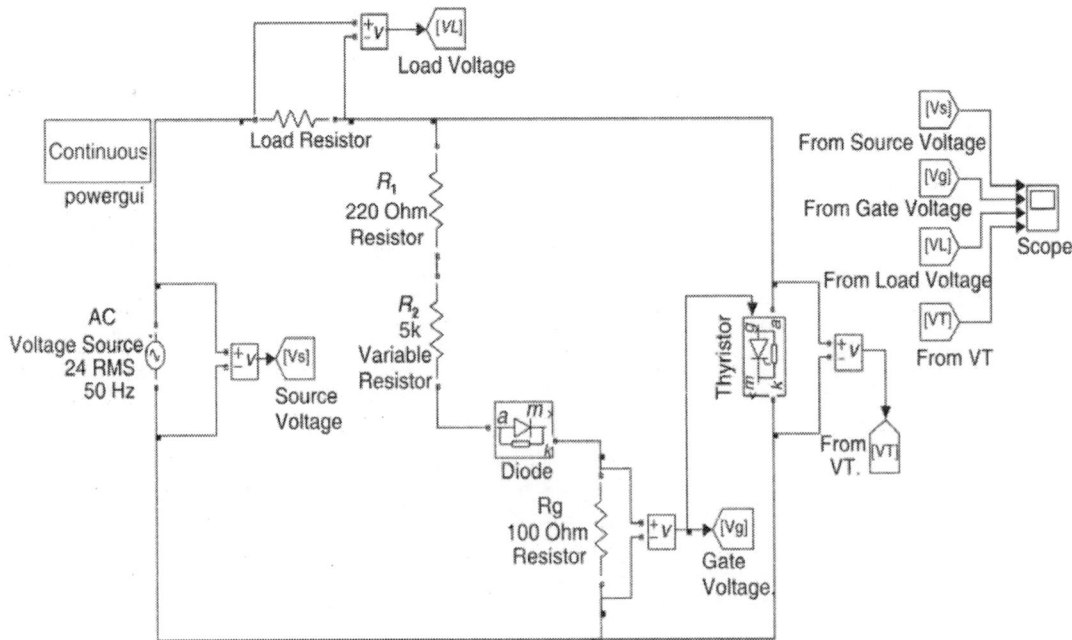

Fig. C1.2: R-Triggering circuit with measurement blocks using MATLAB/Simulink

1. AC voltage source   34 V (24 V RMS), 50 Hz
2. Resistor R_1 220 Ω
3. Resistor R_2 0–5 kΩ
4. Resistor R_g 100 Ω
5. Load resistor R_L 100 Ω
6. Thyristor
7. Diode
8. Goto and From blocks
9. Voltage measurement blocks
10. Scope
11. Powergui block

Step 6: Vary the value of variable resistor R_2 to adjust firing angle, α. Let us select R_2 as 1 kΩ.

Step 7: Save the simulink model. Run the simulation by selecting simulation > start in the model window or use simulation button in the model window toolbar.

The run time can be entered and modified in the space provided in the model window toolbar. For example: run time = 0.06 seconds.

Step 8: Observe and note down the following waveforms by opening the *scope* and save the result.

(a) Source voltage

(b) Gate voltage

(c) Output voltage or voltage across the load and

(d) Voltage across the thyristor

Note: It is recommended to use high values of snubber circuit components for both the diode and thyristor to get better and smooth output voltage waveforms.

Step 9: Adjust the resistor R_2 to keep another value of the firing angle. Continue with steps 6, 7 and 8.

Note:

1. The thyristor is turned OFF or commutated when current through it goes below the holding current. For the sake simplicity usually we consider the holding current of the thyristor is zero and but in practice holding current is not zero. If we closely watch/zoom the simulated load voltage waveform at 180° (negative zero crossing of the supply voltage) as shown in the Fig. C1.3, it can be seen that thyristor is turned OFF just before 180°. This is because the load current or thyristor current becomes less than the holding current at that instant.

2. The firing angle α can be calculated from the instantaneous value of the load voltage waveform. At the instant of turning ON the thyristor, the instantaneous value of the load voltage is noted down. Let it be 14 V.

$$v = V_m \sin \alpha \qquad 14 = 34 \sin \alpha$$
$$\alpha = \sin^{-1}(14/34) = 24.32°$$

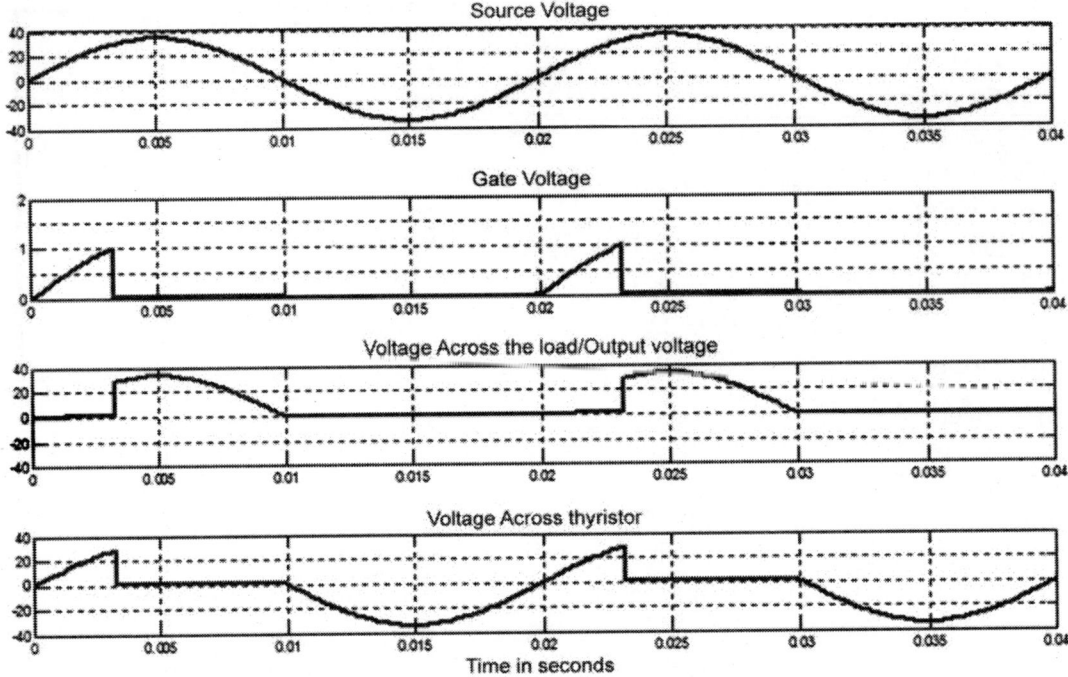

Fig. C1.3: Simulation waveforms of R triggering circuit

C1.4 SIMULATION RESULT

The R triggering circuit was modeled and simulated using MATLAB/Simulink.

Experiment C2

Simulation of an RC Triggering Circuit

C2.1 OBJECTIVE

To simulate a resistance-capacitance (RC) triggering circuit to trigger the gate of a thyristor using MATBLAB/Simulink.

C2.2 RC TRIGGERING CIRCUIT USING MATBLAB/SIMULINK

Theory and design of RC-triggering circuit is discussed in B2, in section B. Figure C2.1 shows the MATLAB/Simulink model of a basic RC triggering circuit for a half-wave rectifier. The basic model of an RC triggering circuit for a full-wave rectifier using MATLAB/Simulink is shown in Fig. C2.4.

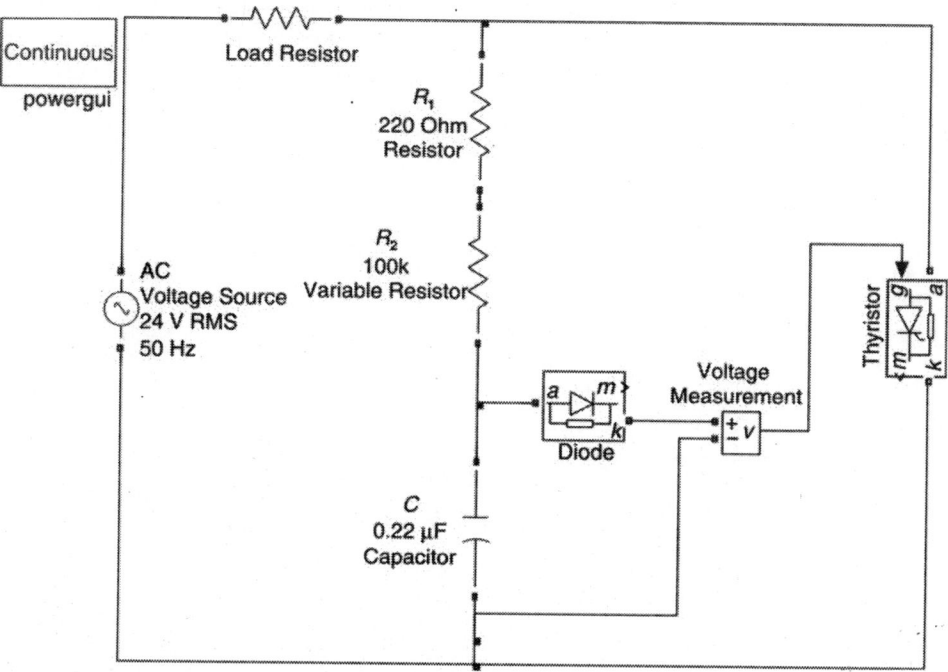

Figure C2.1: Basic model of an RC triggering circuit for Half-wave rectifier using MATLAB/Simulink

C2.3 MODELLING AND SIMULATION OF RC TRIGGERING CIRCUIT USING MATLAB/SIMULINK

Following steps are to be followed for analyzing and modeling the RC-triggering circuit.

Step 1: Enter "simulink" in the MATLAB command window to open the simulink library browser or click simulink button in the MATLAB toolbar.

Step 2: Select **File > New > Model** in the Simulink Library Browser to create a new model or press Ctrl + N. An empty model window is opened and it can be saved by selecting **File>Save** in the model window.

Step 3: Drag and place the following building blocks from *Simulink library* and *Sim Power Systems library* to the new simulink model file

(a) Powergui block [Sim Power Systems]
(b) AC Voltage Source [Sim Power Systems > Electrical Sources]
(c) Series RLC branch (configure it to R) [Sim Power Systems > Elements]
(d) Series RLC branch (configure it to C) [Sim Power Systems > Elements]
(e) Thyristor [Sim Power Systems > Power Electronics]
(f) Diode [Sim Power Systems > Power Electronics]
(g) Voltage Measurement [Sim Power Systems > Measurements]
(h) Scope [Simulink > Sinks]
(i) Display [Simulink > Sinks]
(j) From [Simulink > Signal Routing]
(k) Goto [Simulink > Signal Routing]

Step 4: Connect the blocks according to the circuit diagram of the half wave RC triggering circuit. Also connect required number of voltage measurement blocks to the *scope*, as shown in the Fig. C2.2, to observe necessary waveforms. The *Goto* and *from* blocks need not be necessarily used, but it will reduce the number of connections in the circuit.

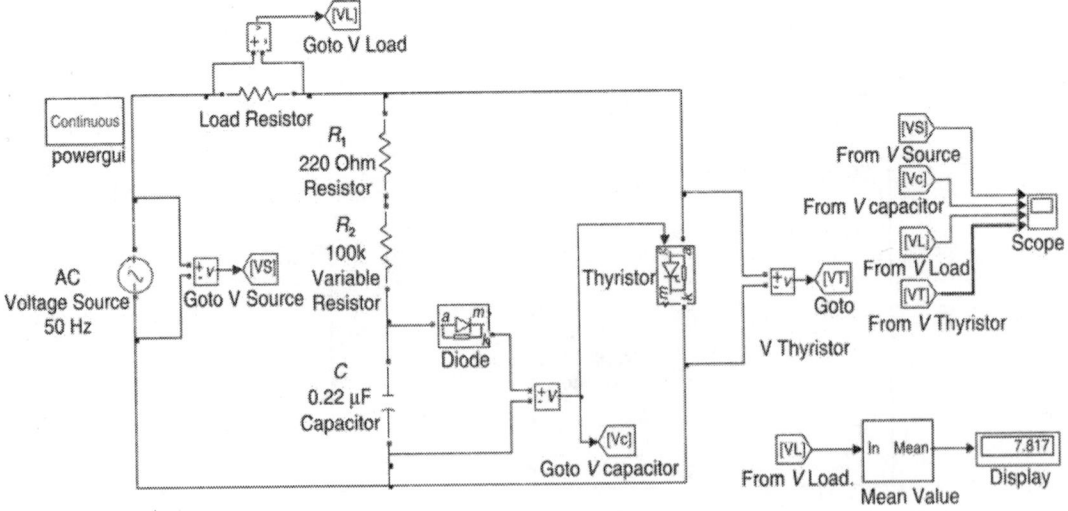

Fig. C2.2: The half-wave RC triggering circuit model with measurement blocks using MATLAB/Simulink

Step 5: Set the parameters of the individual blocks. Components and specifications used in the half wave RC triggering circuit model are:

1.	AC voltage source		34 V (Max. value), 50Hz
2.	Resistor	R_1	220 Ω
3.	Resistor	R_2	0–100 kΩ
4.	Capacitor	C	0.22 μF
5.	Load resistor	R_L	100 Ω
6.	Thyristor		
7.	Diode		
8.	Goto and From blocks		
9.	Voltage measurement blocks		
10.	Scope		
11.	Powergui block		

Step 6: Vary the value of variable resistor R_2 to adjust firing angle, α. Let us select R_2 as 20 kΩ.

Step 7: Save the Simulink model. Run the simulation by selecting **simulation > start** in the model window or use **simulation button** in the model window toolbar. The run time can be entered and modified in the space provided in the model window tool bar.

For example, run time = 0.08 seconds.

Step 8: Observe the following waveforms by opening the *scope*. The waveforms are shown in the Fig. C2.3.

(a) Source voltage

(b) Capacitor voltage [voltage across the capacitor]

(c) Load voltage [Output voltage or voltage across the load resistor R]

(d) Voltage across the thyristor

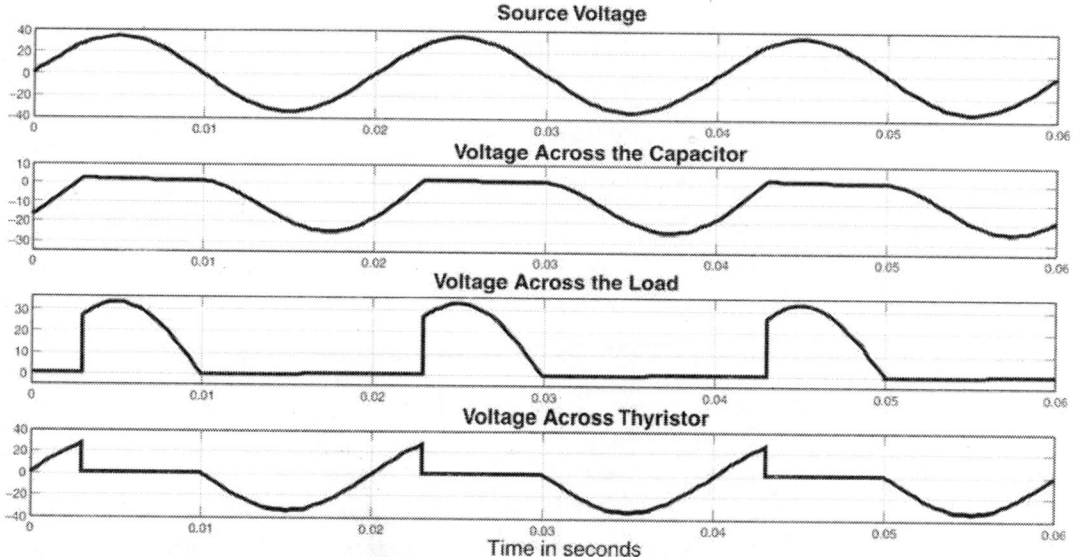

Fig. C2.3: Waveforms - RC triggering circuit for half wave rectifier (a) source voltage (b)Capacitor voltage (c) Voltage across the load (d) Voltage across thyristor

Determine at what firing angle the thyristor is turned ON, by comparing the waveforms of capacitor voltage and load voltage. For the method of calculation, refer to experiment C1.

Next, an RC triggering circuit for a full-wave rectifier is to be modelled and simulated. Connect the blocks according to the circuit diagram of the full-wave RC triggering circuit shown in Fig. C2.4.

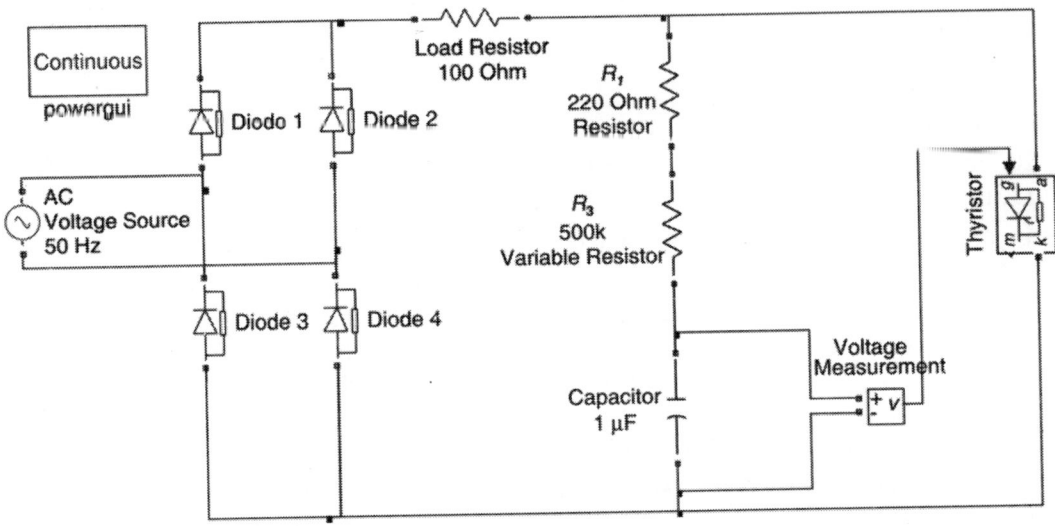

Fig. C2.4: Basic model of an RC triggering circuit for full wave rectifier using MATLAB/Simulink

Now follow steps 1 to 8.

Step 9: Save and Run the Simulink model. Observe the following waveforms by opening the scope. The waveforms are shown in the Fig. C2.5.

(a) Source voltage

(b) Voltage across the diode bridge rectifier

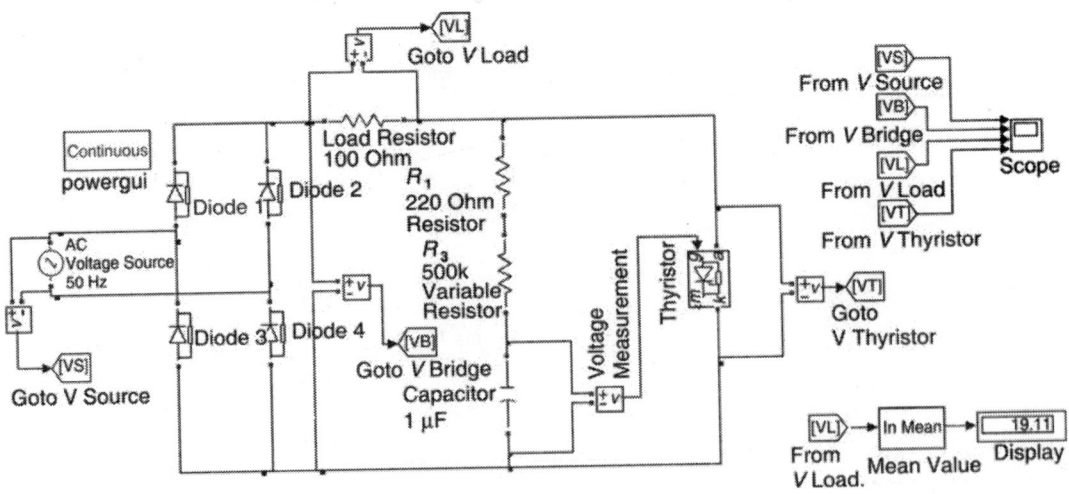

Fig. C2.5: RC triggering circuit model for full-wave rectifier with measurement blocks using MATLAB/Simulink

(c) Load voltage [Output voltage or voltage across the load resistor *R*]

(d) Voltage across thyristor

Determine at what firing angle the thyristor is turned ON, by comparing the waveforms of capacitor voltage and load voltage. For the method of calculation, refer to experiment C1.

Note: It is recommended to use high values of Snubber circuit components for both the diode and thyristor to get better and smooth output voltage waveforms.

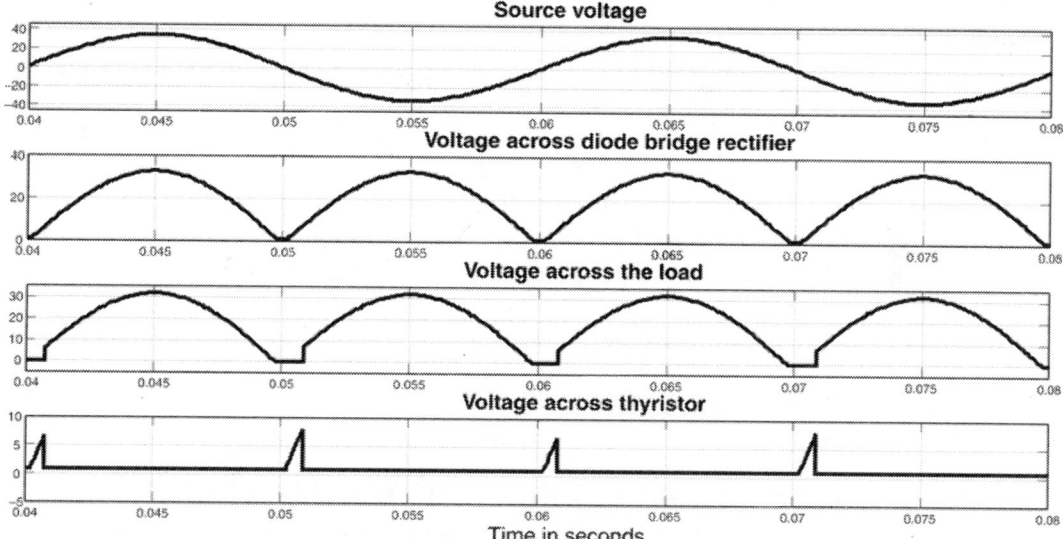

Fig. C2.6: Waveforms related to RC triggering for full-wave rectifier (a) source voltage (b) Voltage across diode bridge rectifier (c) Voltage across the load (d) Voltage across thyristor

C2.4 SIMULATION RESULT

The full-wave and half-wave RC triggering circuits were modelled and simulated using MATLAB/Simulink.

Simulation of a Single Phase Half-wave Rectifier

C3.1 OBJECTIVE

i. To simulate a single phase half-wave rectifier circuit using MATBLAB/Simulink.
ii. To perform the operation of single phase half-wave rectifier under following load condition
   1. Purely resistive load
   2. Resistive-inductive load
   3. Resistive-inductive-back EMF load
   4. Resistive-inductive-back EMF load with freewheeling diode.

C3.2 SINGLE PHASE HALF-WAVE RECTIFIER USING MATLAB/SIMULINK

Theory and working of a single phase half-wave rectifier is discussed in B5 in section B. A basic circuit for the simulation of single phase half-wave rectifier using MATLAB/Simulink is shown in Fig. C3.1.

C3.3 MODELLING AND SIMULATION OF SINGLE PHASE HALF-WAVE RECTIFIER CIRCUIT USING MATLAB/SIMULINK

Following steps are to be followed for analyzing and modelling the single phase half-wave rectifier circuit.

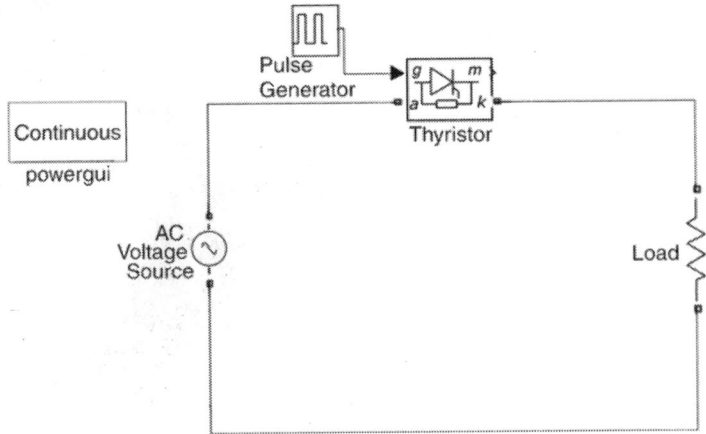

Fig. C3.1: Single phase half-wave rectifier circuit using MATLAB/Simulink

Step 1: Enter "simulink" in the MATLAB Command Window to open the Simulink Library Browser or click Simulink button in the MATLAB toolbar.

Step 2: Select **File > New > Model** in the Simulink Library Browser to create a new model or press Ctrl + N. An empty model window is opened and it can be saved by selecting **File>Save** in the model window.

Step 3: Drag and place the following building blocks from *Simulink library* and *Sim Power Systems library* to the simulink model file, as per the circuit diagram for different loads.

(a) Powergui block [Sim Power Systems]
(b) AC Voltage Source [Sim Power Systems > Electrical Sources]
(c) Series RLC branch (configure it to R/RL) [Sim Power Systems > Elements]
(d) Thyristor [Sim Power Systems> Power Electronics]
(e) Pulse generator [Simulink > Sources]
(f) Diode [Sim Power Systems > Power Electronics]
(g) DC Voltage Source [Sim Power Systems > Electrical Sources]
(h) Voltage Measurement [Sim Power Systems > Measurements]
(i) Current Measurement [Sim Power Systems > Measurements]
(j) Scope [Simulink > Sinks]
(k) Mean value [Sim Power Systems > Extra library > Measurements]
(l) RMS [Sim Power Systems > Extra library > Measurements]
(m) Display [Simulink > Sinks]
(n) From [Simulink > Signal Routing]
(o) Goto [Simulink > Signal Routing]

Half-wave Rectifier with R Load

Step 4: Connect the blocks according to the circuit diagram of the single phase half-wave rectifier circuit with R load. Also connect required number of current and voltage measurement blocks to the scope, as shown in the Fig. C3.2, to observe the necessary waveforms. The *Goto* and From blocks need not be necessarily used, but it will reduce the number of connections in the circuit.

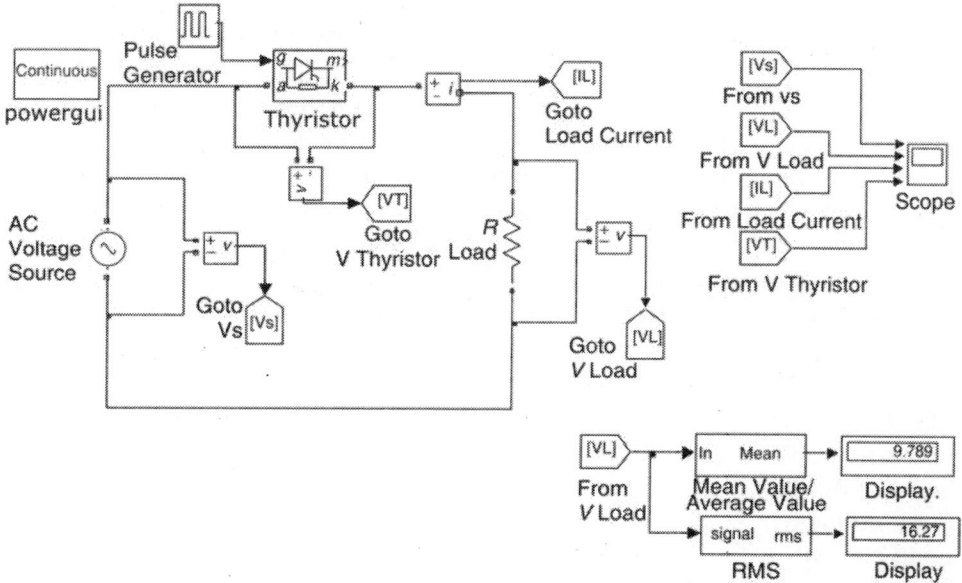

Fig. C3.2: Single phase half-wave rectifier circuit with R load and with measurement blocks

Step 5: Set the parameters of the individual blocks. Components and specifications used in the single phase half-wave rectifier circuit with R load are:

1. AC voltage source	34 V (Max. value), 50 Hz
2. Load Resistor	100 Ω
3. Thyristor	
4. Pulse Generator	
5. Powergui block	
6. Goto and from blocks	
7. Voltage measurement blocks	
8. Current measurement blocks	
9. Scope	
10. Mean value block	
11. RMS block	
12. Display	

The parameters of pulse generator used are:

1. Period (secs): 0.02   (for 50 Hz supply)
2. Pulse Width (% of period): 5
3. Phase delay (secs): 5e –3/3

Phase delay angle can be calculated as explained here. For a firing angle of, $\alpha = 30°$,

$$\text{Phase delay} = \frac{30}{360} \times 20\,ms = \frac{5}{3}ms$$

Note:

1. In simulink, 10^{-3} is entered as $e - 3$. Similarly 1^{-6} is entered as $e - 6$.
   For example
   8 mH is entered as $8e - 3$ 100 μF is entered as $100e - 6$
   1millisecond is entered as $1e - 3$ 1 kΩ is entered as $1e3$
2. Period of source voltage = 1/supply frequency = 1/50 Hz = 0.02 seconds = 20 milli seconds
3. For a thyristor, a gating signal with only a small pulse width is required to latch the device.

Step 6: Since the circuit provides AC to DC conversion, the average and RMS value of the load voltage can be calculated. For that *mean value block* and *RMS block* are used and corresponding values are displayed in the *display block*. Double click on *mean value block and RMS block* and enter *averaging period* as 0.02 seconds and fundamental frequency as 50 Hz respectively, since supply voltage frequency is 50 Hz.

Step 7: Save the simulink model. Run the simulation by selecting **simulation > start** in the model window or use **simulation button** in the model window toolbar.

The run time can be entered and modified in the space provided in the model window toolbar. For example, run time = 0.08 seconds

Step 8: Observe and note down the following waveforms by opening the scope and save the result.

(a) Source voltage
(b) Load voltage
(c) Load current
(d) Voltage across the thyristor

The waveforms are shown in the Fig. C3.3.

Also note down the average and RMS value of the load voltage from the *display blocks*.

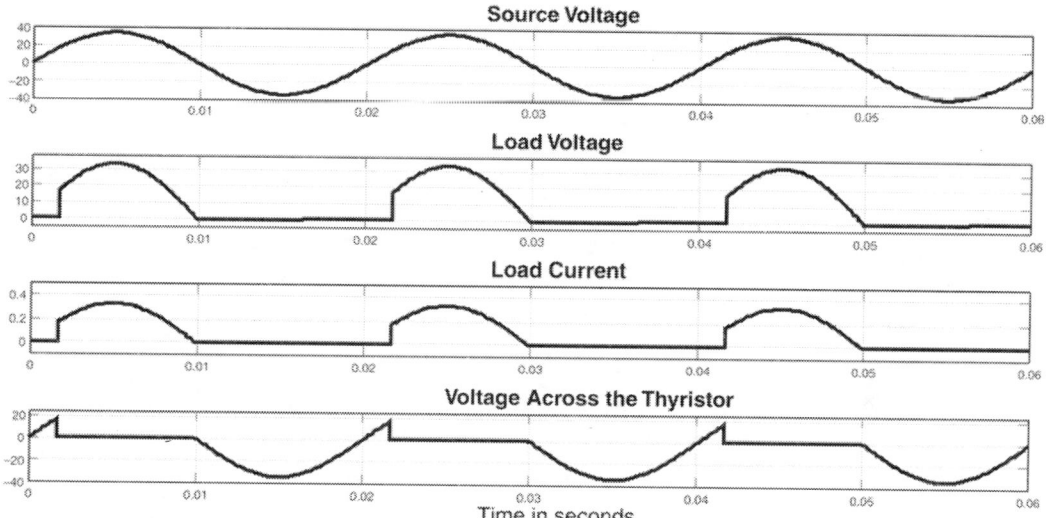

Fig. C3.3: Waveforms of Single phase half wave rectifier with R load (a) Source voltage (b) Voltage across the load (c) Load current (d) Voltage across the thyristor

Half-wave Rectifier with RL Load

Connect the blocks according to the circuit diagram of the single phase half-wave rectifier circuit with RL load as shown in Fig. C3.4. Assume the load current is

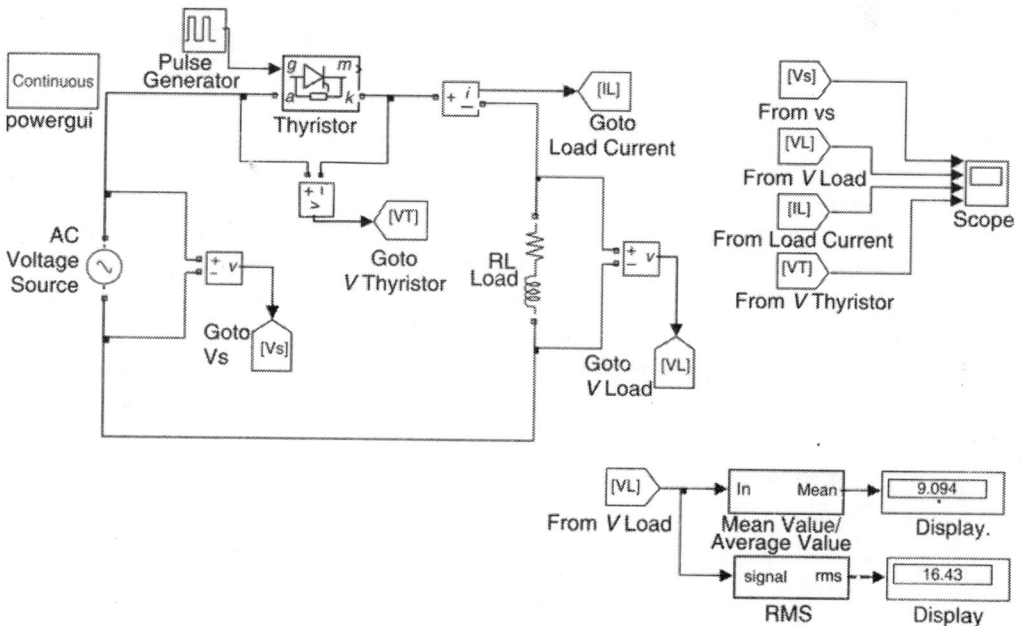

Fig. C3.4: Single phase half-wave rectifier with RL load circuit model with measurement blocks using MATLAB/Simulink

discontinuous, i.e. L/R ratio of the load is low. Double click on R branch and configure to RL. Select the value of inductor L as 150 mH and R as 100 Ω so that L/R ratio is low. Keep the remaining parameters as it is. Follow steps 1 to 8 as in the case of rectifier with R load.

The waveforms are shown in the Fig. C3.5

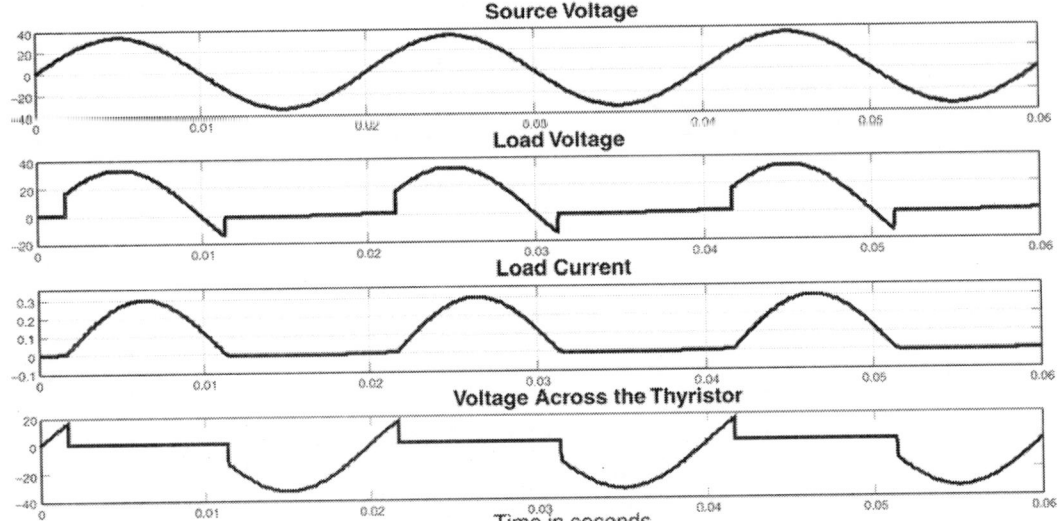

Fig. C3.5: Waveforms of Single phase half-wave rectifier with RL load – Discontinuous conduction mode (a) Source voltage (b) Voltage across the load (c) Load current (d) Voltage across the thyristor

Note down the average and RMS value of the load voltage. It can be seen that the average value is decreased and the RMS value is increased (compared with R load). This is because of the negative excursion of the load voltage shown in Fig. C3.5.

Half-wave Rectifier with RLE Load

Connect the blocks according to the circuit diagram of the single phase half-wave rectifier circuit with RLE load as shown in Fig. C3.6. Assume the load current is discontinuous, i.e. L/R ratio of the load is low. Double click on R branch and configure to RL. Select the value of inductor L = 150 mH, R = 100 Ω and E = 5 V. Keep the remaining parameters as it is. Follow steps 1 to 8 as in the case of rectifier with R load. The waveforms are shown in the Fig. C 3.7

Note down the average and RMS values of the load voltage. It is to be noted that the thyristor can be triggered only when the supply voltage, $v_s = V_m \sin \omega t$ is greater than the back EMF E.

If θ is the minimum firing angle below which the thyristor cannot be turned on, then

$$V_m \sin \theta = E$$

or

$$\theta = \sin^{-1}\left(\frac{E}{V_m}\right)$$

Similarly maximum firing angle is $(\pi - \theta)$

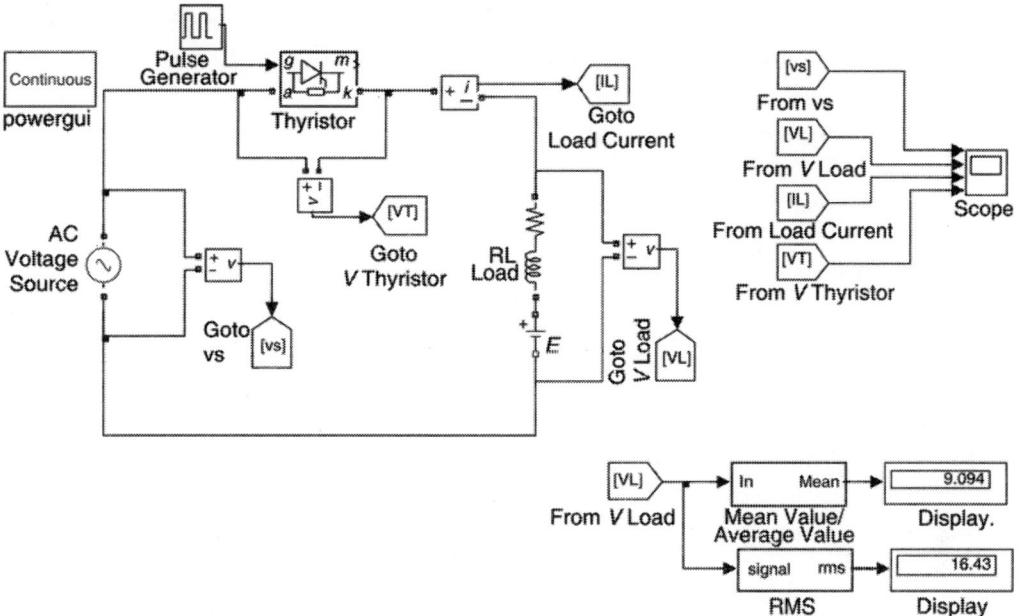

Fig. C3.6: Single phase half-wave rectifier with RLE load circuit model with measurement blocks using MATLAB/Simulink

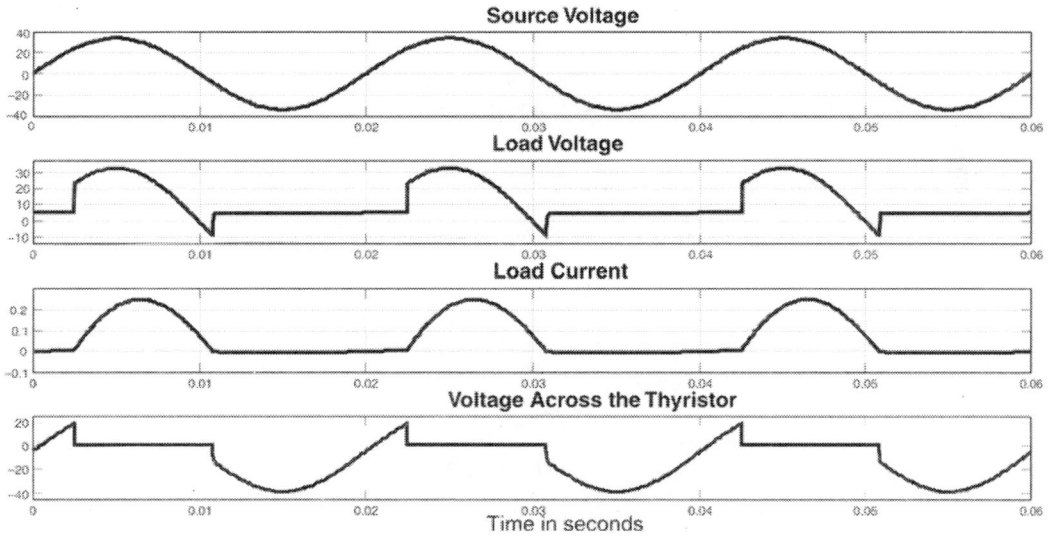

Fig. C3.7: waveforms of single phase half-wave rectifier with RLE load–Discontinuous Conduction mode (a) Source voltage (b) Voltage across the load (c) Load current (d) Voltage across the thyristor

Half-wave Rectifier with RLE Load and Freewheeling Diode (DCM)

Connect the blocks according to the circuit diagram of the single phase half-wave rectifier circuit with RLE load and freewheeling diode as shown in Fig. C3.8. Assume the load current is discontinuous, i.e. L/R ratio of the load is low. Double click on

R branch and configure to *RL*. Select the value of inductor *L* = 150 mH, *R* = 100 Ω and *E* = 5 V. Keep the remaining parameters as it is. Follow steps 1 to 8 as in the case of rectifier with *R* load.

The waveforms are shown in the Fig. C3.9.

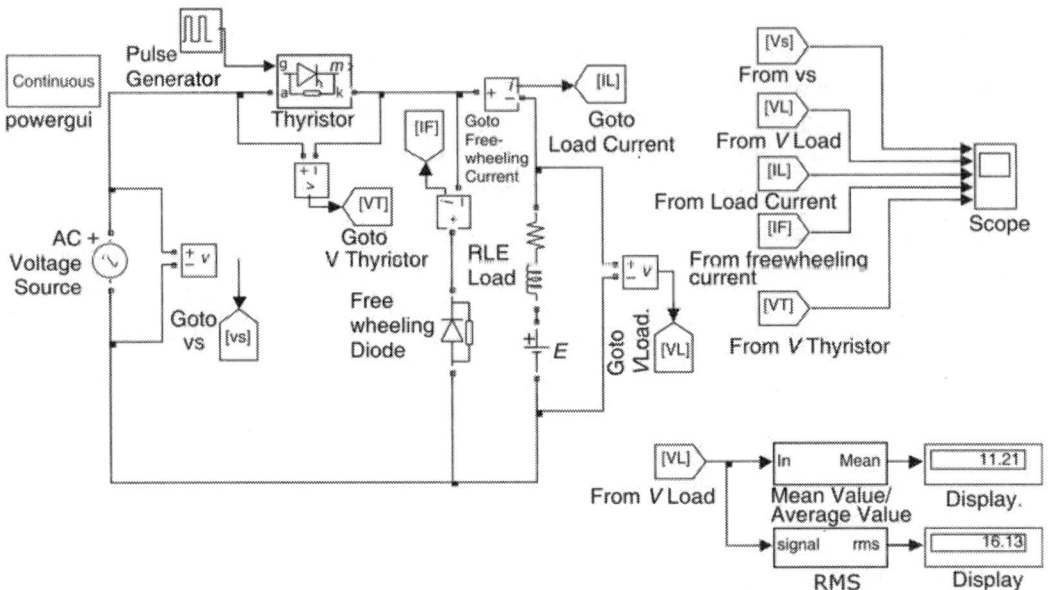

Fig. C3.8: Single phase half-wave rectifier with RLE load and freewheeling diode circuit model with measurement blocks using MATLAB/simulink

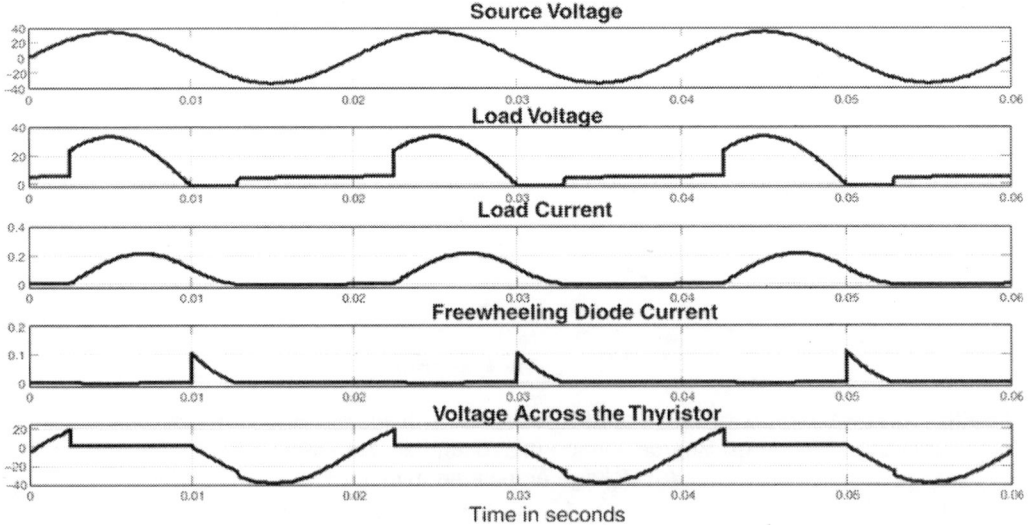

Fig. C3.9: Waveforms of single phase half-wave rectifier with RLE load and freewheeling diode–Discontinuous Conduction Mode (a) Source voltage (b) Voltage across the load (c) Load current (d) Freewheeling diode current (e) Voltage across the thyristor

Note down the average and RMS values of the load voltage. It can be seen that the average value is increased and RMS value is decreased compared with the case of *RLE load without freewheeling diode*. This is because, during the negative half cycle of the supply voltage, the negative excursion of the load voltage is clipped due to the freewheeling diode.

Half-wave Rectifier with RLE Load and Freewheeling Diode (CCM)

Connect the blocks according to the circuit diagram of the single phase half-wave rectifier circuit with RLE load and freewheeling diode as shown in Fig. C3.8. The load current is to be continuous, i.e. the L/R ratio of the load is high. Select a large value of the inductance L say 2500 mH. Keep the remaining parameters as it is. Follow steps 1 to 8 as in the case of rectifier with R load.

The waveforms are shown in the Fig. C3.10.

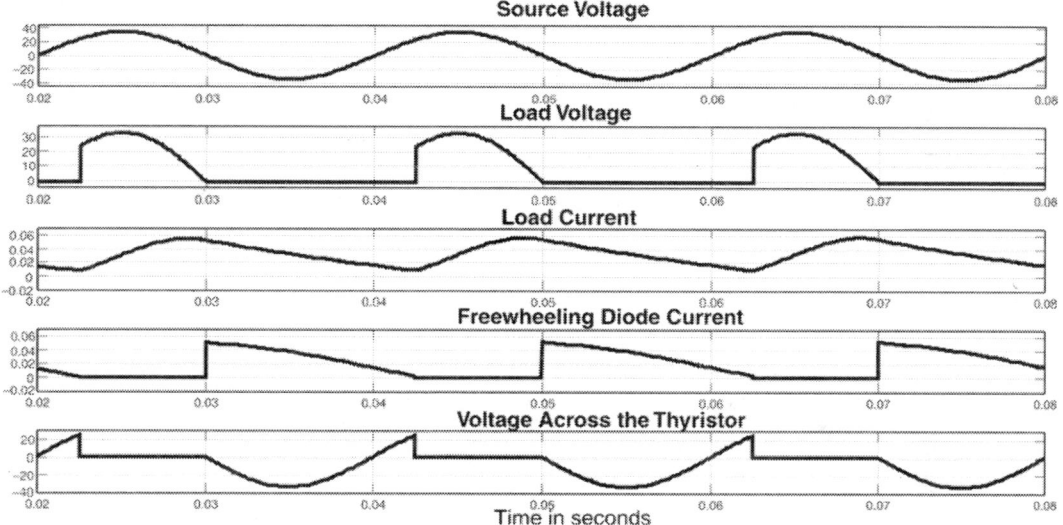

Fig. C3.10: Waveforms of single phase half-wave rectifier with RLE load and freewheeling diode—continuous conduction mode (a) Source voltage (b) Voltage across the load (c) Load current (d) Freewheeling diode current (e) Voltage across the thyristor

Again note down the average and RMS values of the load voltage. It can be seen that both the average value and RMS value of the load voltage is decreased compared with the case of *RLE load discontinuous conduction mode with freewheeling diode*. This is because, during the negative half cycle of the supply voltage, the freewheeling diode starts conduct and which is continuously conducting till thyristor is turned ON again during the positive half cycle of supply voltage.

Note: Different circuits with different load conditions can also be simulated by following the same procedure.

C3.4 SIMULATION RESULT

The single phase half-wave circuit was modelled and simulated using MATLAB/Simulink under the following conditions

1. Purely resistive load
2. Resistive-inductive load-discontinous conduction mode.
3. Resistive-inductive-back EMF load-discontinous conduction mode.
4. Resistive-inductive-back EMF load and with freewheeling diode-discontinous conduction mode.
5. Resistive-inductive-back EMF load and with freewheeling diode-continous conduction mode.

Experiment C4

Simulation of a Single Phase Full-wave Mid-point Rectifier

C4.1 OBJECTIVE

i. To simulate a single phase full-wave rectifier with centre-tapped/mid-point configuration, using MATBLAB/Simulink.

ii. To perform the operation of single phase full-wave rectifier with centre-tapped configuration under following load condition
   1. Purely resistive load
   2. RL load
   3. RL load with freewheeling diode

C4.2 MODELLING AND SIMULATION OF MID-POINT CONFIGURATION OF A SINGLE PHASE FULL-WAVE RECTIFIER CIRCUIT USING MATLAB/SIMULINK

Theory and working of a single phase mid-point converter is discussed in B6 in section B. A basic circuit for the simulation of a single phase mid-point converter using MATLAB/Simulink is shown in Fig. C4.1.

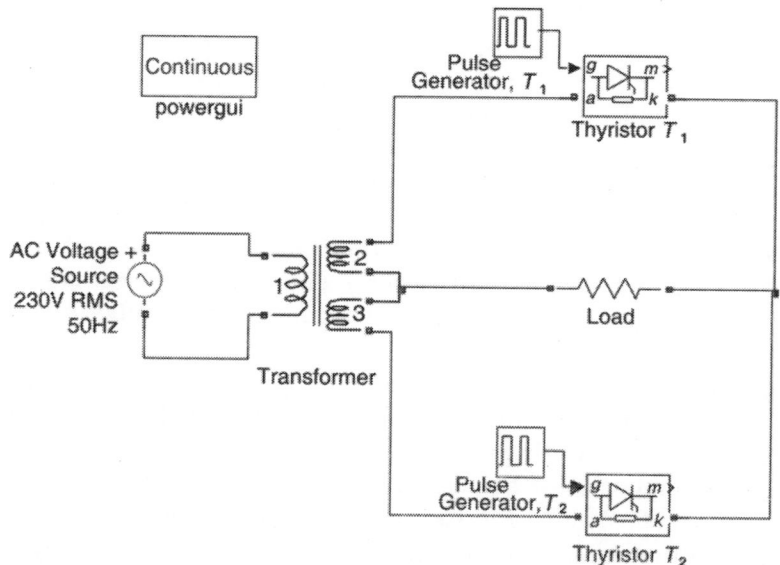

Fig. C4.1: Single phase mid-point converter circuit using MATLAB/Simulink

Following steps are to be followed for analyzing and modelling the single phase mid-point converter circuit.

Step 1: Enter "simulink" in the MATLAB command window to open the simulink library browser or click simulink button in the MATLAB toolbar.

Step 2: Select **File > New > Model** in the simulink library browser to create a new model or press Ctrl + N. An empty model window is opened and it can be saved by selecting **File>Save** in the model window.

Step 3: Drag and place the following building blocks from *simulink library* and *sim power systems library* to the simulink model file according to the circuit diagram with different load conditions.

(a) Powergui block [Sim Power Systems]

(b) AC voltage source [Sim Power Systems > Electrical Sources]

(c) Series RLC branch (configure it to R/RL) [Sim Power Systems > Elements]

(d) Thyristor [Sim Power Systems > Power Electronics]

(e) Pulse generator [Simulink > Sources]

(f) Linear Transformer [Sim Power Systems > Elements > Linear Transformer]

g) Neutral [Sim Power Systems > Elements > Neutral]

(h) Diode [Sim Power Systems > Power Electronics]

(i) Voltage Measurement [Sim Power Systems > Measurements]

(j) Current Measurement [Sim Power Systems > Measurements]

(k) Scope [Simulink > Sinks]

(l) Mean value [Sim Power Systems> Extra library > Measurements > Mean value]

(m) RMS [Sim Power Systems > Extra library > Measurements > RMS]

(n) Display [Simulink > Sinks]

(o) From [Simulink > Signal Routing]

(p) Goto [Simulink > Signal Routing]

Mid-Point Converter with R Load

Step 4: Connect the blocks according to the circuit diagram of a single phase mid-point converter circuit with R load. Also connect required number of current and voltage measurement blocks to the scope, as shown in the Fig. C4.2, to observe the necessary waveforms.

A Neutral block can be used to interconnect two points without drawing a connection line. The *Neutral* block implements a node with a specific node number as shown in Fig. C4.3. Select required numbers of *Neutral* blocks and double click on it and enter corresponding node numbers, as shown in the Fig. C4.3. The *Neutral*, *Goto* and *From* blocks need not be necessarily used, but it will reduce the number of connections in the circuit.

A *linear transformer* model can be used as a transformer with mid-point configuration. A linear transformer model has three windings, namely, winding1 (primary winding), winding–2 (secondary winding–1) and winding–3 (secondary winding–2).The magnetizing characteristics of the core of the *linear transformer* is modelled by a magnetizing branch consisting of inductance, L_m and resistance, R_m.

Double click on the transformer model and select winding–1 voltage, $V_1 = 230V$, winding–2 voltage, $V_2 =12$ V and winding–3 voltage, $V_3 =12V$.Choose a very low value

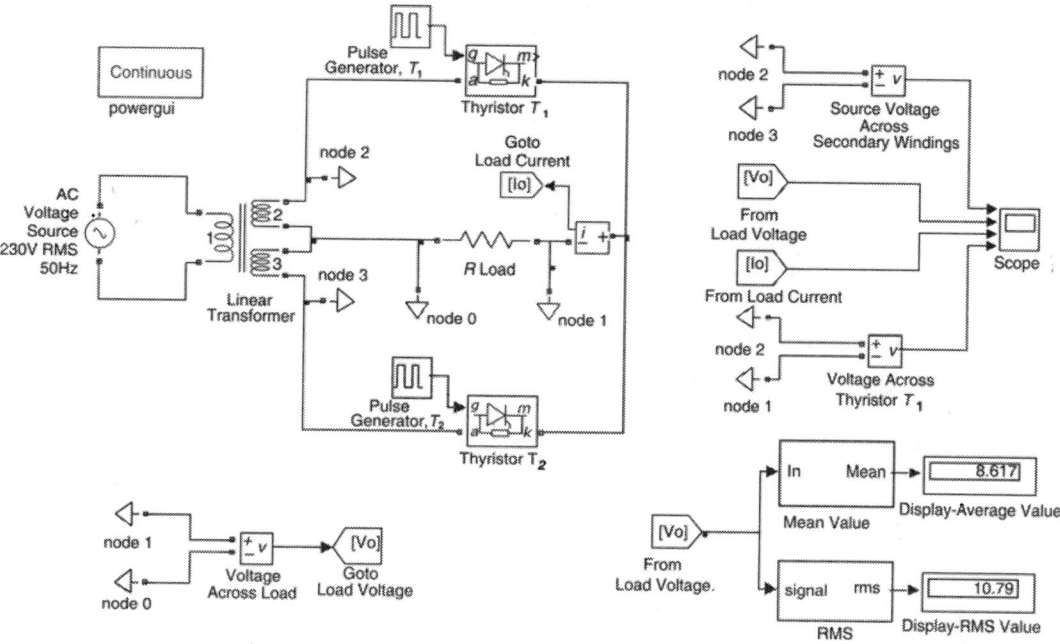

Fig. C4.2: Single phase mid-point converter with R load circuit model with measurement blocks

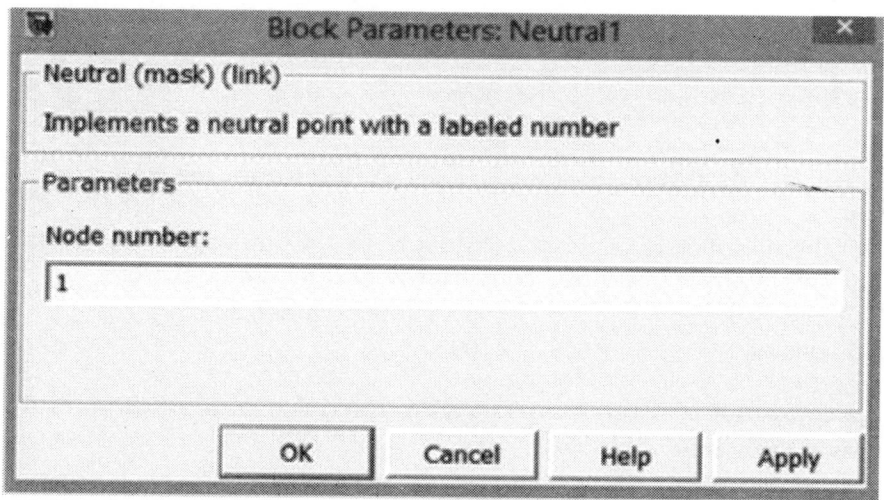

Fig. C4.3: Neutral block-dialog box and parameters

of transformer secondary resistance and secondary leakage inductance values, say $R_2 = R_3 = 0.002$ and $L_2 = L_3 = 0$, as shown in the Fig. C4.4. Select parameter in per unit (pu) and frequency as 50 Hz.

To implement an ideal transformer model, set the winding resistances and inductances to 0, and the magnetization resistance and inductance (R_m L_m) to *inf*.

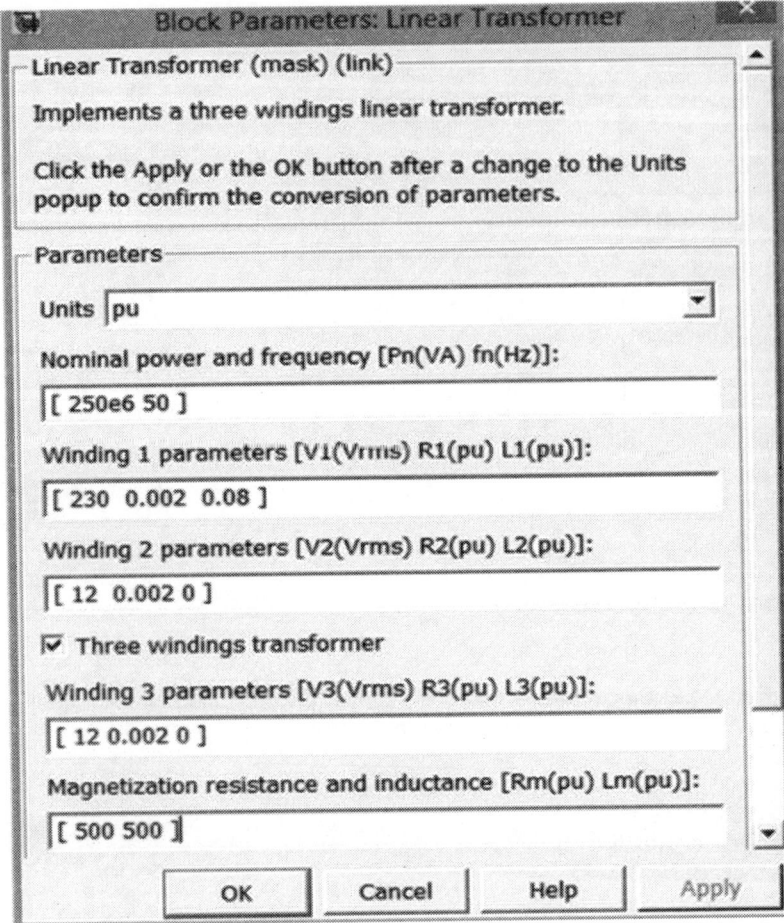

Fig. C4.4: Linear transformer-dialog box and sample parameters

Step 5: Set the parameters of the individual blocks. Components and specifications used in the single phase mid-point converter with R load model are:

1. AC voltage source   325 V (Max. value), 50 Hz
2. Load resistor   100 Ω
3. Thyristor
4. Pulse generator
5. Linear transformer
6. Powergui block
7. Voltage measurement blocks
8. Current measurement blocks
9. Scope
10. Mean value block
11. RMS block
12. Display
13. Neutral blocks
14. Goto and from blocks

The parameters of pulse generator used are:

Pulse generator T_1:
1. Period (secs): 0.02 (for 50 Hz supply)
2. Pulse Width (% of period): 5
3. Phase delay (secs): 2.5e –3

Phase delay angle can be calculated as explained here. For a firing angle of, $\alpha_{T_1} = 45°$,

$$\text{Phase delay} = \frac{45}{360°} \times 20\,\text{ms} = 2.5\,\text{ms}$$

Pulse generator T_2:
1. Period (secs): 0.02 (for 50 Hz supply)
2. Pulse Width (% of period): 5
3. Phase delay (secs): 12.5e –3

Phase delay angle can be calculated for a firing angle of, $\alpha_{T_2} = 180° + 45°$,

$$\text{Phase delay} = \frac{(180° + 45°)}{360°} \times 20\,\text{ms} = 12.5\,\text{ms}$$

Step 6: Since the circuit provides AC to DC conversion, the average and RMS value of the load voltage can be calculated. For that *mean value block* and *RMS block* are used and corresponding values are displayed in the *display block*. Double click on *mean value block and RMS block* and enter *averaging period* as 0.02 seconds and fundamental frequency as 50 Hz respectively, since supply voltage frequency is 50 Hz.

Step 7: Save the Simulink model. Run the simulation by selecting **simulation > start** in the model window or use **simulation button** in the model window toolbar.

The run time can be entered and modified in the space provided in the model window tool bar. For example, run time = 0.08 seconds

Step 8: Observe and note down the following waveforms by opening the scope and save the result.

(a) Transformer secondary voltage
(b) Load voltage
(c) Load current
(d) Voltage across the thyristor T_1

The waveforms are shown in the Fig. C4.5.

Also note down the average and RMS value of the load voltage from the *display blocks*.

Mid-Point Converter with RL Load

Connect the blocks according to the circuit diagram of the single phase mid-point converter circuit with *RL* load as shown in Fig. C4.6. Assume the load current is discontinuous, i.e. L/R ratio of the load is low. Double click on *R* branch and configure to *RL*. Select the value of inductor *L* as 100 mH and *R* as 100 Ω so that L/R ratio is low. Keep the remaining parameters as it is. Follow steps 1 to 8 as in the case of mid-point converter with *R* load.

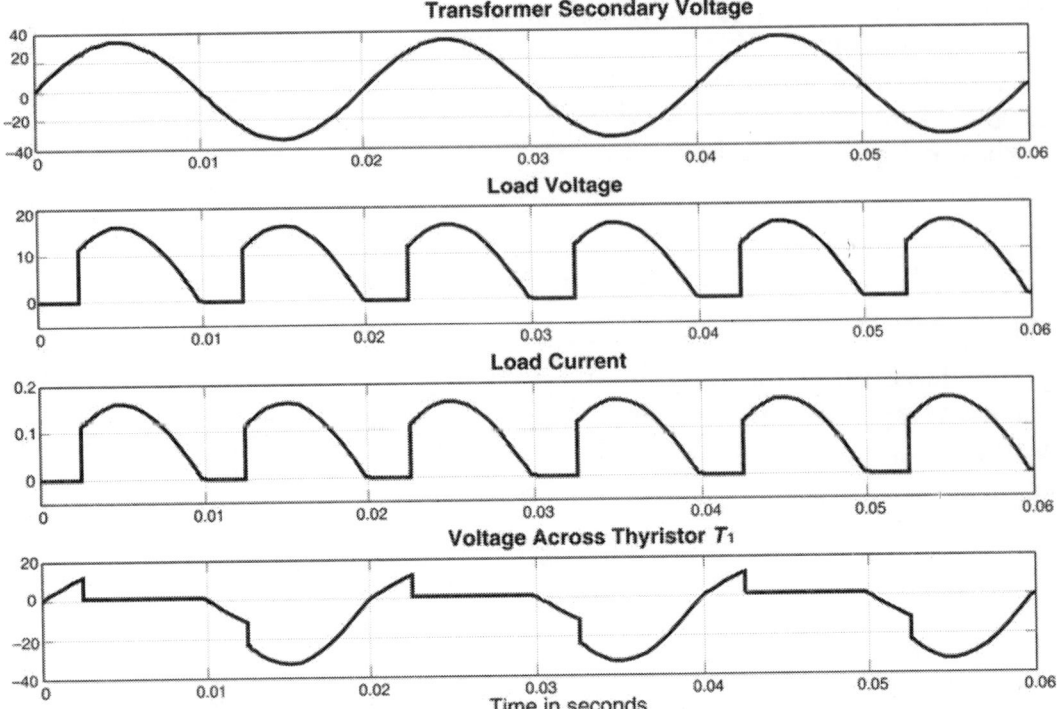

Fig. C4.5: Waveforms of Single phase mid-point converter with R load (a) Transformer secondary voltage (b) Voltage across the load (c) Load current (d) Voltage across the thyristor T_1

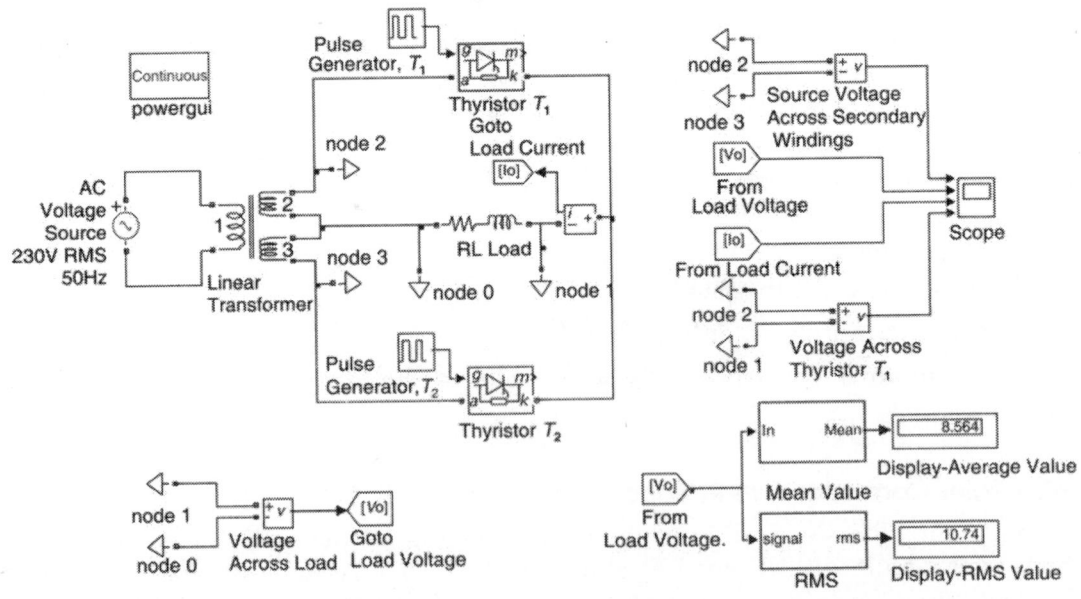

Fig. C4.6: Single phase mid-point converter with RL load circuit model with measurement blocks

The waveforms are shown in the Fig. C4.7.

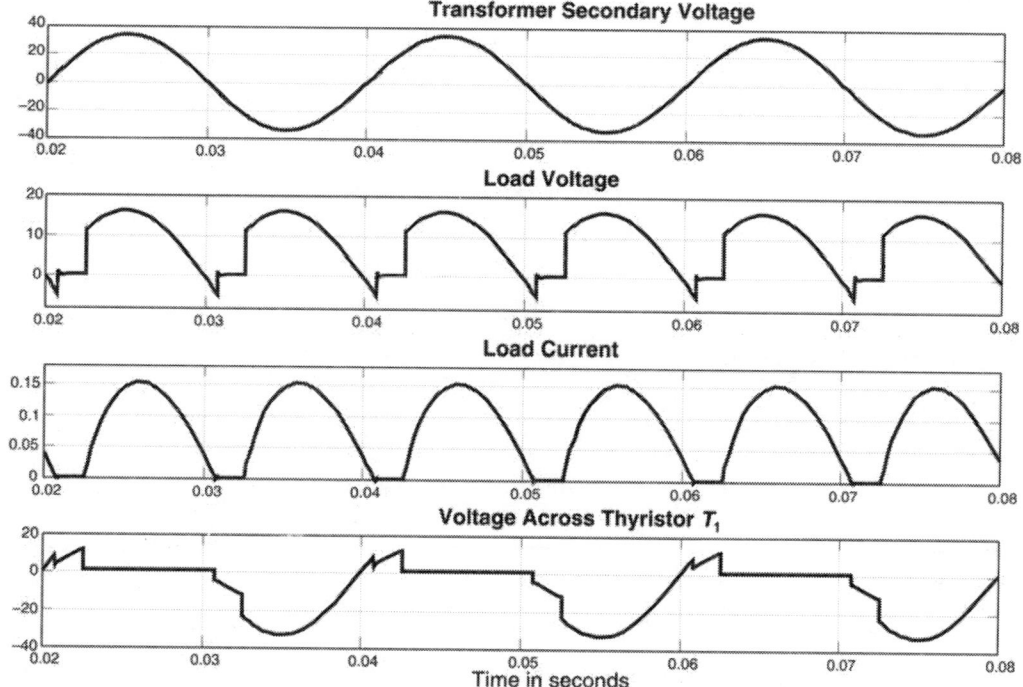

Fig. C4.7: Waveforms of Single phase mid-point converter with RL load-Discontinuous
Conduction Mode - (a) Transformer secondary voltage (b) Voltage across the load
(c) Load current (d) Voltage across the thyristor T_1

Note down the average and RMS value of the load voltage. It can be seen that the
average value is decreased and the RMS value is increased (compared with R load).
This is because of the negative excursion of the load voltage shown in Fig. C4.7.

Now, consider L/R ratio of the load is high, i.e. the load current is continuous.
Double click on RL branch and select the value of inductor L as 500 mH and R as 100 Ω
so that L/R ratio is high. Keep the remaining parameters as it is. Follow steps 1 to 8 as
in the case of mid-point converter with R load.

The waveforms are shown in the Fig. C4.8.

Note down the average and RMS value of the load voltage.

Mid-Point Converter with RL Load and Freewheeling Diode

Connect the blocks according to the circuit diagram of a single phase mid-point
converter circuit with RL load and freewheeling diode as shown in Fig. C4.9. Assume
the load current is discontinuous, i.e. L/R ratio of the load is low. Follow steps 1 to 8.
The waveforms are shown in the Fig. C4.10.

Note down the average and RMS values of the load voltage. It can be seen that the
average value is increased and RMS value is decreased compared with the case of RL
load without freewheeling diode. This is because, during the negative half cycle of the
supply voltage, the negative excursion of the load voltage is clipped by the freewheeling
diode.

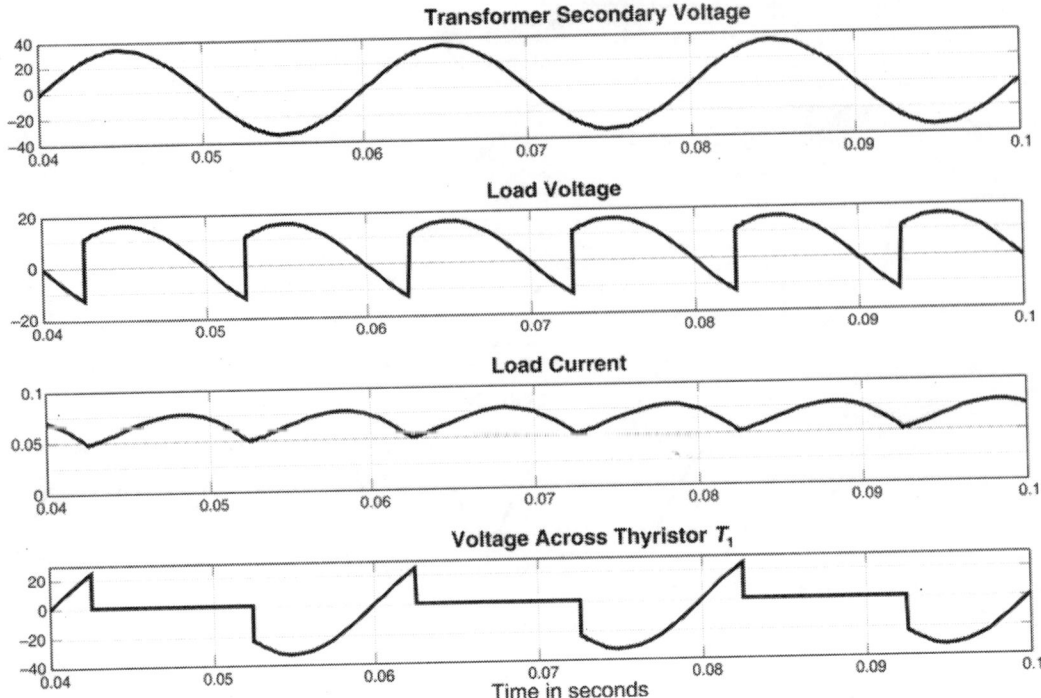

Fig. C4.8: Waveforms of Single phase mid-point converter with RL load-continuous conduction Mode: (a) Transformer secondary voltage (b) Voltage across the load (c) Load current (d) Voltage across the thyristor T_1

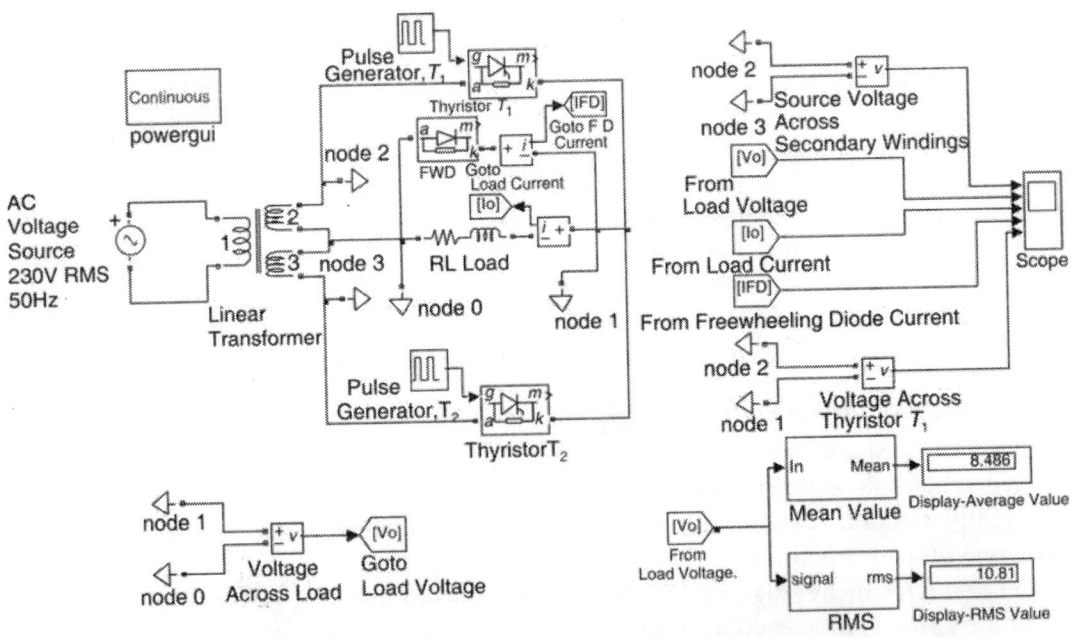

Fig. C4.9: Single phase mid-point converter with *RL* loadand freewheeling diode circuit model with measurement blocks

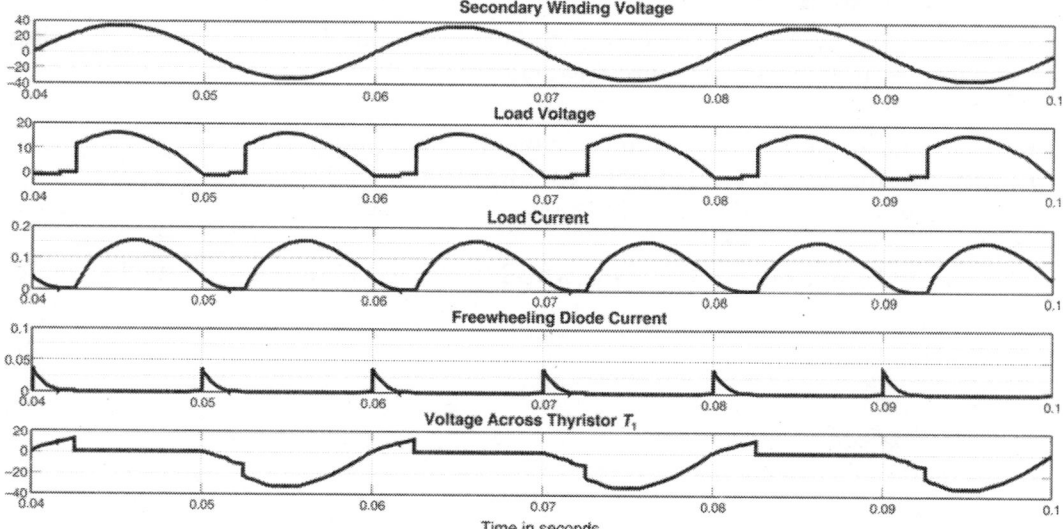

Fig. C4.10: Waveforms of single phase mid-point converter with *RL* load and Freewheeling diode-discontinuous conduction mode: (a) Transformer secondary voltage (b) Voltage across the load (c) Load current (d) Freewheeling diode current (e) Voltage across the thyristor T_1

Now, make the load current is continuous by adjusting the L/R ratio of the load. Follow steps 1 to 8. The waveforms are shown in the Fig. C4.11. Again note down the average and RMS values of the load voltage.

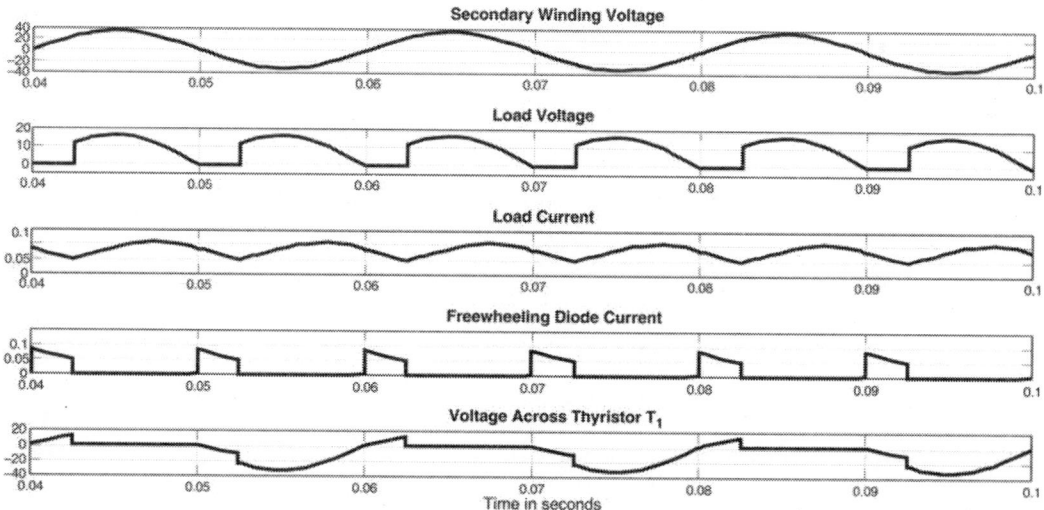

Fig. C4.11: Waveforms of single phase mid-point converter with *RL* load and freewheeling diode-continuous conduction Mode: (a) Transformer secondary voltage (b) Voltage across the load (c) Load current (d) Freewheeling diode current (e)Voltage across the thyristor T_1

C4.3 SIMULATION RESULT

The single phase mid-point converter circuit was modelled and simulated using MATLAB/simulink under the following conditions

1. Purely resistive load
2. Resistive-inductive load-discontinuous conduction mode.
3. Resistive-inductive load-continuous conduction mode.
4. Resistive-inductive load and with freewheeling diode-discontinuous conduction mode.
5. Resistive-Inductive load and with freewheeling diode -continuous conduction mode.

C5.1 OBJECTIVE

i. To simulate a single phase full-wave half-controlled bridge rectifier (semi converter) using MATBLAB/Simulink.

ii. To perform the operation of single phase half-controlled bridge rectifier/converter under following load condition
1. Purely resistive load
2. Resistive-inductive load - DCM
3. Resistive-inductive-back EMF load with freewheeling diode-CCM

C5.2 MODELLING AND SIMULATION OF SINGLE PHASE FULL-WAVE HALF-CONTROLLED BRIDGE CONVERTER CIRCUIT USING MATLAB/SIMULINK

Theory and working of a single phase full-wave half-controlled bridge rectifier circuit is discussed in B7.

A single phase semi converter circuit model using MATLAB/simulink is shown in Fig. C5.1.

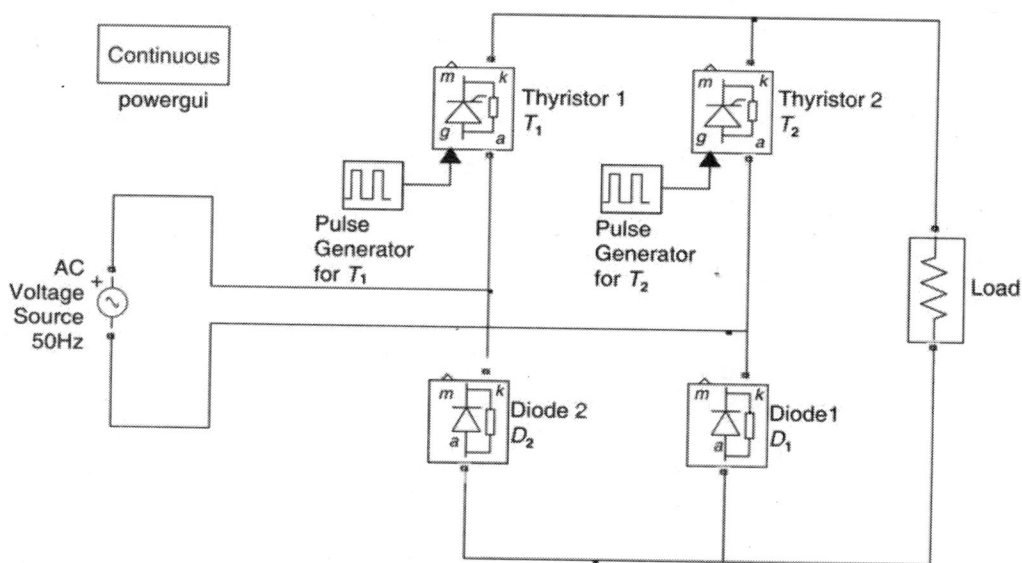

Figure C5.1: A single phase full-wave half-controlled bridge converter with symmetrical connection–circuit model using MATLAB/simulink

Following steps are to be followed for analyzing and modelling the single phase half-wave rectifier circuit.

Step 1: Enter "simulink" in the MATLAB command window to open the simulink library browser or click simulink button in the MATLAB toolbar.

Step 2: Select **File > New > Model** in the simulink library browser to create a new model or press Ctrl + N. An empty model window is opened and it can be saved by selecting **File > Save** in the model window.

Step 3: Drag and place the following building blocks from *Simulink library* and *Sim Power Systems library* to the simulink model file, as per the circuit diagram for different loads.

   (a) Powergui block [Sim Power Systems]
   (b) AC Voltage Source [Sim Power Systems > Electrical Sources]
   (c) Series RLC branch (configure it to R/RL) [Sim Power Systems > Elements]
   (d) Thyristor [Sim Power Systems > Power Electronics]
   (e) Pulse generator [Simulink > Sources]
   (f) Diode [Sim Power Systems > Power Electronics]
   (g) DC Voltage Source [Sim Power Systems > Electrical Sources]
   (h) Voltage Measurement [Sim Power Systems > Measurements]
   (i) C urrent Measurement [Sim Power Systems > Measurements]
   (j) Scope [Simulink > Sinks]
   (k) Mean value [Sim Power Systems > Extra library > Measurements]
   (l) RMS [Sim Power Systems > Extra library > Measurements]
  (m) Display [Simulink > Sinks]
   (n) From [Simulink > Signal Routing]
   (o) Goto [Simulink > Signal Routing]

Semiconverter with R Load

Step 4: Connect the blocks according to the circuit diagram of the single phase semi-converter circuit with *R* load. Also connect required number of current and voltage measurement blocks to the scope, as shown in the Fig. C5.2, to observe the necessary waveforms. The *Goto* and *From* blocks need not be necessarily used, but it will reduce the number of connections in the circuit.

Step 5: Set the parameters of the individual blocks. Components and specifications used in the single phase semi converter circuit with *R* load are:

   1. AC voltage source   34 V (Max. value), 50 Hz
   2. Load Resistor   100 Ω
   3. Thyristor
   4. Pulse Generator
   5. Diode
   6. Powergui block
   7. Voltage measurement blocks
   8. Current measurement blocks
   9. Scope
 10. Mean value block
 11. RMS block
 12. Display
 13. Goto block
 14. From block

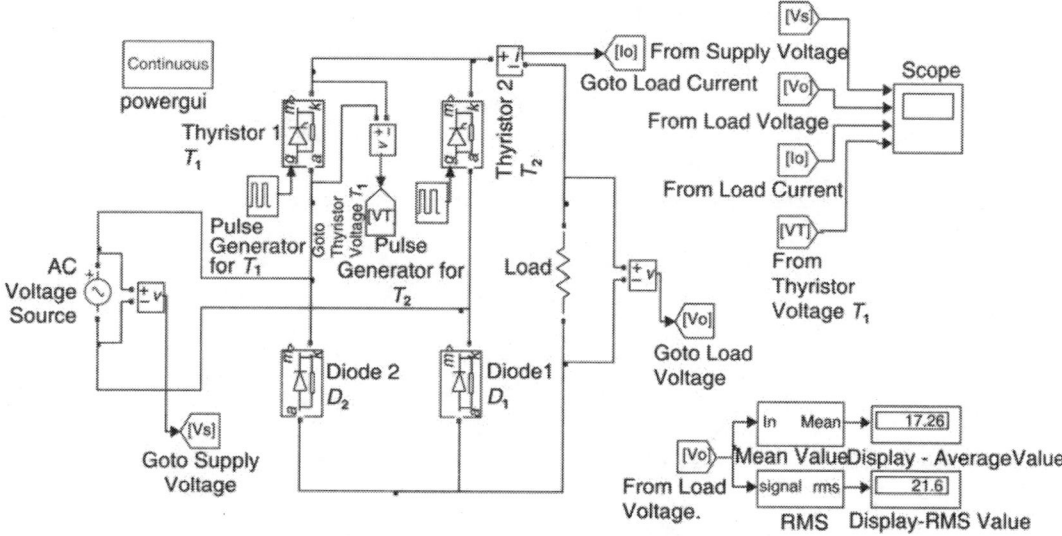

Fig. C5.2: The single phase full-wave half-controlled bridge converter circuit model feeding *R* load and with measurement blocks

Parameters of pulse generator used are:

Pulse generator for T_1:

1. Period (secs): 0.02 (i.e., 50 Hz)
2. Pulse width (% of period): 5
3. Phase delay (secs): 5e –3/2

Pulse generator for T_2:

1. Period (secs): 0.02 (i.e., 50 Hz)
2. Pulse width (% of period): 5
3. Phase delay (secs): 12.5e –3

Note: Firing angle of Thyristor T_2 = 180° + Firing angle of thyristor T_1

Step 6: Since the circuit provides AC to DC conversion, the average and RMS value of the load voltage can be calculated. For that *mean value block* and *RMS block* are used and corresponding values are displayed in the *display block*. Double click on *mean value block* and *RMS block* and enter *averaging period* as 0.02 seconds and fundamental frequency as 50 Hz respectively since supply voltage frequency is 50 Hz.

Consider that the firing angle for thyristor T_1 is 45° and thyristor T_2 is 180° + 45°.

Step 7: Save the Simulink model. Run the simulation by selecting **simulation > start** in the model window or use **simulation button** in the model window toolbar.

The run time can be entered and modified in the space provided in the model window tool bar. For example, run time = 0.08 seconds.

Step 8: Observe and note down the following waveforms by opening the scope and save the result.

  (a) Source voltage
  (b) Load voltage [Output voltage or voltage across the load resistor *R*]
  (c) Current through the load.
  (d) Voltage across the thyristor T_1

The waveforms are shown in the Fig. C5.3.

Note down the average and RMS value of the load voltage from the *display blocks*.

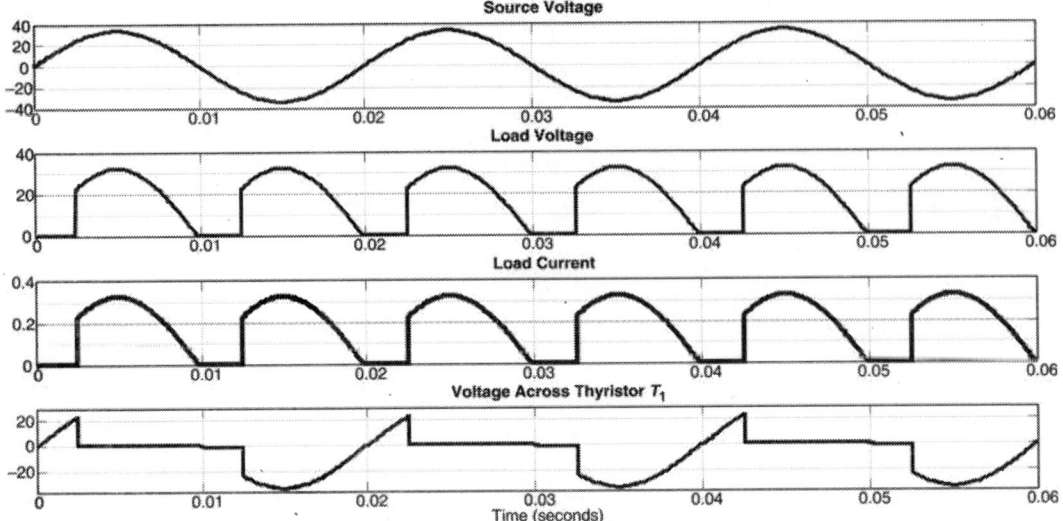

Fig. C5.3: Waveforms of single phase full-wave half-controlled bridge converter with R load:
(a) Source voltage (b) Voltage across the load (c) Load current
(d) Voltage across the thyristor T_1

Semi Converter with RL Load

Step 9: Now change the load of semiconverter to resistive-inductive (RL) load as shown in Fig. C5.4. Assume the load current is discontinuous, i.e., L/R ratio of the load is low. Double click on R branch and configure to RL. Select the value of inductor L and R, say $L = 80$ mH and $R =$ as $100\ \Omega$, so that L/R ratio is low. Consider that the firing

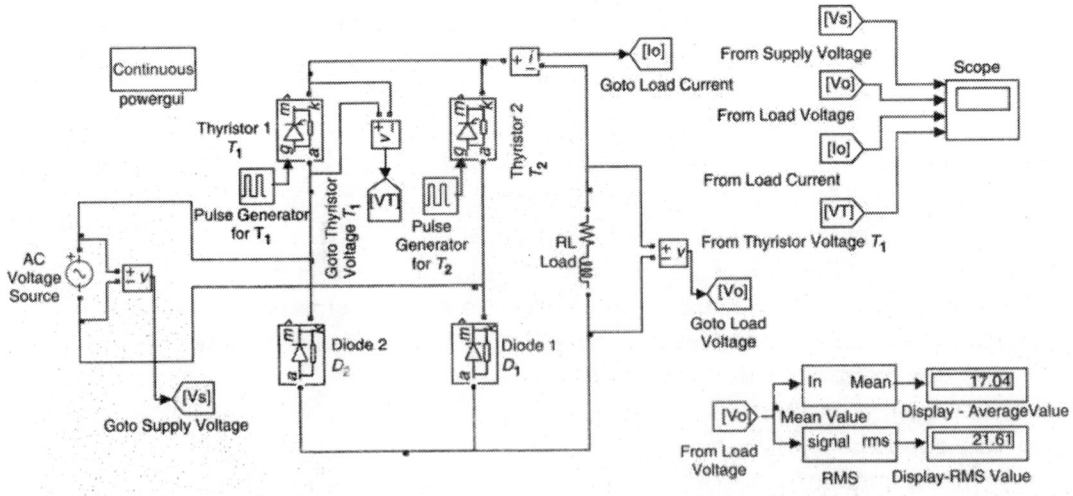

Fig. C5.4: The single phase full-wave half-controlled bridge converter circuit model having *RL* load with measurement blocks using MATLAB/Simulink

angle α is greater than the load impedance angle of the load ϕ, i.e., $\alpha > \phi$ where, $\phi = \tan^{-1}(wL/R)$, so that the load current is discontinuous. Keep the remaining parameters as it is. Follow steps 1 to 8 as in the case of semiconverter with R load. Waveforms are shown in the Fig. C5.5.

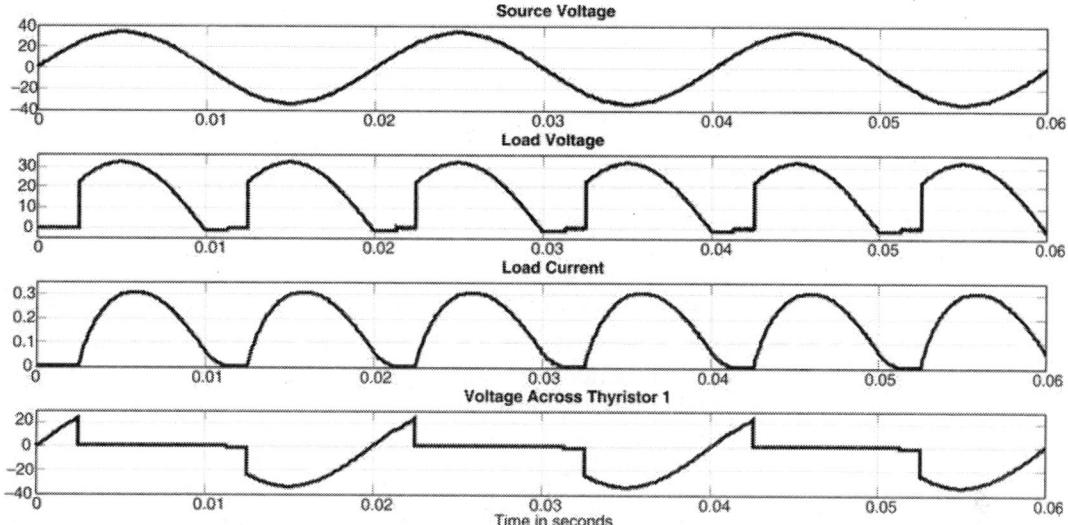

Fig. C5.5: Waveforms of single phase full-wave half-controlled bridge converter with RL load – Discontinuous Conduction Mode (a) Source voltage (b) Voltage across the load (c) Load current (d) Voltage across the thyristor T_1

Note down the average and RMS value of the load voltage from the *display blocks*. It can be seen that the average value is decreased and the RMS value is increased (compared with R load values). This is because, during the negative half cycle of the supply voltage, freewheeling takes place through the thyristor T_1 and diode D_2 till the energy stored in the inductor reaches zero. Similarly during the next positive half cycle, devices T_2 and D_1 carries freewheeling current till the energy in the inductor reaches zero.

Semi Converter with RLE Load with Freewheeling Diode

Step 10: Now change the load to resistive-inductive-back EMF (RLE) load with freewheeling diode as shown in the Fig. C5.6. Consider a large value of L/R ratio so that the load current is continuous. For example take $R = 100\ \Omega$, $L = 400$ mH and back EMF $E = 5$ V. Keep remaining parameters as it is. Also select the firing angle α to be less than the load impedance angle ϕ, i.e., $\alpha < \phi$, where $\phi = \tan^{-1}(\omega L/R)$

Follow steps 1 to 8 as in the case of semiconverter with R load. Waveforms are shown in the Fig. C5.7. Note down the average and RMS values of the load voltage from the display blocks.

It is to be noted that the thyristor can be triggered only when the supply voltage, $v_s = V_m \sin \omega t$ is greater than the back EMF E. If α is the minimum firing angle below which the thyristor cannot be turned on, then,

$$V_m \sin \theta = E \quad \text{or} \quad \theta = \sin^{-1}(E/V_m)$$

Similarly maximum firing angle is $(\pi - \theta)$

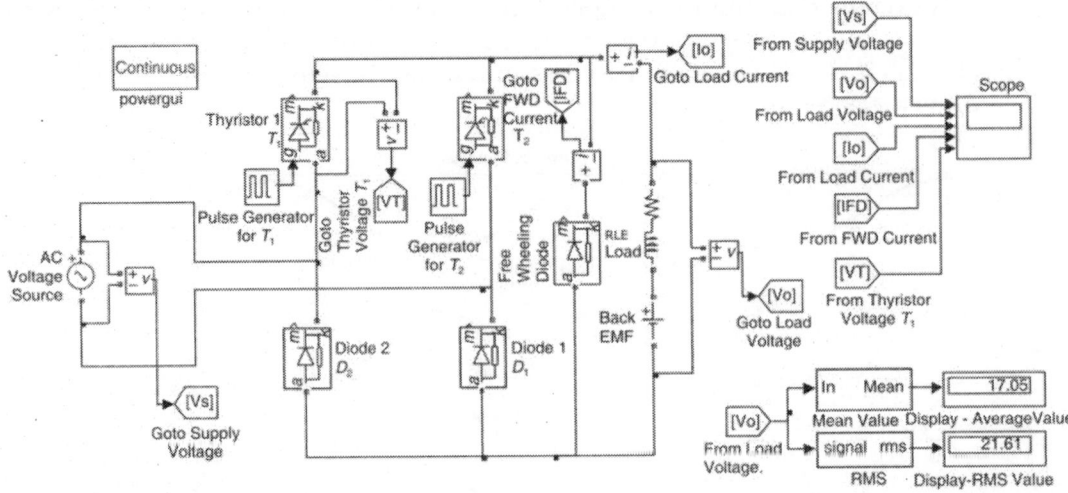

Fig. C5.6: Single phase full-wave half-controlled bridge converter circuit model having RLE load and freewheeling diode with measurement blocks using MATLAB/Simulink

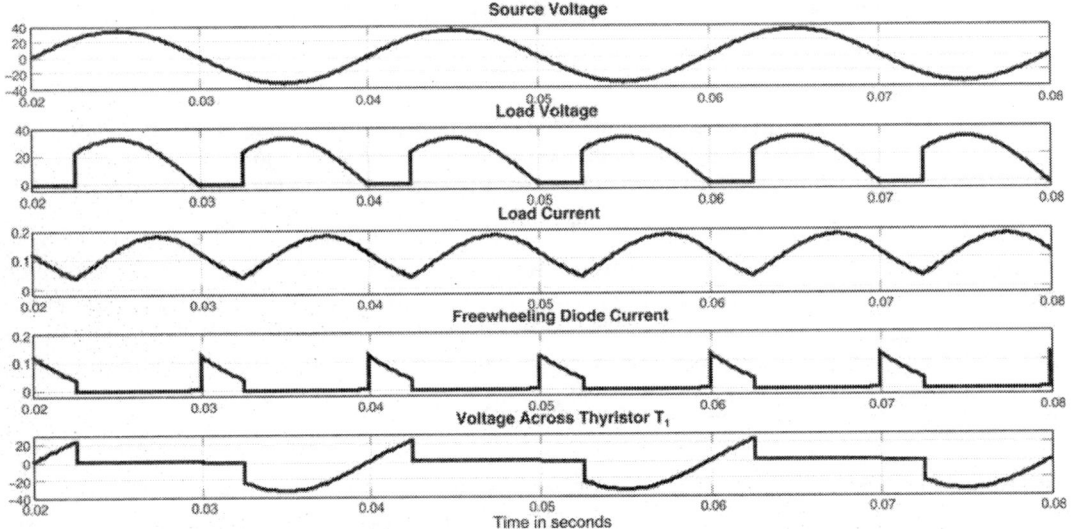

Fig. C5.7: Waveforms of single phase full-wave half-controlled bridge converter with RLE load and freewheeling diode–continuous conduction mode (a) Source voltage (b) Voltage across the load (c) Load current (d) Freewheeling diode current (e) Voltage across the thyristor T_1

Note: Different circuits with different load conditions can also be simulated by following the same procedure.

C5.3 SIMULATION RESULT

The single phase full-wave half-controlled bridge converter circuit was modelled and simulated using MATLAB/Simulink under the following conditions
1. Purely resistive load
2. Resistive-inductive load–discontinous conduction mode
3. Resistive-inductive-back EMF load with freewheeling diode-continuous conduction mode

Simulation of a Single Phase Full-wave Fully Controlled Bridge Rectifier

C6.1 OBJECTIVE

To simulate a single phase full-wave fully controlled bridge rectifier using MATBLAB/Simulink, under the following load conditions:

1. Resistive load
2. Resistive-inductive load-DCM
3. Resistive-inductive-back EMF load with freewheeling diode-CCM

C6.2 MODELLING AND SIMULATION OF A SINGLE PHASE FULL-WAVE FULLY CONTROLLED BRIDGE RECTIFIER CIRCUIT USING MATLAB/SIMULINK

Theory and working of a single phase full-wave fully controlled bridge rectifier circuit is discussed in B8.

A single phase full converter circuit model using MATLAB/Simulink is shown in Fig. C6.1.

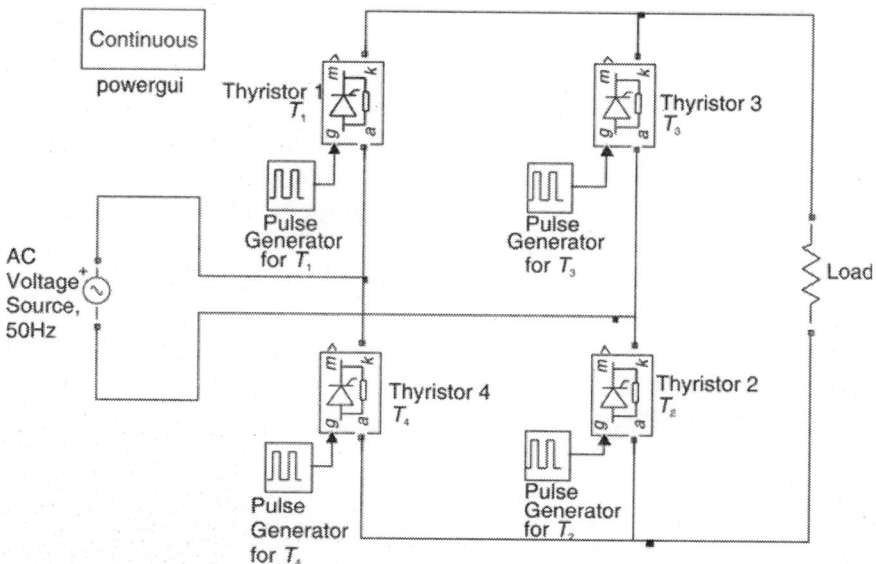

Fig. C6.1: A single phase full-wave fully controlled bridge converter circuit model using MATLAB/Simulink

Following steps are to be followed for analyzing and modelling the single phase half-wave rectifier circuit.

Step 1: Enter "simulink" in the MATLAB Command Window to open the Simulink Library Browser or click Simulink button in the MATLAB toolbar.

Step 2: Select **File > New > Model** in the Simulink Library Browser to create a new model or press Ctrl + N. An empty model window is opened and it can be saved by selecting **File>Save** in the model window.

Step 3: Drag and place the following building blocks from *Simulink library* and *Sim Power Systems library* to the simulink model file, as per the circuit diagram for different loads.

   (a) Powergui block [Sim Power Systems]
   (b) AC Voltage Source [Sim Power Systems > Electrical Sources]
   (c) Series RLC branch (configure it to RL) [Sim Power Systems > Elements]
   (d) Thyristor [Sim Power Systems> Power Electronics]
   (e) Pulse generator [Simulink > Sources]
   (f) Diode [Sim Power Systems > Power Electronics]
   (g) DC Voltage Source [Sim Power Systems> Electrical Sources]
   (h) Voltage Measurement [Sim Power Systems > Measurements]
   (i) Current Measurement [Sim Power Systems > Measurements]
   (j) Scope [Simulink > Sinks]
   (k) Mean value [Sim Power Systems> Extra library > Measurements]
   (l) RMS [Sim Power Systems> Extra library > Measurements]
 (m) Display [Simulink > Sinks]
  (n) From [Simulink > Signal Routing]
  (o) Goto [Simulink > Signal Routing]

Full Converter with Pure Resistive Load

Step 4: Connect the blocks according to the circuit diagram of the single phase semi-converter circuit with *R* load. Also connect required number of current and voltage measurement blocks to the *scope*, as shown in the Fig. C6.2, to observe the necessary waveforms. The *Goto* and *From* blocks need not be necessarily used, but it will reduce the number of connections in the circuit.

Step 5: Set the parameters of the individual blocks. Components and specifications used in the single phase semi converter circuit with *R* load are:

   1. AC voltage source   34 V (Max. value), 50 Hz
   2. Load resistor   100 Ω
   3. Thyristors
   4. Pulse generators
   5. Powergui block
   6. Voltage measurement blocks
   7. Current measurement blocks
   8. Scope
   9. Mean value block
 10. RMS block
 11. Display
 12. Goto block
 13. From block

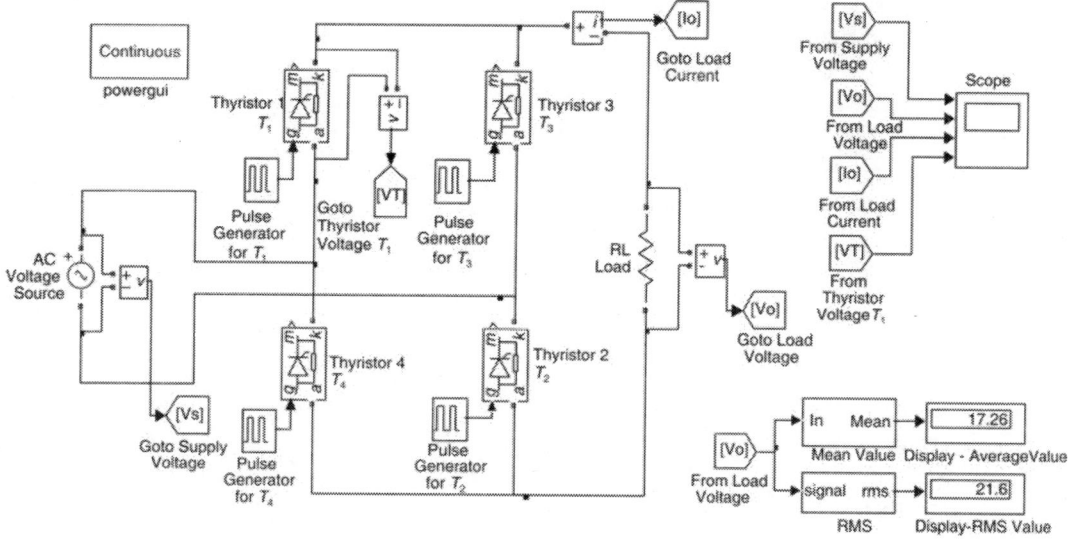

Fig. C6.2: A single phase full-wave fully controlled bridge converter circuit model with *R* Load and with measurement blocks

Parameters of pulse generator used are:

Pulse generator for T_1 and T_2

1. Period (secs): 0.02 (i.e., 50 Hz)
2. Pulse Width (% of period): 5
3. Phase delay (secs): 5e –3/2

Pulse generator for T_3 and T_4

1. Period (secs): 0.02 (i.e., 50 Hz)
2. Pulse Width (% of period): 5
3. Phase delay (secs): 12. 5e –3

Note: Firing angle of Thyristor T_3 = 180° + Firing angle of Thyristor T_1

Step 6: Since the circuit provides AC to DC conversion, the average and RMS value of the load voltage can be calculated. For that *mean value block* and *RMS block* are used and corresponding values are displayed in the *display block*. Double click on *mean value block* and *RMS block* and enter *averaging period* as 0.02 seconds and fundamental frequency as 50 Hz respectively since supply voltage frequency is 50 Hz.

Step 7: Save the Simulink model. Run the simulation by selecting **simulation > start** in the model window or use **simulation button** in the model window toolbar.

The run time can be entered and modified in the space provided in the model window toolbar. For example, run time = 0.08 seconds.

Step 8: Observe and note down the following waveforms by opening the scope and save the result.

(a) Source voltage

(b) Load voltage [Output voltage/voltage across the load resistor *R*]

(c) Current through the load

(d) Voltage across the thyristor T_1 or T_2

The waveforms are shown in the Fig. C6.3. Note down the average and RMS value of the load voltage from the *display blocks*.

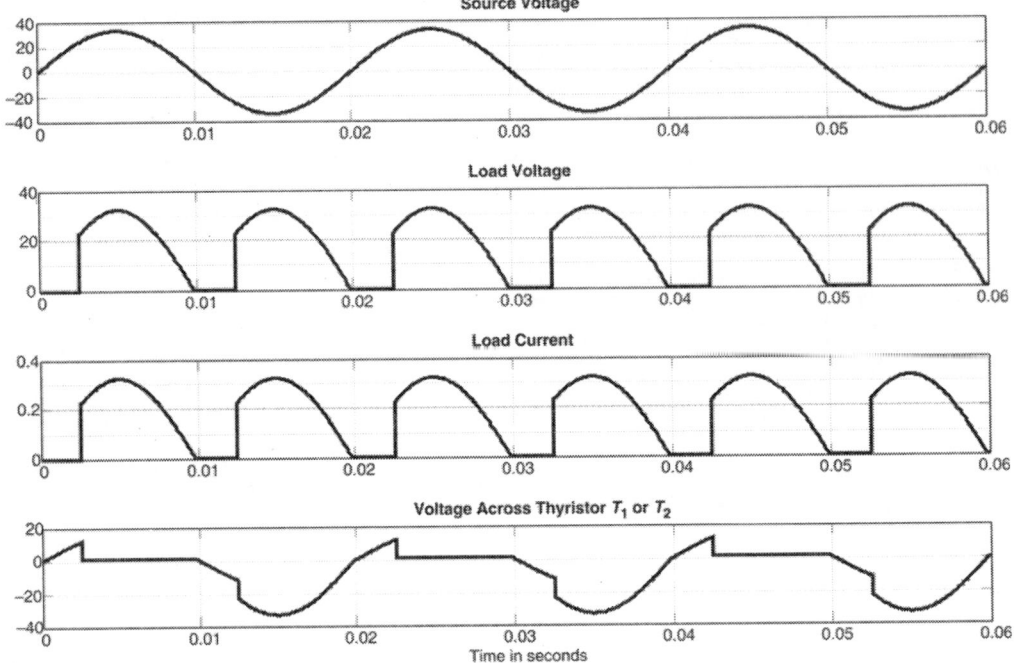

Fig. C6.3: Waveforms of single phase full-wave fully controlled bridge converter with R load (a) Source voltage (b) Voltage across the load (c) Load current (d) Voltage across the thyristor T_1 or T_2

Full Converter with RL Load

Step 9: Now change the load of full converter to resistive-inductive (*RL*) load as shown in Fig. C6.4. Assume the load current is discontinuous, i.e. L/R ratio of the load is low.

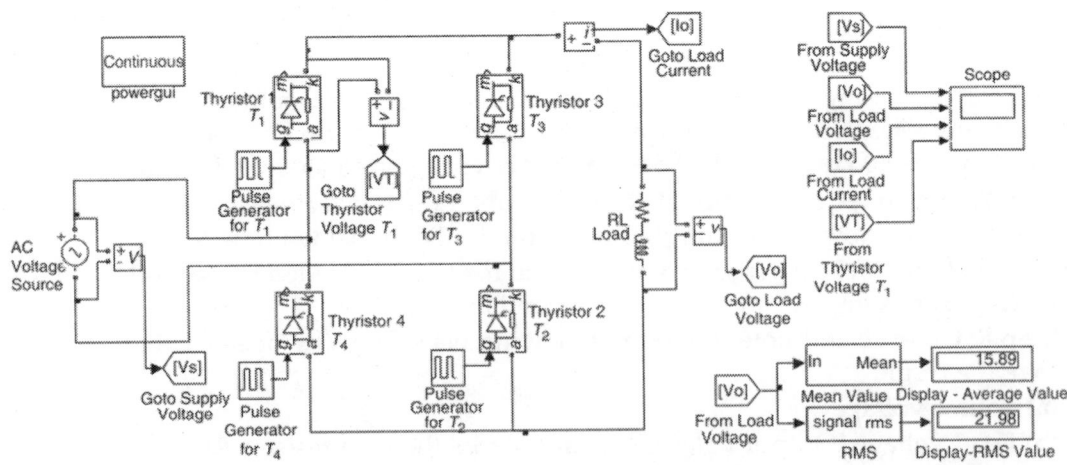

Fig. C6.4: The single phase full-wave fully controlled bridge converter circuit model feeding RL load and with measurement blocks

Double click on *R* branch and configure to *RL*. Select the value of inductor *L* and *R*, say *L* = 180 mH and *R* = as 100 Ω. Choose a proper value for the firing angle so that the load current is discontinuous. Keep the remaining parameters as it is. Follow steps 1 to 8 as in the case of full converter with *R* load. Waveforms are shown in the Fig. C6.5. Note down the average and RMS value of the load voltage from the *display blocks*. Here average value of the output voltage is less compared to the average value of the output voltage in a semi converter. This is because of negative excursion of output voltage waveform. This is absent in a semi converter because of the freewheeling action.

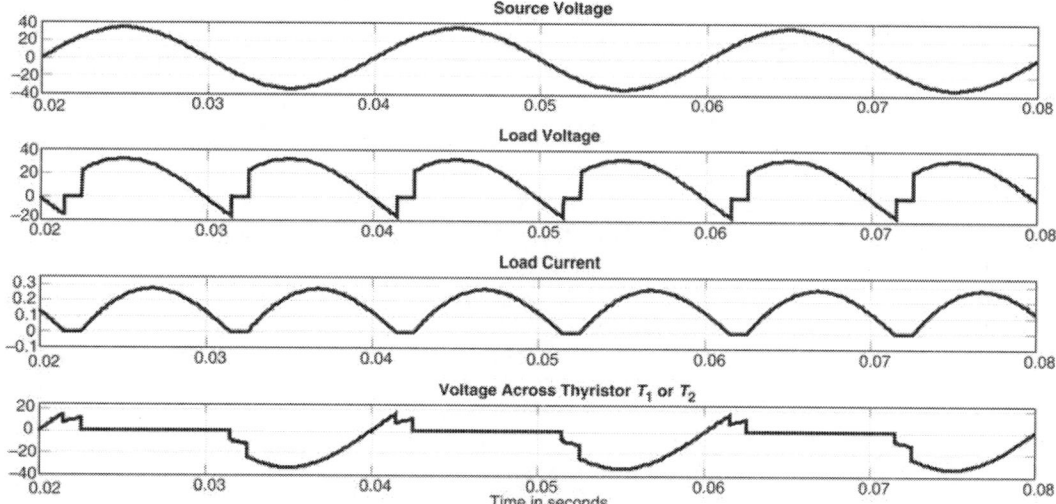

Fig. C6.5: Waveforms of single phase full-wave fully controlled bridge converter with RL load–discontinuous conduction mode (a) Source voltage (b) Voltage across the load (c) Load current (d) Voltage across the thyristor T_1 or T_2

Full Converter with RLE Load with Freewheeling Diode

Step 10: Now change the load to resistive-inductive-back EMF (RLE) load with freewheeling diode as shown in the Fig. C6.6.

Consider a large value of *L/R* ratio so that the load current is continuous. Let us choose inductance *L* = 400 mH, resistance *R* = 100 Ω and back EMF = 5 V. Keep remaining parameters as it is. Consider that the firing angle α is less than the load impedance angle ϕ, i.e., α < ϕ where, ϕ = tan⁻¹ (ω*L*/*R*), so that the load current is continuous.

Follow steps 1 to 8 as in the case of semiconverter with *R* load. Waveforms are shown in the Fig. C6.7. Note down the average and RMS values of the load voltage from the *display* blocks.

It can be noticed that the average value is increased and RMS value is decreased compared with the case of without FWD. This is because of the presence of freewheeling diode which prevent negative excursion of the output voltage during the both half cycle of the supply voltage.

It is to be noted that the thyristor can be triggered only when the supply voltage, $v_s = V_m \sin \omega t$ is greater than the back EMF E.

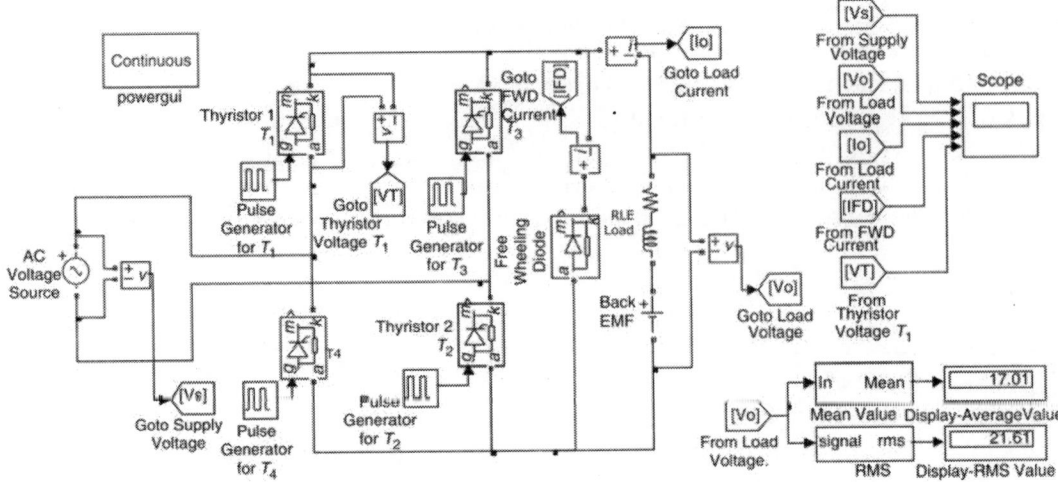

Fig. C6.6: Single phase full-wave fully controlled bridge converter circuit model with RLE load and freewheeling diode with measurement blocks

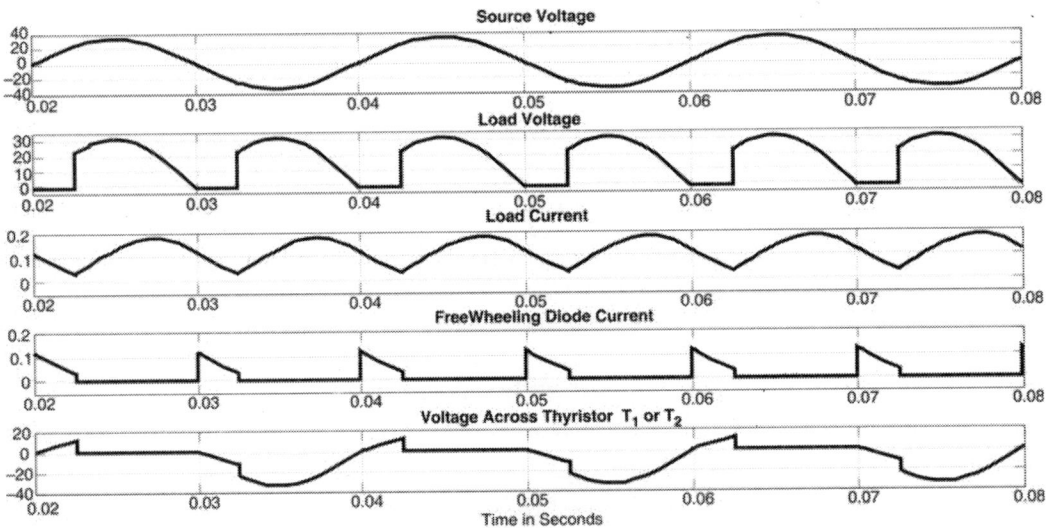

Fig. C6.7: Waveforms of single phase full-wave fully controlled bridge converter with RLE load and freewheeling diode-continuous conduction mode: (a) Source voltage (b) Voltage across the load (c) Load current (d) Freewheeling Diode current (e) Voltage across thyristor T_1 or T_2

If θ is the minimum firing angle below which the thyristors cannot be turned on, then,

$$V_m \sin \theta = E \quad \text{or} \quad \theta = \sin^{-1}\left(\frac{E}{V_m}\right)$$

Similarly maximum firing angle is $(\pi - \theta)$.

C6.3 SIMULATION RESULT

The single phase full-wave fully controlled bridge converter circuit was modelled and simulated using MATLAB/Simulink under the following conditions.

1. Pure resistive load.
2. Resistive-inductive load-discontinuous conduction mode.
3. Resistive-inductive-back EMF load with Freewheeling diode-continuous conduction mode.

Experiment C7

Simulation of Speed Control of a DC Motor Using a Controlled Rectifier

C7.1 OBJECTIVE

To model and simulate the speed control of DC motor using a single phase semi-converter circuit with freewheeling diode using MATBLAB/Simulink and plot following graphs

(a) Speed of the DC motor versus firing angle

(b) Armature voltage versus firing angle

(c) Speed versus armature voltage

C7.2 MODELING AND SIMULATION OF SPEED CONTROL OF DC MOTOR USING MATLAB/SIMULINK

Theory and working of speed contorl of a DC motor using single phase semi-converter circuit with symmetrical configuration is discussed in B9 in section B.

The speed of a DC motor can be controlled by varying the armature voltage. The armature voltage can be varied by the varying the firing angle of converter. The field of DC motor is assumed to be of permanent magnet type and hence field flux is constant.

A DC motor can be driven in the continuous current mode using a freewheeling diode which is connected across the armature terminals. Select a high value of armature inductance so that the armature current is continuous.

The circuit of speed control of a DC motor having single phase semiconverter with freewheeling diode using MATLAB/Simulink is shown in Fig C7.1.

Following steps are to be adopted for analyzing and modelling the speed control of DC motor using single phase semi converter.

Step 1: Enter "simulink" in the MATLAB command window to open the simulink library browser or click simulink button in the MATLAB toolbar.

Step 2: Select **File > New > Model** in the simulink library browser to create a new model or press Ctrl+N. An empty model window is opened and it can be saved by selecting **File>Save** in the model window.

Step 3: Drag and place the following building blocks from *simulink library* and *Sim Power Systems library* to the simulink model file.

(a) Powergui block [Sim Power Systems]

(b) AC voltage source [Sim Power Systems> Electrical Sources]

(c) Thyristor [Sim Power Systems > Power Electronics]

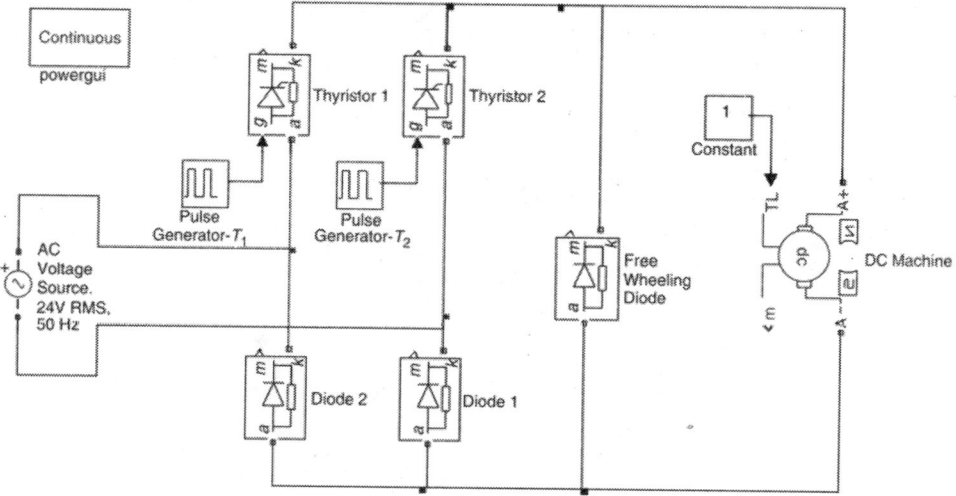

Fig. C7.1: Circuit model for speed control of a DC motor with freewheeling diode using single phase semiconverter

(d) Pulse generator [Simulink > Sources]

(e) Diode [Sim Power Systems > Power Electronics]

(f) DC machine [Sim Power Systems > Machines]

(g) Constant [Simulink > Sources]

(h) Bus selector [Simulink > Signal Routing]

(i) Product [Simulink > Commonly Used Blocks]

(j) Scope [Simulink > Sinks]

(k) Mean value [Sim Power Systems > Extra library > Measurements]

(l) Voltage Measurement [Sim Power Systems > Measurements]

(m) Display [Simulink > Sinks]

(n) From [Simulink > Signal Routing]

(o) Goto [Simulink > Signal Routing]

Step 4: Connect the blocks according to the circuit diagram of speed control of DC motor using single phase semiconverter and freewheeling diode.

Also connect required measurement blocks to the *scope*, as shown in the Fig. C7.2, to observe the necessary waveforms. The *Goto* and *From* blocks need not be necessarily used, but it will reduce the number of connections in the circuit.

Step 5: Set the parameters of the individual blocks. Components and specifications used in the circuit model are:

1. AC voltage source   34 V (Max. value), 50 Hz

2. Constant   1

3. Product   180/pi

4. Thyristors

5. Pulse Generators

6. DC machine

7. Diodes

8. Powergui block

9. Scope

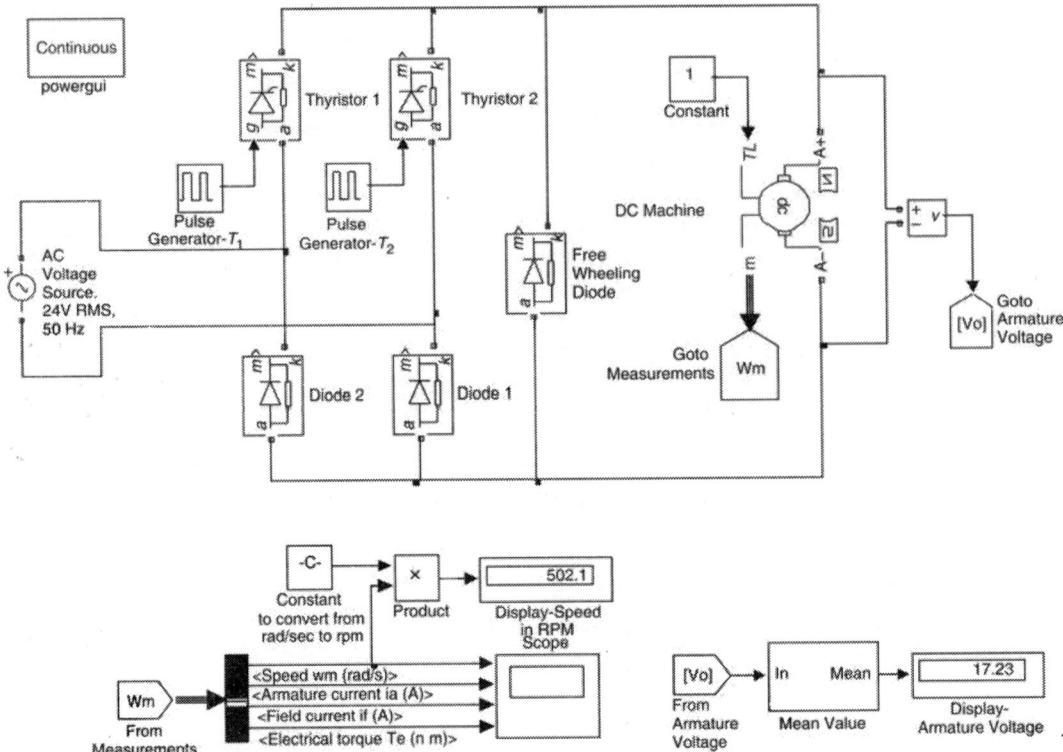

Fig. C7.2: Speed control of a DC motor using single phase semiconverter
with measurement blocks

10. Mean value block
11. Goto and From blocks
12. Display
13. Voltage measurement block
14. Bus selector

The parameters of pulse generator used are:

Pulse generator T_1:

1. Period (secs): 0.02 (for 50 Hz supply)
2. Pulse Width (% of period): 5
3. Phase delay (secs): 2.5e-3

   Phase delay angle can be calculated as explained here. For a firing angle $\alpha_{T_1} = 45°$,

$$\text{Phase delay} = \frac{45°}{360°} \times 20\,\text{ms} = 2.5\,\text{ms}$$

Pulse generator T_2:

1. Period (secs): 0.02 (for 50 Hz supply)
2. Pulse Width (% of period): 5
3. Phase delay (secs): 12. 5e-3

Phase delay angle can be calculated for a firing angle $\alpha_{T_2} = 180° + 45°$,

$$\text{Phase delay} = \frac{(180° + 45°)}{360°} \times 20\,\text{ms} = 12.5\,\text{ms}$$

The voltage measurement block is connected across armature terminals, A^+ and A^-, to measure its voltage which is varied by the firing angle of the converter and the average value of armature voltage is displayed in the display block.

The measurement terminal "m" of the DC machine provides the following four signals and these signals can be demultiplexed by a **Bus Selector** block.

Signal	Definition	Units
1	Speed, ω_m	rad/sec
2	Armature current, i_a	Ampere
3	Field current, i_f	Ampere
4	Electrical torque, T_e	N.m

Now, double click on DC machine and select *configuration* tab. Mechanical output to the DC motor can be selected as either torque TL or rotor speed ω. Select load torque as *Torque TL* and field type as *permanent magnet*, as shown in the Fig. C7.3. Then select *Parameter* tab where armature resistance and armature inductance values can be entered. Since the armature current is assumed to be continuous, choose a large value of armature inductance. Let $R_a = 2.6\,\Omega$ and $L_a = 0.212$ H. Select *torque constant* in Nm/A as shown in the Fig. C7.4.

A *constant* block is added which acts as external load torque on the DC motor. Double click on *constant* block and enter a constant value. For full load torque, enter 1. The measurement terminal "m" of the DC machine provides speed in rad/sec. To convert the speed from rad/sec to RPM, multiply it with $180/\pi$. For this purpose, *constant* and *product* blocks are used.

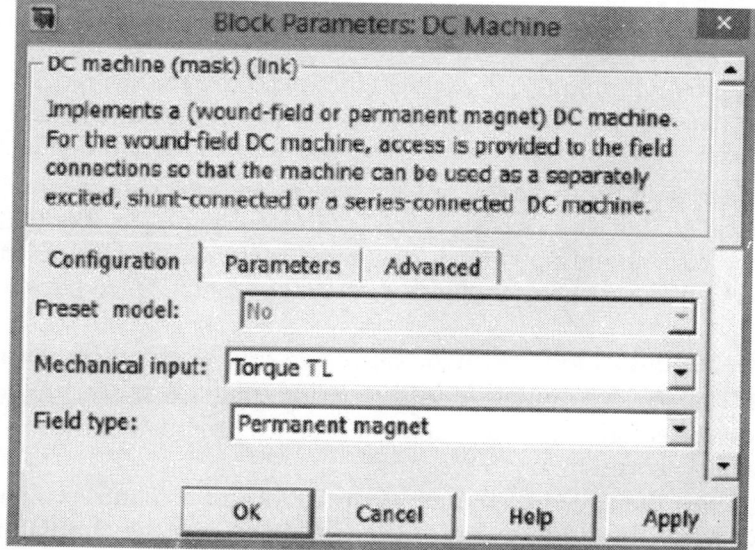

Fig. C7.3: Block parameters of permanent magnet type DC machine – Configuration tab

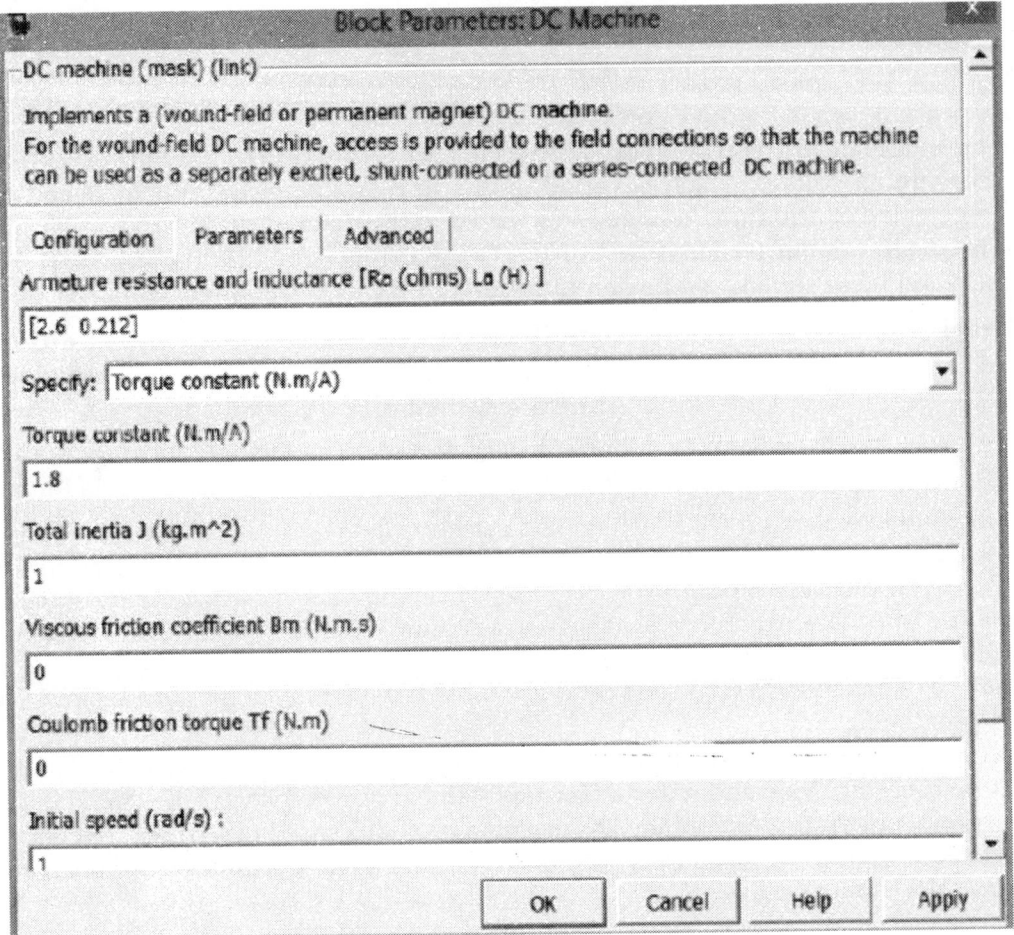

Fig. C7.4: Block parameters of permanent magnet type DC machine – Parameters tab

From measurement terminal *"m"*, connect the speed to a *product* block which is also connected to a *constant* block. Double click on *constant* block and enter 180/pi. The speed of the motor in RPM is shown in *display* block as shown in the Fig. C7.2.

Step 6: Since the circuit provides AC to DC conversion, the average value of the load voltage can be calculated. In this case, it acts as armature voltage. For that *mean value block* is used and corresponding value is displayed in the *display block*. Double click on *mean value block* and enter *averaging period* as 0.02 seconds, since supply frequency is 50 Hz.

Step 7: Save the Simulink model. Run the simulation by selecting **simulation > start** in the model window or use **simulation button** in the model window toolbar.

The run time can be entered and modified in the space provided in the model window tool bar. For example, run time = 5 seconds.

Step 8: Note down the speed of the motor in RPM and armature voltage from the display blocks. Vary the firing angle of the converter, run the simulation and note down the corresponding changes in the speed and armature voltage.

Sample readings are:

Firing angle of T_1	Firing angle of T_2	Speed in RPM	Armature voltage
(a) 1.66 ms ($\alpha_1 = 30°$)	11.66 ms ($\alpha_2 = 210°$)	550.2	18.74 V
(b) 2.5 ms ($\alpha_1 = 45°$)	12.5 ms	502.1	17.23 V
(c) 3.33 ms ($\alpha_1 = 60°$)	13.33 ms	435.3	15.13 V
(d) 5 ms ($\alpha_1 = 90°$)	15 ms	272.4	10 V
(e) 6.66 ms ($\alpha_1 = 120°$)	16.66 ms	109.4	4.88 V

Plot following graph:

(1) Speed of the motor in RPM versus firing angle of converter
(2) Armature voltage of DC motor versus firing angle of the converter
(3) Speed versus armature voltage

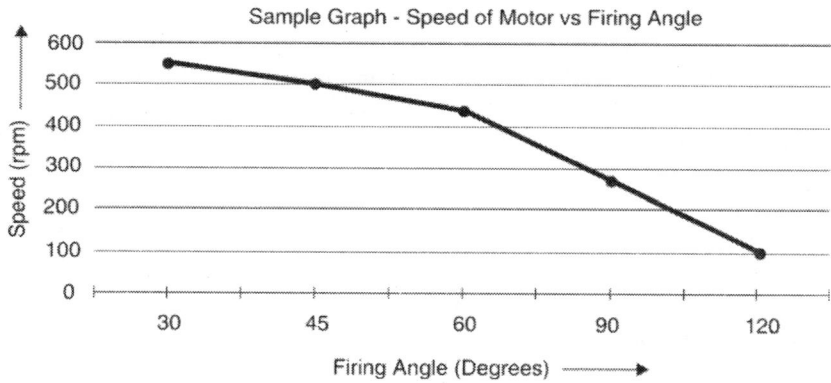

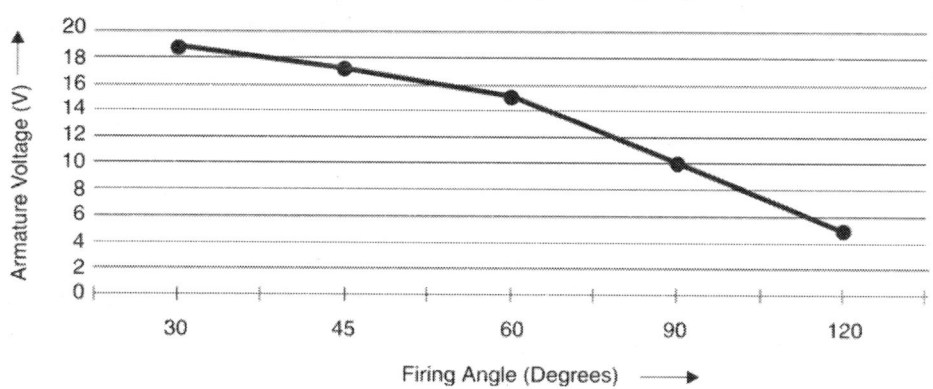

C7.3 SIMULATION RESULT

A single phase thyristor controlled semi converter for the speed control of a DC motor was modelled and simulated using MATLAB/Simulink. The variations of speed versus the firing angle and the variation of armature voltage versus firing angle were observed.

Experiment C8

Simulation of a Single Phase AC Voltage Regulator Using Thyristors

C8.1 OBJECTIVE

To simulate and analyse the opertion of a single phase AC voltage regulator with two thyristors connected in anti parallel using MATLAB/Simulink and to observe the following waveforms:
1. Load voltage
2. Load current
3. Voltage across the thyristor

C8.2 MODELLING AND SIMULATION OF AC VOLTAGE REGULATOR CIRCUIT USING MATLAB/SIMULINK

The AC voltage regulator/controller is a step down variable magnitude fixed frequency AC–AC converter. The AC voltage regulator circuit using MATLAB/Simulink is shown in Fig. C8.1. Theory and working of the AC voltage regulator is discussed in section B10. The circuit consists of two thyristors which are connected in anti-parallel. The load can be a pure resistive or resistive-inductive load.

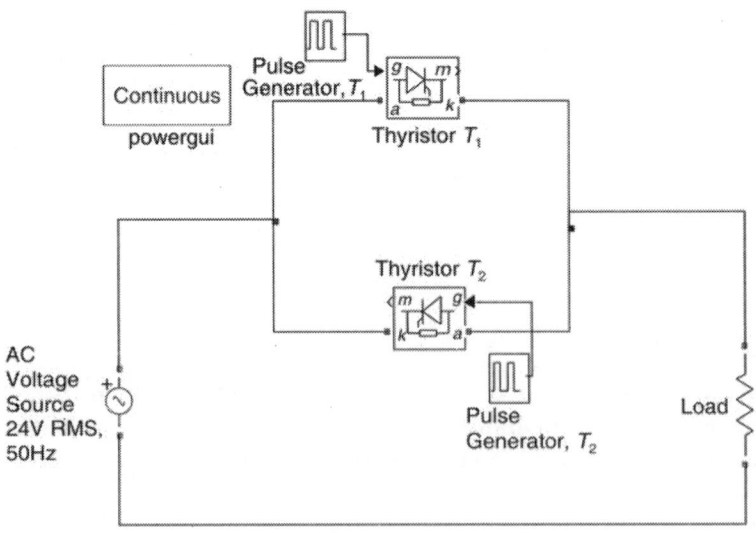

Fig. C8.1: The AC voltage regulator circuit model using MATLAB/Simulink

Following steps are to be followed for analyzing and modelling the single phase AC voltage regulator.

Step 1: Enter "simulink" in the MATLAB command window to open the simulink library browser or click simulink button in the MATLAB toolbar.

Step 2: Select **File > New > Model** in the simulink library browser to create a new model or press Ctrl+N. An empty model window is opened and it can be saved by selecting **File>Save** in the model window.

Step 3: Drag and place the following building blocks from *Simulink library* and *Sim Power Systems library* to the simulink model file, as per the circuit diagram for different loads.

(a) Powergui block [Sim Power Systems]
(b) AC Voltage Source [Sim Power Systems> Electrical Sources]
(c) Series RLC branch (configure it to R/RL) [Sim Power Systems> Elements]
(d) Thyristor [Sim Power Systems> Power Electronics]
(e) Pulse generator [Simulink > Sources]
(f) Voltage Measurement [Sim Power Systems > Measurements]
(g) Scope [Simulink > Sinks]
(h) RMS [Sim Power Systems > Extra library > Measurements]
(i) Display [Simulink > Sinks]
(j) From [Simulink > Signal Routing]
(k) Goto [Simulink > Signal Routing]

AC Voltage Regulator with R Load

Step 4: Connect the blocks according to the circuit diagram of the single phase AC–AC voltage regulator circuit with R load. Also connect required number of current and voltage measurement blocks to the *scope*, as shown in the Fig. C8.2, to observe the

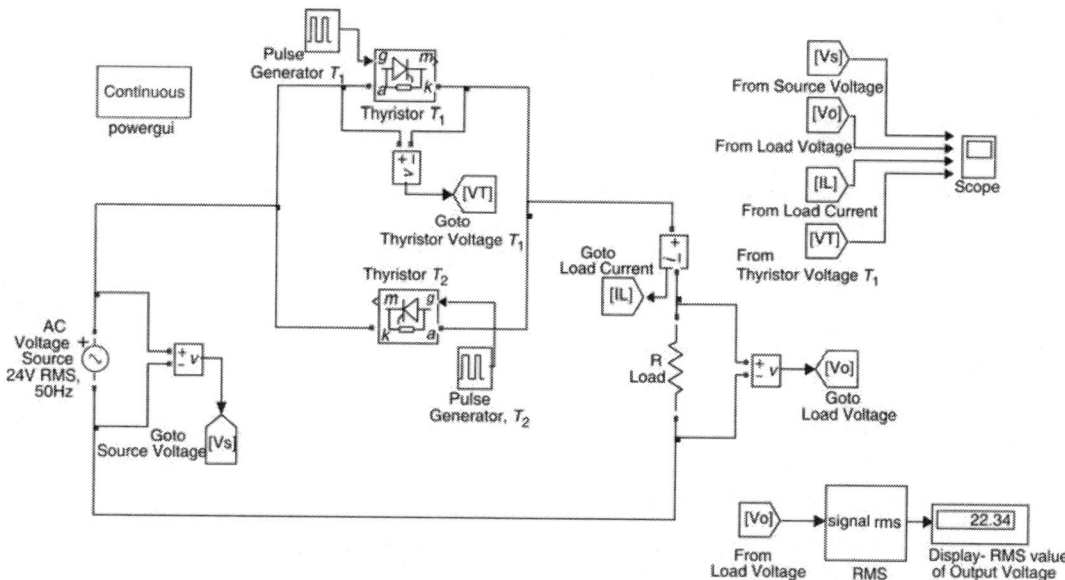

Fig. C8.2: AC voltage regulator circuit model with R load and measurements blocks

necessary waveforms. The *Goto* and *From* blocks need not be necessarily used, but it will reduce the number of connections in the circuit.

Step 5: Set the parameters of the individual blocks. Components and specifications used in the single phase AC voltage regulator circuit with R load are:

1. AC voltage source	34 V (Max. value), 50 Hz
2. Load resistor	100 Ω
3. Load inductor	60 mH

4. Thyristor
5. Pulse generator
6. Powergui block
7. Voltage measurement blocks
8. Current measurement blocks
9. Scope
10. RMS block
11. Display
12. Goto block
13. From block

Parameters of pulse generator used are:

Pulse generator for T_1:

1. Period (secs): 0.02 (i.e. 50 Hz)
2. Pulse width (% of period): 5
3. Phase delay (secs): 2.5e-3

Pulse generator for T_2:

1. Period (secs): 0.02 (i.e. 50 Hz)
2. Pulse width (% of period): 5
3. Phase delay (secs): 12.5e-3

Note: Firing angle of Thyristor $T_2 = 180°$ + Firing angle of Thyristor T_1

Step 6: Since the circuit provides AC to AC conversion, the RMS value of the load voltage can be calculated. For that *RMS block* is used and corresponding value is displayed in the *display block*. Double click on *RMS block* and enter fundamental frequency as 50 Hz.

Consider that the firing angle for thyristor T_1 is 45° and thyristor T_2 is 180° + 45°.

Step 7: Save the Simulink model. Run the simulation by selecting **Simulation > start** in the model window or use **simulation button** in the model window toolbar.

The run time can be entered and modified in the space provided in the model window tool bar. For example, run time = 0.08 seconds.

Step 8: Observe and note down the following waveforms by opening the scope and save the result.

(a) Source voltage
(b) Load voltage
(c) Current through the load
(d) Voltage across the thyristor T_1

The waveforms are shown in the Fig. C8.3.

Note down the RMS value of the load voltage from the *display block*.

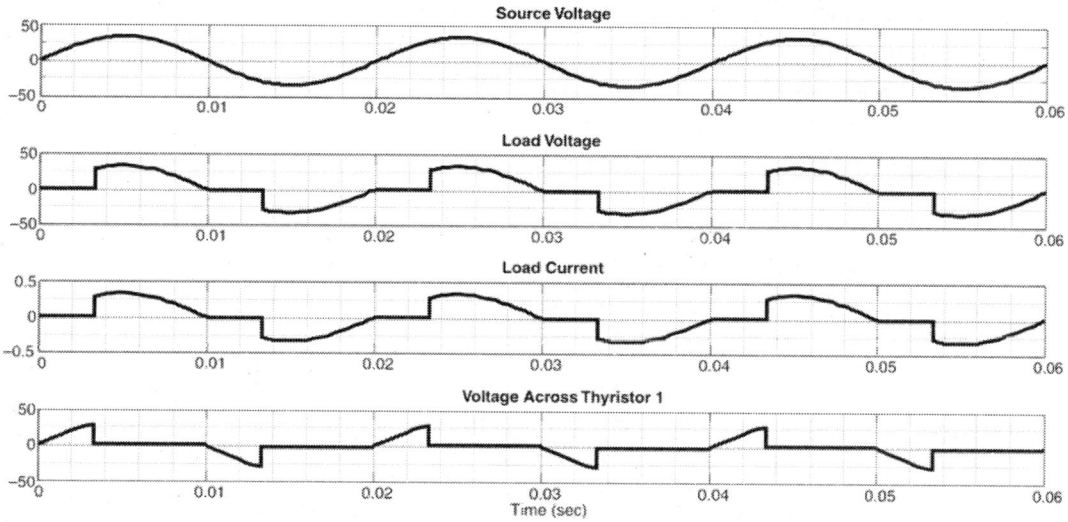

Fig. C8.3: AC voltage regulator with R load - waveforms (a) source voltage (b) voltage across the load (c) load current (d) voltage across thyristor 1

AC Voltage Regulator with RL Load

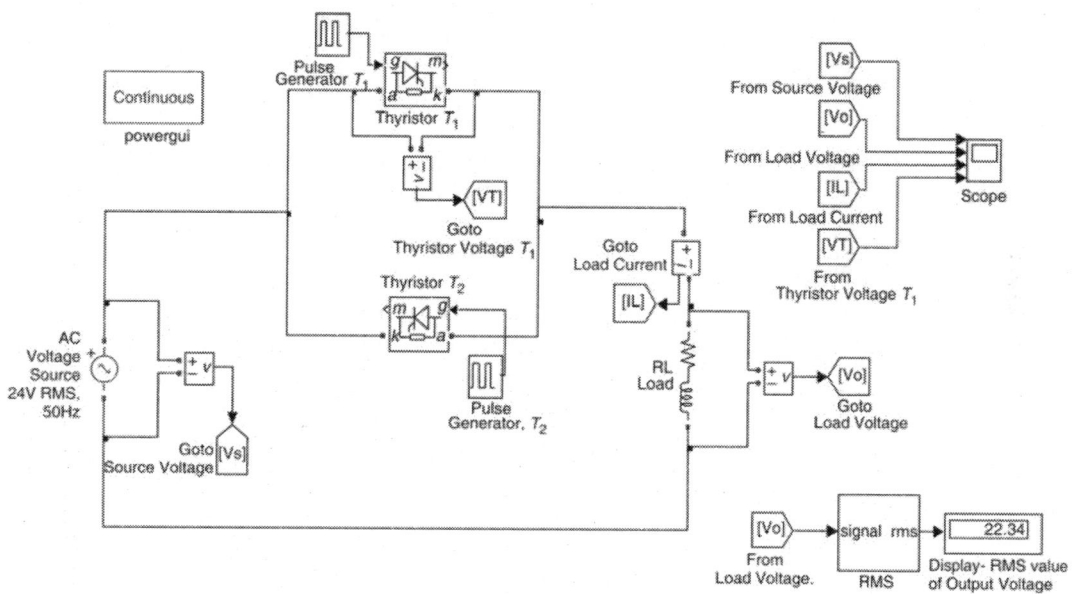

Fig. C8.4: AC voltage regulator circuit model with RL load and measurements blocks

Step 9: Now change the load of the voltage regulator to resistive-inductive (*RL*) load as shown in Fig. C8.4. Double click on *R* branch and configure to *RL*. Select the value of inductor *L* and *R*, say *L* = 80 mH and *R* = 100 Ω so that *L/R* ratio is low. Consider that the firing angle α is greater than the load impedance angle of the load ϕ, i.e. $\alpha > \phi$ where, $\phi = \tan^{-1}(\omega L/R)$, so that the load current is discontinuous. Keep the remaining parameters as it is. Follow steps 1 to 8 as in the case of R load. Waveforms are shown in the Fig. C8.5.

Note down the RMS value of the load voltage from the *display block*.

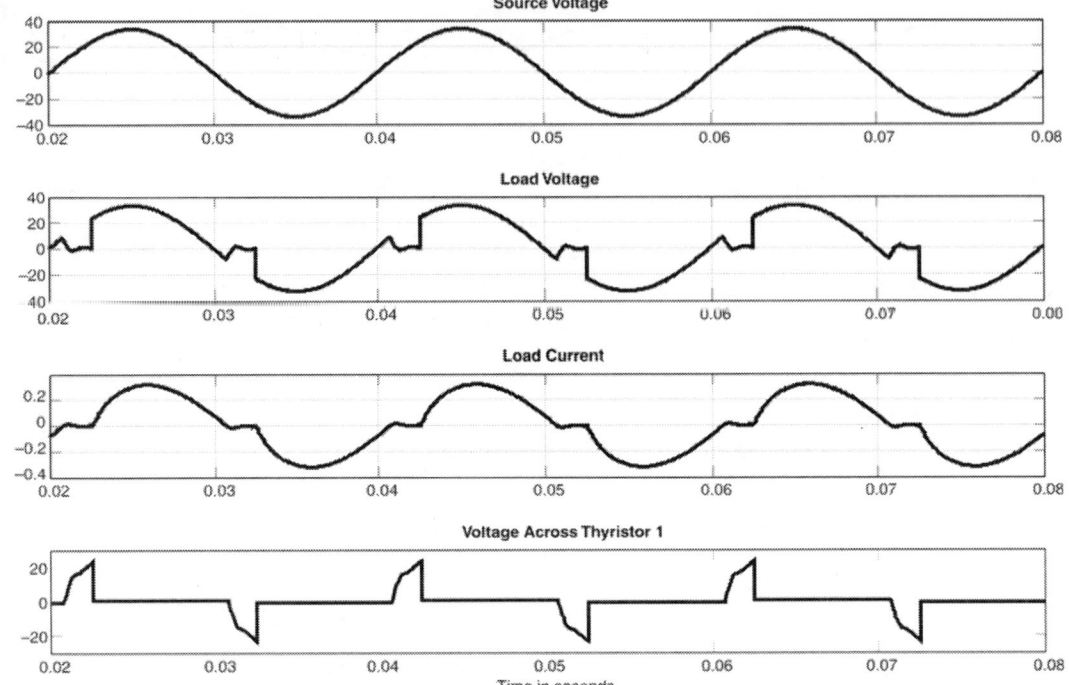

Fig. C8.5: AC voltage regulator with RL load - waveforms (a) source voltage (b) Voltage across the load (c) Load current (d) Voltage across thyristor

C8.3 SIMULATION RESULT

The AC voltage regulator circuit was modelled and simulated using MATLAB/ Simulink. The waveforms of the circuit were analysed with R and RL load.

C9.1 OBJECTIVE

To simulate a complementary commutation circuit, using MATBLAB/Simulink and to observe the following waveforms:
1. Capacitor voltage
2. Voltage across main thyristor
3. Voltage across complementary thyristor
4. Load voltage.

C9.2 MODELLING AND SIMULATION OF COMPLEMENTARY COMMUTATION CIRCUIT USING MATLAB/SIMULINK

Theory and design of a complementary commutation circuit is discussed in section B12. A basic circuit for the simulation of complementary commutation circuit using MATLAB/Simulink is shown in Fig. C9.1.

Following steps are to be followed for analyzing and modelling the complementary commutation circuit.

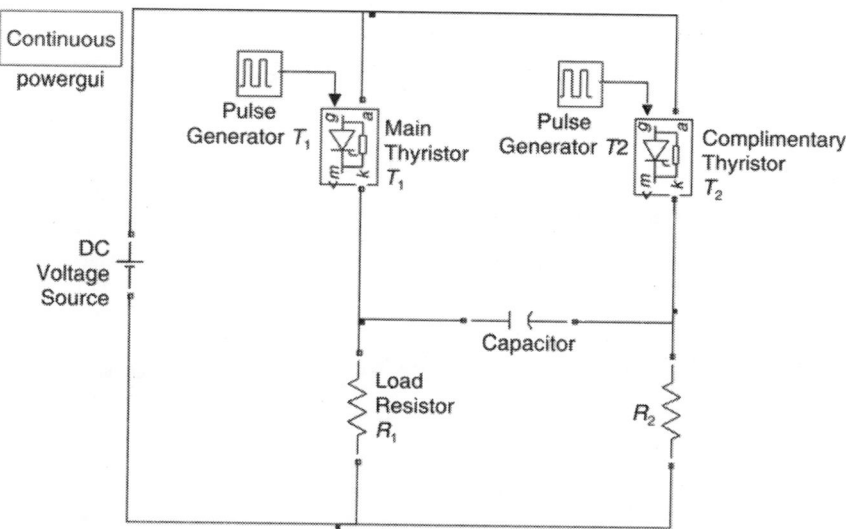

Fig. C9.1: Complementary commutation circuit using MATLAB/Simulink

Step 1: Enter "simulink" in the MATLAB command window to open the simulink library browser or click simulink button in the MATLAB toolbar.

Step 2: Select **File > New > Model** in the simulink library browser to create a new model or press Ctrl+N. An empty model window is opened and it can be saved by selecting File>**Save** in the model window.

Step 3: Drag and place the following building blocks from *simulink library* and *sim power systems library* to the simulink model file according to the circuit diagram.

(a) Powergui block [Sim Power Systems]
(b) DC Voltage Source [Sim Power Systems > Electrical Sources]
(c) Thyristor [Sim Power Systems > Power Electronics]
(d) Pulse generator [Simulink > Sources]
(e) Series RLC branch (configure it to R) [Sim Power Systems > Elements]
(f) Series RLC branch (configure it to C) [Sim Power Systems > Elements]
(g) Voltage measurement [Sim Power Systems > Measurements]
(h) Scope [Simulink > Sinks]
(i) From [Simulink > Signal Routing]
(j) Goto [Simulink > Signal Routing]

Step 4: Connect the blocks according to the circuit diagram of the complementary commutation circuit. Also connect required number of voltage measurement blocks to the *scope*, as shown in the Fig. C9.2, to observe the necessary waveforms. The *Goto* and *From* blocks need not be necessarily used, but it will reduce the number of connections in the circuit.

Step 5: Set the parameters of the individual blocks. Components and specifications used in the complementary commutation circuit model are:

1. DC voltage source	12 V
2. Load resistor, R_1	100 Ω
3. Resistor, R_2	220 Ω
4. Capacitor, C	3.3 μF

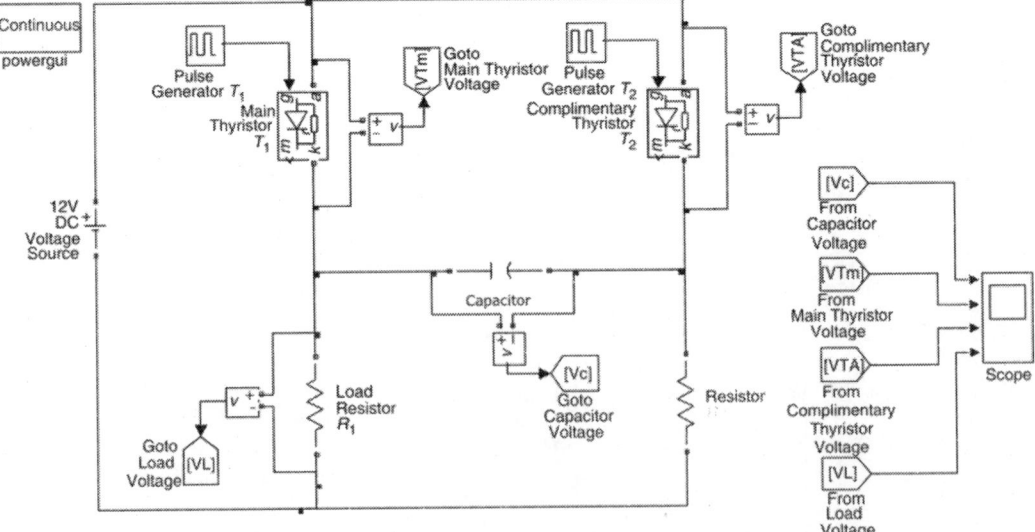

Fig. C9.2: Complementary commutation circuit with measurement blocks using MATLAB/Simulink

5. Thyristor
6. Pulse generator
7. Powergui block
8. Goto and from blocks
9. Voltage measurement blocks
10. Scope

The parameters of pulse generator used are:
Pulse generator for main thyristor T_1:
1. Period (secs): 0.006 (Select period = 6 ms)
2. Pulse width (% of period): 5
3. Phase delay (secs): 2e-3

Pulse generator for complementary thyristor T_2:
1. Period (secs): 0.006
2. Pulse width (% of period): 5
3. Phase delay (secs): 4e-3

Step 6: Save the simulink model. Run the simulation by selecting **simulation > start** in the model window or use **simulation button** in the model window toolbar.
The run time can be entered and modified in the space provided in the model window tool bar. For example, run time = 0.018 seconds. Since the switching period is 6 ms, this will give three cycles of the waveforms in the *scope*.

Step 7: Observe and note down the following waveforms by opening the scope and save the result.
1. Capacitor voltage
2. Voltage across main thyristor
3. Voltage across complementary thyristor
4. Load voltage.

The waveforms are shown in the Fig. C9.3.

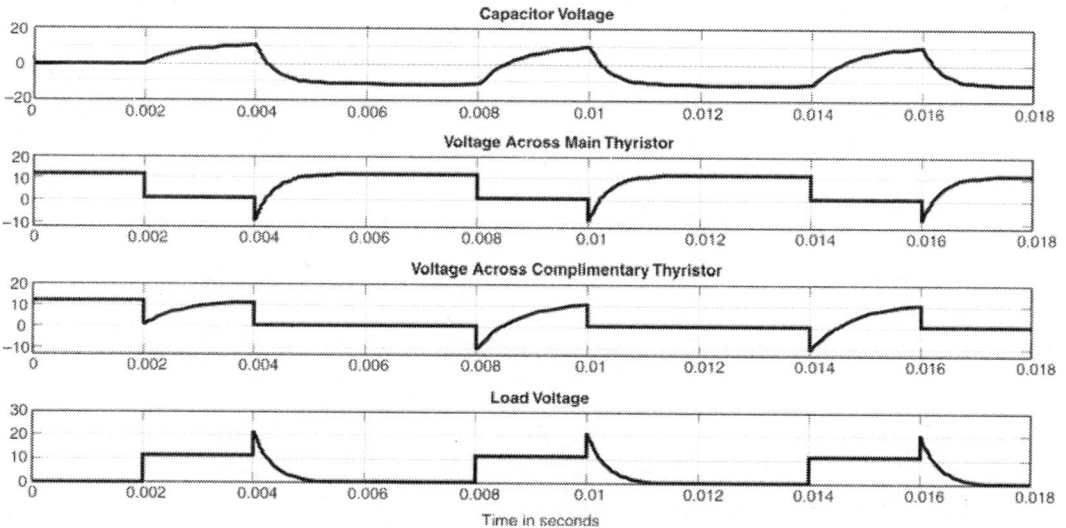

Fig. C9.3: Complementary commutation circuit waveforms

C9.3 SIMULATION RESULT

The complementary commutation circuit was modelled and simulated using MATLAB/Simulink and the following waveforms are observed:

1. Capacitor voltage
2. Voltage across main thyristor
3. Voltage across complementary thyristor
4. Load voltage.

Experiment C10

Simulation of A Resonance Commutation Circuit

C10.1 OBJECTIVE

To simulate the parallel resonance commutation circuit for a thyristor using MATLAB/Simulink and to observe the following waveforms:

(a) Capacitor current
(b) Load current
(c) Current through the thyristor
(d) Load voltage
(e) Voltage across the thyristor

C10.2 MODELLING AND SIMULATION OF PARALLEL RESONANCE COMMUTATION CIRCUIT USING MATLAB/SIMULINK

Theory and design of a parallel resonance commutation circuit is discussed in *section B13*. A basic parallel resonance commutation circuit model using MATLAB/Simulink is shown in Fig. C10.1.

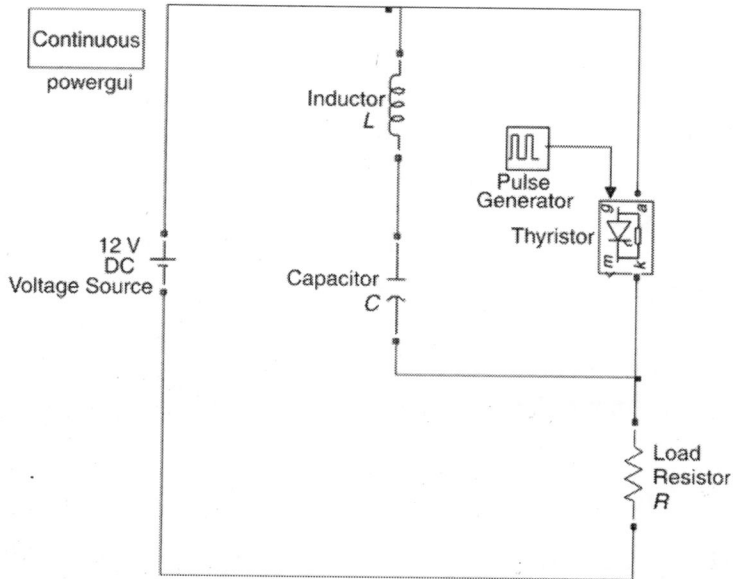

Fig. C10.1: A parallel resonance commutation circuit model using MATLAB/Simulink

Following steps are to be followed for analyzing and modelling a parallel resonance commutation.

Step 1: Enter "simulink" in the MATLAB command window to open the simulink library browser or click simulink button in the MATLAB toolbar.

Step 2: Select **File > New > Model** in the simulink library browser to create a new model or press Ctrl+N. An empty model window is opened and it can be saved by selecting **File > Save** in the model window.

Step 3: Drag and place the following building blocks from *Simulink library* and *Sim Power Systems library* to the simulink model file, as per the circuit diagram.

  (a) Powergui block [Sim Power Systems]
  (b) DC Voltage Source [Sim Power Systems > Electrical Sources]
  (c) Series RLC branch (configure it to R) [Sim Power Systems > Elements]
  (d) Series RLC branch (configure it to L) [Sim Power Systems > Elements]
  (e) Series RLC branch (configure it to C) [Sim Power Systems > Elements]
  (f) Thyristor [Sim Power Systems > Power Electronics]
  (g) Pulse generator [Simulink > Sources]
  (h) Voltage Measurement [Sim Power Systems> Measurements]
  (i) Current Measurement [Sim Power Systems > Measurements]
  (j) Scope [Simulink > Sinks]
  (k) Display [Simulink > Sinks]
  (l) From [Simulink > Signal Routing]
  (m) Goto [Simulink > Signal Routing]

Step 4: Connect the blocks according to the circuit diagram of the parallel resonance commutation circuit. Also connect required number of current and voltage measurement blocks to the *scope*, as shown in the Fig. C10.2, to observe the necessary waveforms. The *Goto* and *From* blocks need not be necessarily used, but it will reduce the number of connections in the circuit.

Step 5: Set the parameters of the individual blocks. Components and specifications used in the parallel resonance commutation circuit are:

1.	DC voltage source	12 V
2.	Load resistor	100 Ω
3.	Inductor	6 mH
4.	Capacitor	2.7 μF
5.	Thyristor	
6.	Pulse Generator	
7.	Powergui block	
8.	Voltage measurement blocks	
9.	Current measurement blocks	
10.	Scope	
11.	Display	
12.	Goto block	
13.	From block	

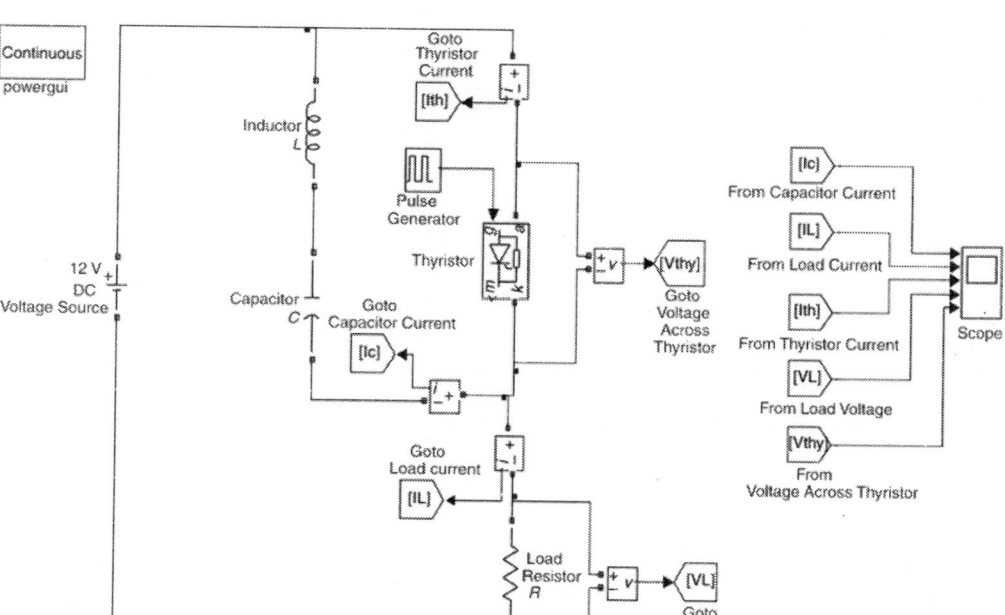

Fig. C10.2: Parallel resonance commutation circuit model using MATLAB/Simulink
with measurement blocks

Parameters of pulse generator used are:

1. Period (secs): 0.01
2. Pulse Width (% of period): 5
3. Phase delay (secs): 1.25e-3

Switching period or chopping period = 10 milliseconds

Step 6: Save the simulink model. Run the simulation by selecting **simulation > start** in the model window or use **simulation button** in the model window toolbar.

The run time can be entered and modified in the space provided in the model window tool bar. For example, run time = 0.03 seconds. Since the switching period is 10 ms, this will give three cycles of the waveforms in the scope.

Step 7: Observe and note down the following waveforms by opening the scope and save the result.

(a) Capacitor current

(b) Load current

(c) Current through the thyristor

(d) Load voltage

(e) Voltage across the thyristor

The waveforms are shown in the Fig. C10.3.

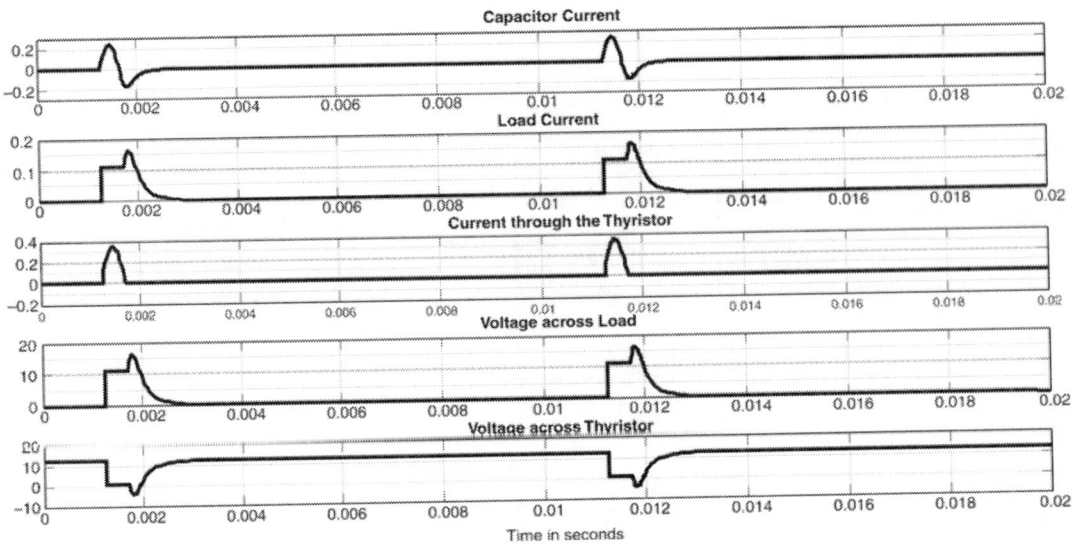

Fig. C10.3: Waveforms-parallel resonance commutation

C10.3 SIMULATION RESULT

The resonance commutation circuit was modelled and simulated using MATLAB/Simulink and the following waveforms are observed.

(a) Capacitor current
(b) Load current
(c) Current through the thyristor
(d) Load voltage
(e) Voltage across the thyristor.

Simulation of an Impulse Commutation Circuit

C11.1 OBJECTIVE

To simulate the impulse (auxiliary) commutation circuit for a thyristor using MATLAB/Simulink and to observe the following waveforms:

(a) Capacitor current

(b) Voltage across the capacitor

(c) Current through the main thyristor

(d) Voltage acrossmain thyristor

(e) Voltage across auxiliary thyristor

(f) Load voltage

C11.2 MODELLING AND SIMULATION OF AN IMPULSE COMMUTATION CIRCUIT USING MATLAB/SIMULINK

The basic circuit model of an impulse commutation circuit using MATLAB/Simulink is shown in Fig. C11.1. Theory and design of an impulse commutation circuit is discussed in section B14.

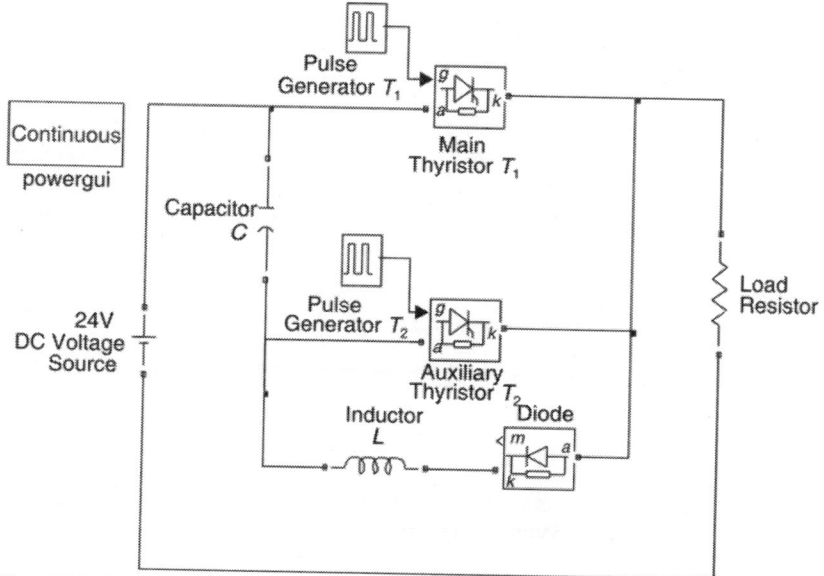

Fig. C11.1: An auxiliary commutation circuit model using MATLAB/Simulink

Following steps are to be followed for analyzing and modelling an auxiliary commutation circuit.

Step 1: Enter "simulink" in the MATLAB command window to open the simulink library browser or click simulink button in the MATLAB toolbar.

Step 2: Select **File > New > Model** in the simulink library browser to create a new model or press Ctrl+N. An empty model window is opened and it can be saved by selecting **File > Save** in the model window.

Step 3: Drag and place the following building blocks from *Simulink library* and *Sim Power Systems library* to the simulink model file, as per the circuit diagram.

(a) Powergui block [Sim Power Systems]
(b) DC voltage source [Sim Power Systems > Electrical Sources]
(c) Series RLC branch (configure it to R) [Sim Power Systems > Elements]
(d) Series RLC branch (configure it to L) [Sim Power Systems > Elements]
(e) Series RLC branch (configure it to C) [Sim Power Systems > Elements]
(f) Thyristor [Sim Power Systems> Power Electronics]
(g) Pulse generator [Simulink > Sources]
(h) Voltage measurement [Sim Power Systems> Measurements]
(i) Current measurement [Sim Power Systems > Measurements]
(j) Scope [Simulink > Sinks]
(k) From [Simulink > Signal Routing]
(l) Goto [Simulink > Signal Routing]

Step 4: Connect the blocks according to the circuit diagram of the auxiliary commutation circuit. Also connect required number of current and voltage measurement blocks to the *scope*, as shown in the Fig. C11.2, to observe the necessary waveforms. The *Goto*

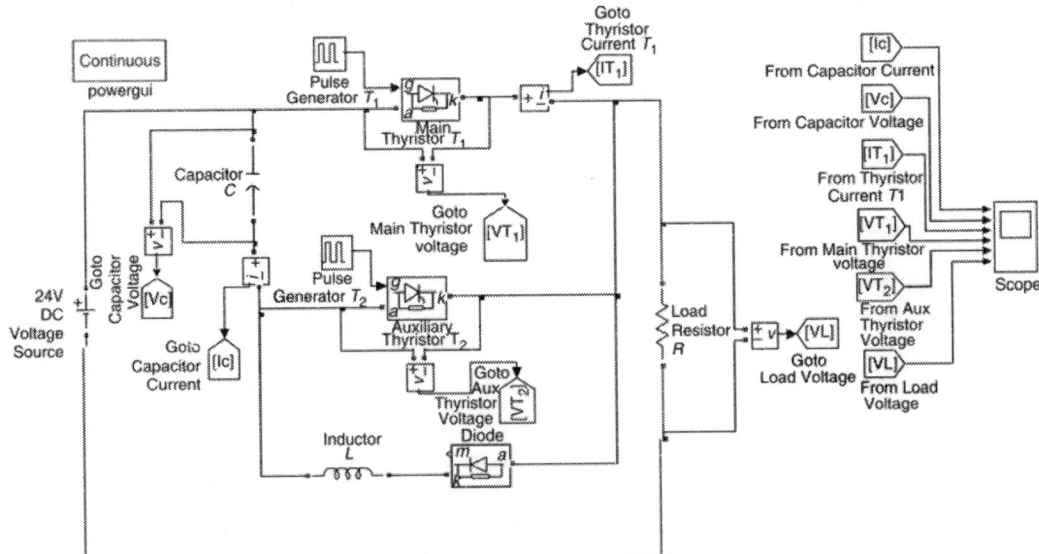

Fig. C11.2: Auxiliary commutation circuit model using MATLAB/Simulink
With measurements blocks

and *From* blocks need not be necessarily used, but it will reduce the number of connections in the circuit.

Step 5: Set the parameters of the individual blocks. Components and specifications used in the auxiliary commutation circuit are:

1. DC voltage source   24 V
2. Load resistor   100 Ω
3. Inductor   5 mH
4. Capacitor   3.3 µF
5. Thyristor
6. Pulse generator
7. Powergui block
8. Voltage measurement blocks
9. Current measurement blocks
10. Scope
11. Goto block
12. From block

Parameters of pulse generator used are:
Pulse generator for main thyristor
1. Period (secs): 0.01 (10 milliseconds)
2. Pulse width (% of period): 5
3. Phase delay (secs): 0.001
Pulse generator for auxiliary thyristor
1. Period (secs): 0.01 (10 milliseconds)
2. Pulse Width (% of period): 5
3. Phase delay (secs): 0.006

Choose the switching period as 10 milli seconds (i.e. switching frequency = 100 Hz), since the thyristor is a low switching frequency device.

Step 6: Save the simulink model. Run the simulation by selecting **simulation > start** in the model window or use **simulation button** in the model window toolbar.

The run time can be entered and modified in the space provided in the model window tool bar. For example, run time = 0.03 seconds. Since the switching period is 10 ms, this will give three cycles of the waveforms in the scope.

Step 7: Observe and note down the following waveforms by opening the scope and save the result.

(a) Capacitor current
(b) Capacitor voltage
(c) Main thyristor current
(d) Voltage across main thyristor
(e) Voltage across the auxiliary thyristor
(f) Load voltage

The waveforms are shown in the Fig. C11.3.

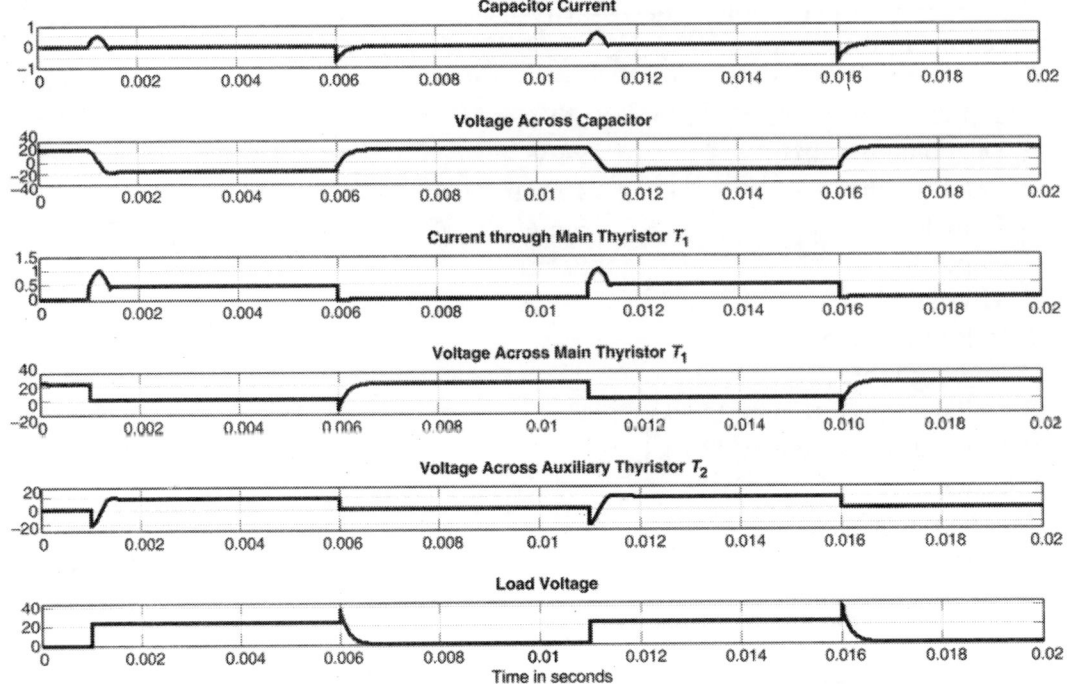

Fig. C11.3: Waveforms: Impulse commutation circuit

C11.3 SIMULATION RESULT

The impulse commutation circuit was modelled and simulated using MATLAB/ Simulink and the following waveforms are observed.

(a) Capacitor current
(b) Voltage across the capacitor
(g) Current through the main thyristor
(h) Voltage acrossmain thyristor
(i) Voltage across auxiliary thyristor
(j) Load voltage.

Experiment C12

Simulation of A Current Commutated Chopper

C12.1 OBJECTIVE

To perform variable frequency and constant frequency operation of current commutated chopper (oscillation chopper) to control the load voltage.

C12.2 THEORY

Theory, operation and analysis are explained in section B17.

C12.3 MODELLING AND SIMULATION OF OSCILLATION CHOPPER CIRCUIT USING MATLAB/SIMULINK

The output voltage or load voltage of the oscillation chopper can be controlled by two different schemes, namely, *constant frequency operation* and *variable frequency operation*.

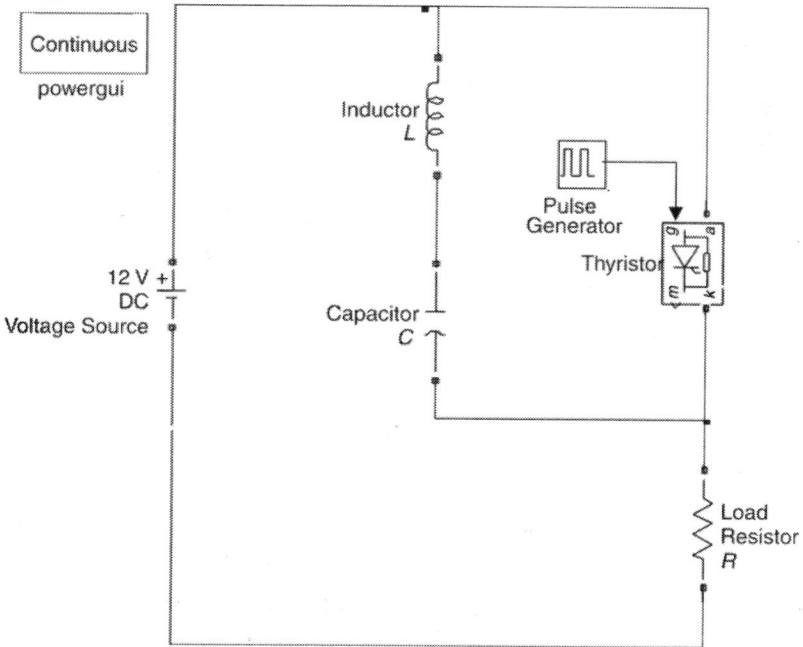

Fig. C12.1: A current commutated chopper circuit model using MATLAB/Simulink

205

Following steps are to be followed for analyzing and modelling a current commutated chopper.

Step 1: Enter "simulink" in the MATLAB command window to open the simulink library browser or click simulink button in the MATLAB toolbar.

Step 2: Select **File > New > Model** in the simulink library browser to create a new model or press Ctrl+N. An empty model window is opened and it can be saved by selecting **File > Save** in the model window.

Step 3: Drag and place the following building blocks from *simulink library* and *Sim Power Systems library* to the simulink model file, as per the circuit diagram.

(a) Powergui block [Sim Power Systems]

(b) DC voltage source [Sim Power Systems > Electrical Sources]

(c) Series RLC branch (configure it to R) [Sim Power Systems > Elements]

(d) Series RLC branch (configure it to L) [Sim Power Systems > Elements]

(e) Series RLC branch (configure it to C) [Sim Power Systems > Elements]

(f) Thyristor [Sim Power Systems > Power Electronics]

(g) Pulse generator [Simulink > Sources]

(h) Voltage measurement [Sim Power Systems > Measurements]

(i) Current measurement [Sim Power Systems > Measurements]

(j) Scope [Simulink > Sinks]

(k) Mean value [Sim Power Systems > Extra library > Measurements]

(l) Display [Simulink > Sinks]

(m) From [Simulink > Signal Routing]

(n) Goto [Simulink > Signal Routing]

Step 4: Connect the blocks according to the circuit diagram of the current commutated chopper. Also connect required number of current and voltage measurement blocks to the *scope*, as shown in the Fig. C12.2, to observe the necessary waveforms. The *Goto* and *From* blocks need not be necessarily used, but it will reduce the number of connections in the circuit.

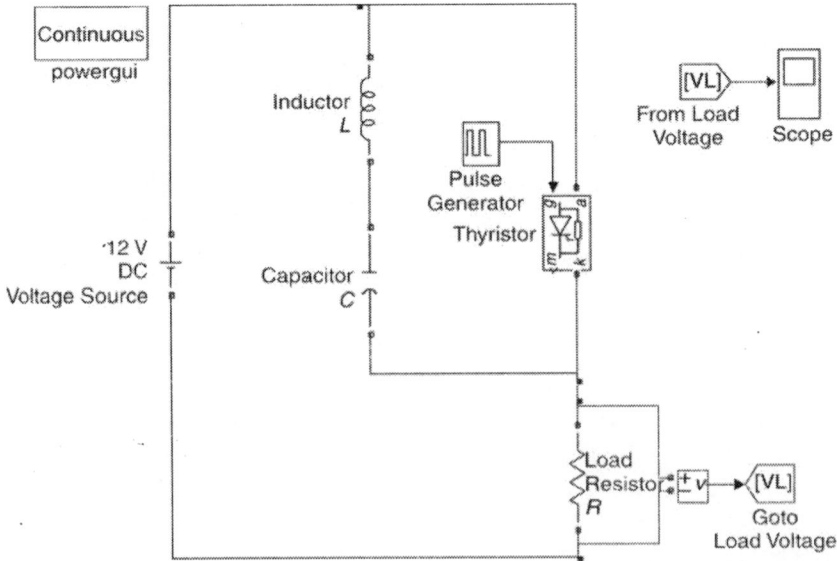

Fig. C12.2: Current commutated chopper circuit model using MATLAB/Simulink with voltage measurement block

Step 5: Set the parameters of the individual blocks. Components and specifications used in the current commuted chopper circuit are:

1. DC voltage source   12 V
2. Load resistor   100 Ω
3. Inductor   6 mH
4. Capacitor   2.7 µF
5. Thyristor
6. Pulse generator
7. Powergui block
8. Voltage measurement blocks
9. Scope
10. Display
11. Goto block
12. From block
13. Mean value block

Parameters of pulse generator used are:

1. Period (secs): 0.01
2. Pulse Width (% of period): 5
3. Phase delay (secs): 1.25e-3

Switching period or chopping period = 10 milliseconds

Step 6: The average value of the load voltage can be measured. For that *mean value block* is used here and the average value is displayed in *display block*. Double click on *mean value block* and enter *averaging period* as 0.01 seconds. Whenever switching frequency is changed, the averaging period in the mean value block should also be changed.

Step 7: Save the Simulink model. Run the simulation by selecting **simulation > start** in the model window or use **simulation button** in the model window toolbar.

The run time can be entered and modified in the space provided in the model window tool bar. For example, run time = 0.03 seconds. Since the switching period is 10 ms, this will give three cycles of the waveforms in the scope.

Step 8: Observe the load voltage waveform by opening the scope and save the result. Note down the mean value of the load voltage from the *display block*.

C12.4. CONSTANT FREQUENCY OPERATION

C12.4.1 Time Ratio Control

The output voltage or load voltage of the voltage commutated chopper can be controlled by two different schemes, namely, *constant frequency operation* and *variable frequency operation*.

Step 9: In *constant frequency operation*, the chopping period T (or chopping frequency) is kept constant and the thyristor ON time, t_{ON} (time for which the thyristor is ON) is varied. The ON time can be varied by adjusting the values of the commutating elements L and C. Select the design values of $L = 6$ mH and $C = 2.7$ µF and $T = 0.01$ seconds. Now run the simulation. Fig. C12.4(a) shows the waveform of load voltage using the design values of L and C. Fig. C12.4(b) shows the waveform with t_{ON} varied and T is constant, for a different set of values of L and C ($L = 3$ mH and $C = 2.7$ µF). Simulating the circuit model shown in Fig. 12.3, the load voltage waveforms for both the cases can be observed simultaneously.

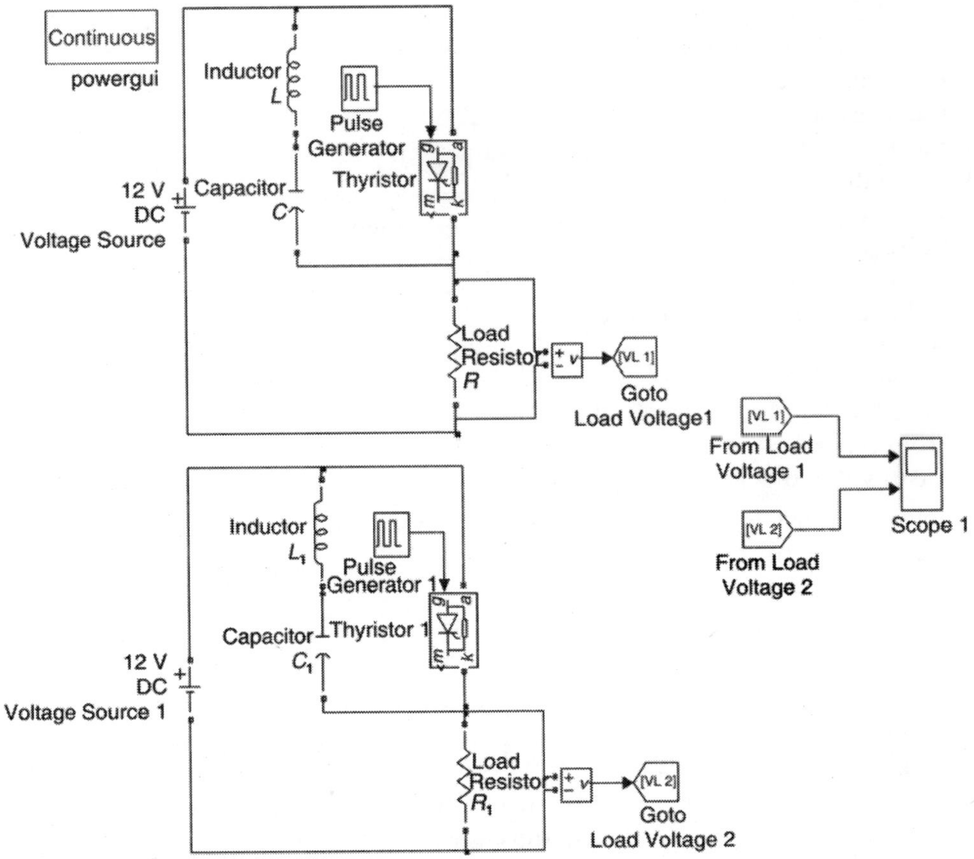

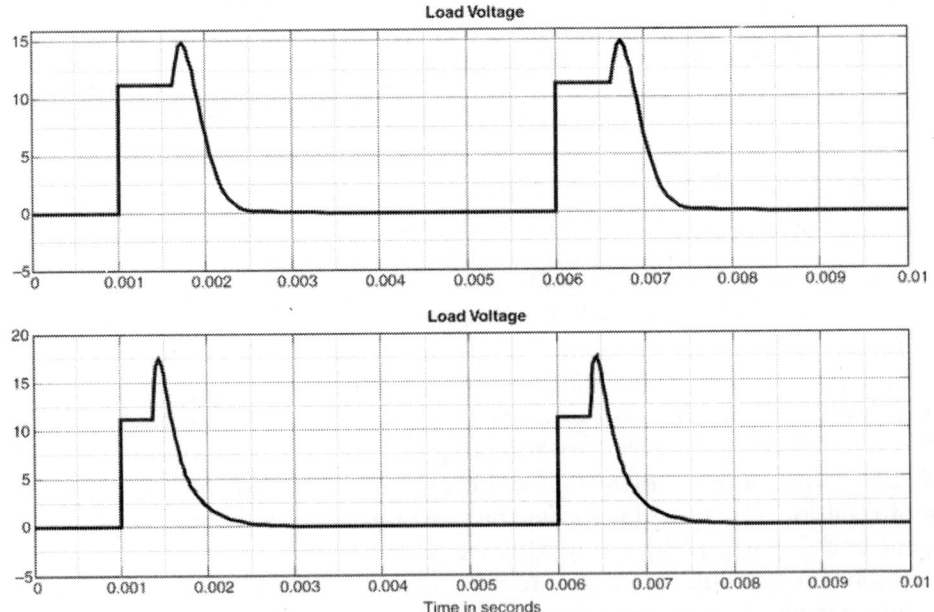

Fig C12.3: Time ratio control of current commutated chopper

Fig C12.4: Time ratio control of current commutated chopper-constant frequency

C12.4.2 Variable Frequency Operation

Step 10: In *variable frequency operation*, the chopping period T is varied by adjusting the switching frequency of the thyristor. In this case commutating elements L and C are of design values. Hence t_{ON} is constant. For simulation, keep **period** of *Pulse Generator* as 0.01 seconds and keep remaining parameters as it is.

Figure C12.5 shows the variable frequency operation of oscillation chopper. In Fig. C12.5(a) the chopping period is 0.01 seconds where as in Fig. C12.5(b) chopping period is selected as 0.005 seconds. Hence in the latter case the chopping period is twice as that of the former.

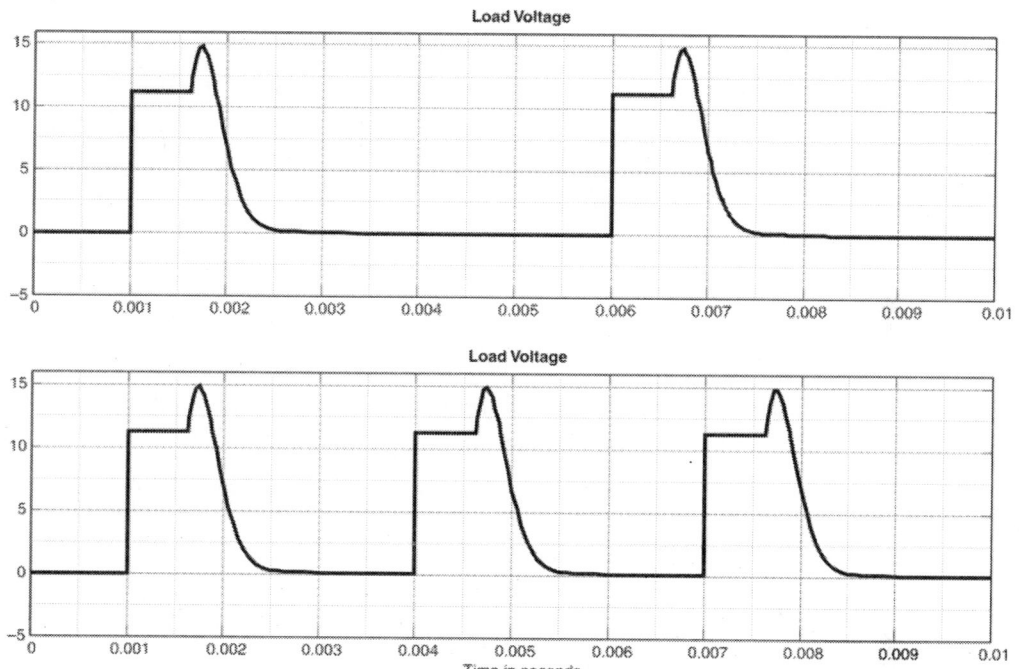

Fig. C12.5: Time ratio control of current commutated chopper-variable frequency

C12.5 SIMULATION RESULT

The oscillation chopper circuit was modelled and simulated using MATLAB/Simulink. The waveforms of the circuit were analysed. The constant and variable frequency operation of the oscillation chopper was performed.

Experiment C13

Simulation of a Voltage Commutated Chopper

C13.1 OBJECTIVE

To perform variable frequency and constant frequency operation of voltage commutated chopper to control the load voltage.

C13.2 THEORY

Theory, operation and analysis are explained in section B18.

C13.3 MODELLING AND SIMULATION OF VOLTAGE COMMUTATED CHOPPER CIRCUIT USING MATLAB/SIMULINK

The output voltage or load voltage of a voltage commutated chopper can be controlled by two different schemes, namely, *constant frequency operation* and *variable frequency operation*.

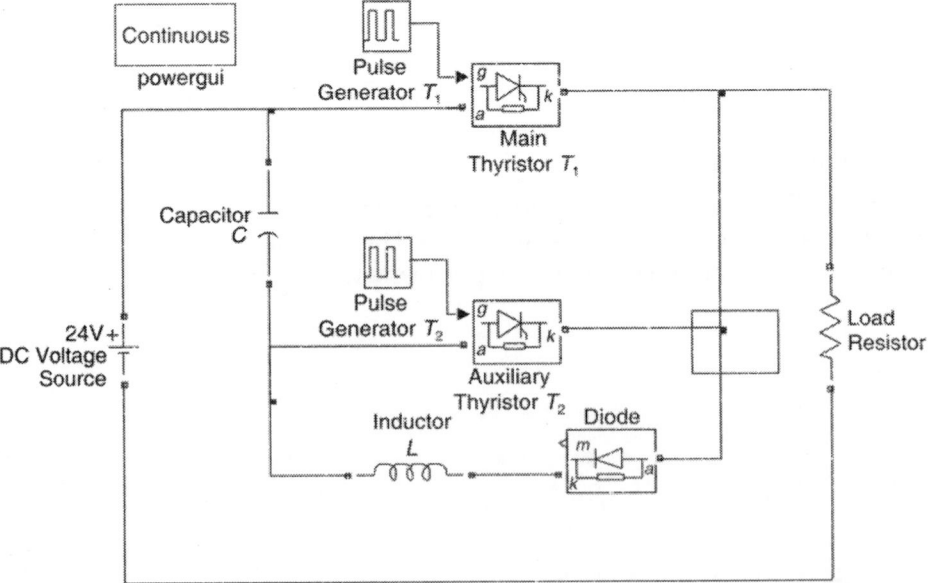

Fig. C13.1: A voltage commutated chopper circuit model using MATLAB/Simulink

210

Following steps are to be followed for analyzing and modelling a voltage commutated chopper.

Step 1: Enter "simulink" in the MATLAB command window to open the Simulink Library Browser or click Simulink button in the MATLAB toolbar.

Step 2: Select **File > New > Model** in the simulink library browser to create a new model or press Ctrl+N. An empty model window is opened and it can be saved by selecting **File > Save** in the model window.

Step 3: Drag and place the following building blocks from *Simulink library* and *Sim Power Systems library* to the simulink model file, as per the circuit diagram.

(a) Powergui block [Sim Power Systems]
(b) DC voltage source [Sim Power Systems > Electrical Sources]
(c) Series RLC branch (configure it to R) [Sim Power Systems > Elements]
(d) Series RLC branch (configure it to L) [Sim Power Systems > Elements]
(e) Series RLC branch (configure it to C) [Sim Power Systems > Elements]
(f) Thyristor [Sim Power Systems > Power Electronics]
(g) Pulse generator [Simulink > Sources]
(h) Voltage measurement [Sim Power Systems > Measurements]
(i) Current measurement [Sim Power Systems > Measurements]
(j) Scope [Simulink > Sinks]
(k) Mean value [Sim Power Systems > Extra library > Measurements]
(l) Display [Simulink > Sinks]
(m) From [Simulink > Signal Routing]
(n) Goto [Simulink > Signal Routing]

Step 4: Connect the blocks according to the circuit diagram of the voltage commutated chopper. Also connect required number of current and voltage measurement blocks to the *scope*, as shown in the Fig. C13.2, to observe the necessary waveforms. The *Goto*

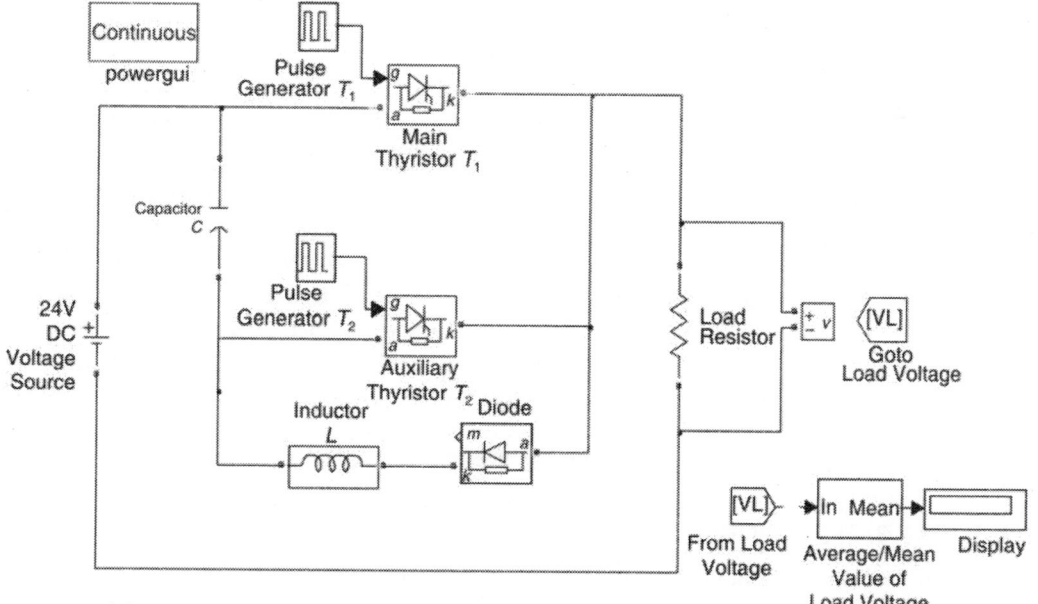

Fig. C13.2: Voltage commutated chopper circuit model using MATLAB/Simulink with voltage measurement block and *From* blocks need not be necessarily used, but it will reduce the number of connections in the circuit.

Step 5: Set the parameters of the individual blocks. Components and specifications used in the current commuted chopper circuit are:

1. DC voltage source   24 V
2. Load resistor   100 Ω
3. Inductor   5 mH
4. Capacitor   3.3 μF
5. Thyristor
6. Pulse generator
7. Powergui block
8. Voltage measurement blocks
9. Scope
10. Display
11. Goto block
12. From block
13. Mean value block

Parameters of pulse generator used are:

Pulse generator for main thyristor
1. Period (secs): 0.003 (3 milliseconds)
2. Pulse Width (% of period): 5
3. Phase delay (secs): 0.001

Pulse generator for auxiliary thyristor
1. Period (secs): 0.003 (3 milliseconds)
2. Pulse Width (% of period): 5
3. Phase delay (secs): 0.002

Choose the switching period as 3 milli seconds.

Step 6: The average value of the load voltage can be measured. For that *mean value block* is used here and the average value is displayed in *display block*. Double click on *mean value block* and enter *averaging period* as 0.003 seconds. Whenever switching frequency is changed, the averaging period in the mean value block should also be changed.

Step 7: Save the Simulink model. Run the simulation by selecting **simulation > start** in the model window or use **simulation button** in the model window toolbar.

The run time can be entered and modified in the space provided in the model window tool bar. For example, run time = 0.03 seconds. Since the switching period is 3 ms, this will give three cycles of the waveforms in the scope.

Step 8: Observe the load voltage waveform by opening the *scope* and save the result. Note down the mean value of the load voltage from the *display block*.

C13.4 TIME RATIO CONTROL

The output voltage or load voltage of the voltage commutated chopper can be controlled by two different schemes, namely, *constant frequency operation* and *variable frequency operation*.

C13.4.1 Constant Frequency Operation

Step 9: In **constant frequency operation** or pulse width modulation control (PWM), the time period T (or switching frequency) is kept constant and main thyristor ON time, t_{ON} (time for which the thyristor is ON) is varied. The ON time of main thyristor can be varied by adjusting the switching instant of auxiliary thyristor.

By keeping commutating elements L and C values kept constant, turn ON instant of auxiliary thyristor is changed to new value. For that, adjust the value of *phase delay (secs)* of pulse generator of the auxiliary thyristor. (For example, phase delay (secs): = 0.0025).

Simulating the circuit model shown in Fig. C13.3, the load voltage waveforms for both the cases can be observed simultaneously.

Waveforms are shown in Fig. C13.4 for the constant frequency operation of voltage commutated chopper.

C13.4.2 Variable Frequency Operation

Step 10: In **variable frequency operation**, the switching period T (or switching frequency) is varied. Here ON time, t_{ON} is kept constant. In this case commutation elements (L and C) are also kept constant. When the switching period of the main thyristor is changed to a new value, the switching period of the auxiliary thyrisor must also be changed to the same value.

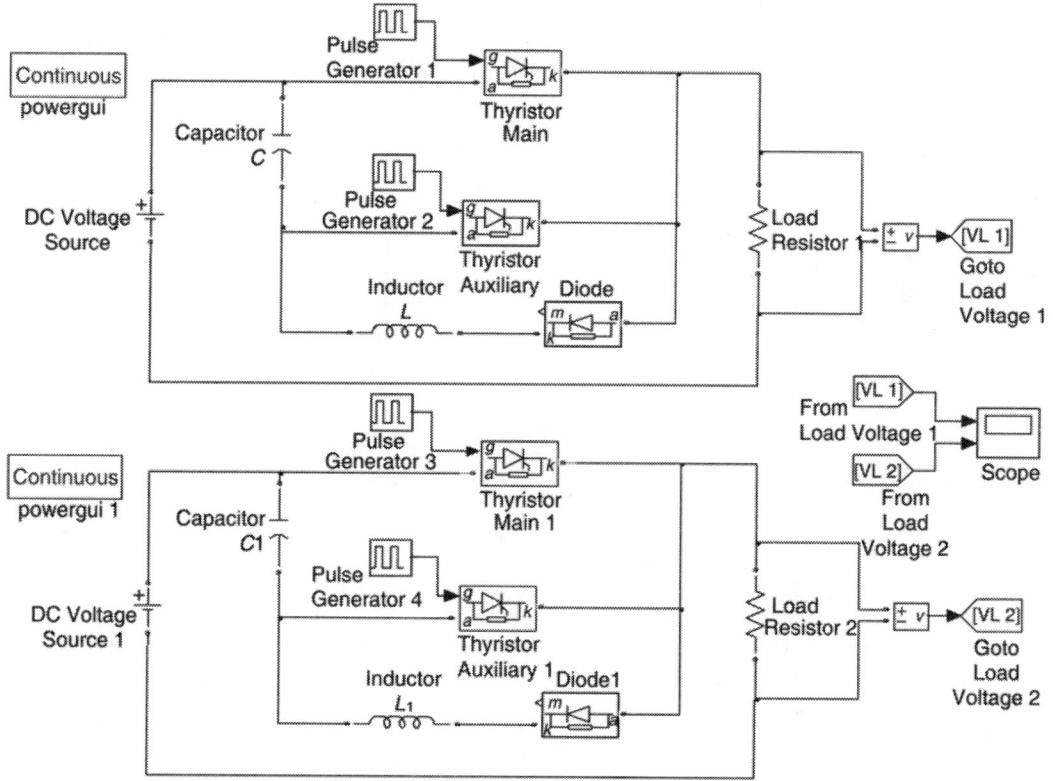

Fig C13.3: Time ratio control of voltage commutated chopper

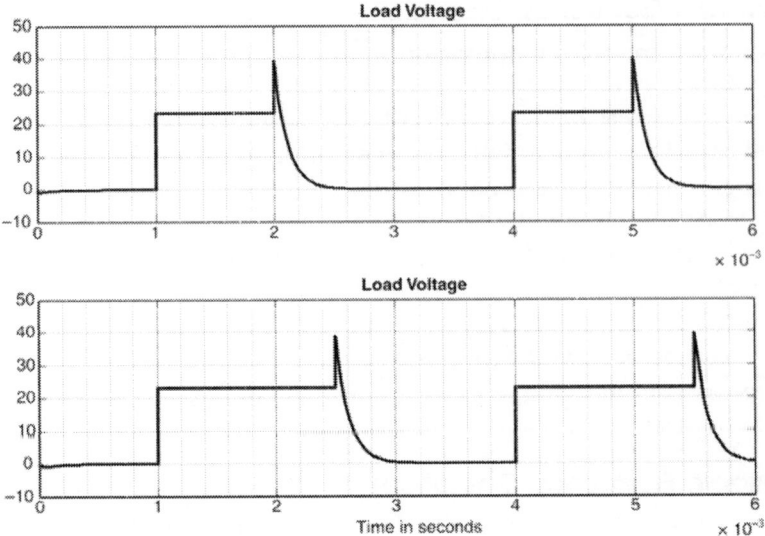

Fig. C13.4: Time ratio control of voltage commutated chopper-constant frequency

In the *pulse generator* of both main and auxiliary thyristor, change *period (secs)* value to new value and keep the phase delay of both the main and auxiliary thyristors the same. Keep the remaining parameters as it is.

(For example, for both main and auxiliary thyristors: period (secs): = 0.004). Phase delay = 0.002. Waveforms are shown in Fig. C13.5 for the variable frequency operation of voltage commutated chopper.

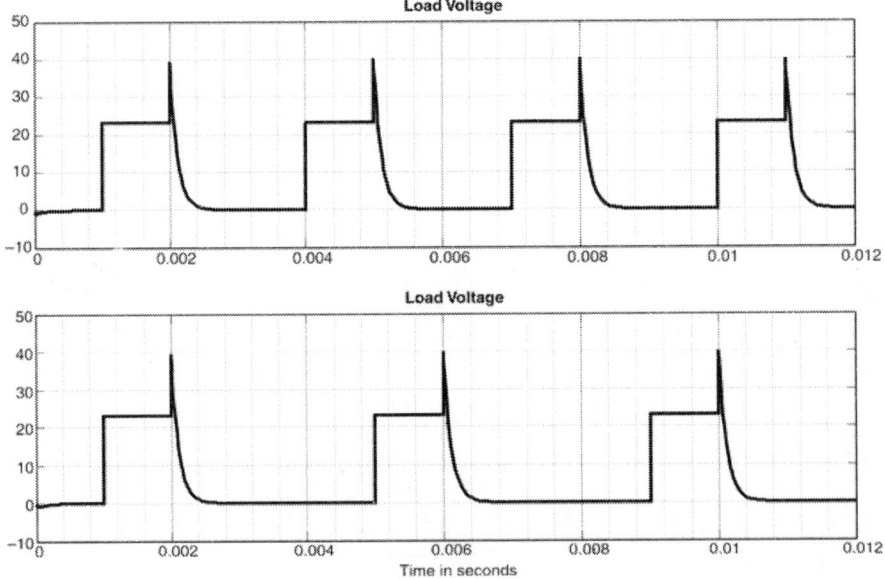

Fig. C13.5: Time ratio control of voltage commutated chopper-variable frequency

C13.5 SIMULATION RESULT

The voltage commutated chopper circuit was modelled and simulated using MATLAB/Simulink. The waveforms of the circuit were analysed. The constant and variable frequency operation of the voltage commutated chopper was performed.

Experiment C14

Simulation of a Buck Regulator

C14.1 OBJECTIVE

To model and simulate the buck regulator using MATLAB/Simulink and observe following waveforms

(a) Voltage across inductor

(b) Current through inductor

(c) Current through capacitor

(d) Load voltage

C14.2 MODELLING AND SIMULATION OF BUCK REGULATOR CIRCUIT USING MATLAB/SIMULINK

Theory and design of a buck regulator circuit is discussed in section B15. A basic circuit for the simulation of buck regulator circuit using MATLAB/Simulink is shown in Fig. C14.1.

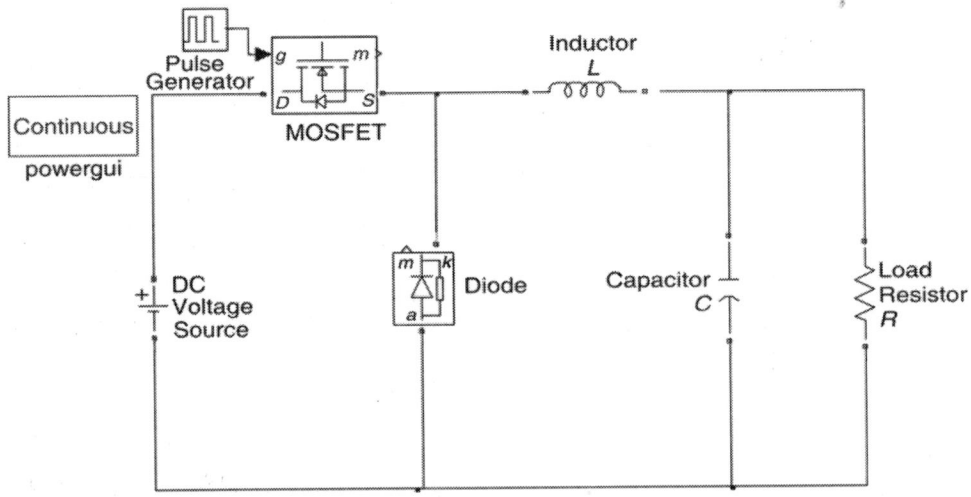

Fig. C14.1: Buck regulator circuit model using MATLAB/Simulink

Following steps are to be followed for analyzing and modelling the buck regulator circuit.

Step 1: Enter "simulink" in the MATLAB command window to open the simulink library browser or click simulink button in the MATLAB toolbar.

Step 2: Select **File > New > Model** in the simulink library browser to create a new model or press Ctrl+N. An empty model window is opened and it can be saved by selecting **File > Save** in the model window.

Step 3: Drag and place the following building blocks from *Simulink library* and *Sim Power Systems library* to the simulink model file according to the circuit diagram.

(a) Powergui block [Sim Power Systems]

(b) DC Voltage Source [Sim Power Systems > Electrical Sources]

(c) Series RLC branch (configure it to R) [Sim Power Systems > Elements]

(d) Series RLC branch (configure it to L) [Sim Power Systems > Elements]

(e) Series RLC branch (configure it to C) [Sim Power Systems > Elements]

(f) Diode [Sim Power Systems > Power Electronics]

(g) Mosfet [Sim Power Systems > Power Electronics]

(h) Pulse generator [Simulink > Sources]

(i) Voltage measurement [Sim Power Systems > Measurements]

(j) Current measurement [Sim Power Systems > Measurements]

(k) Scope [Simulink > Sinks]

(l) Mean value [Sim Power Systems > Extra library > Measurements]

(m) Display [Simulink > Sinks]

Step 4: Connect the blocks according to the circuit diagram of the buck regulator circuit. Also connect required number of voltage and current measurement blocks to the *scope*, as shown in the Fig. C14.2, to observe the necessary waveforms. The *Goto* and *From* blocks need not be necessarily used, but it will reduce the number of connections in the circuit.

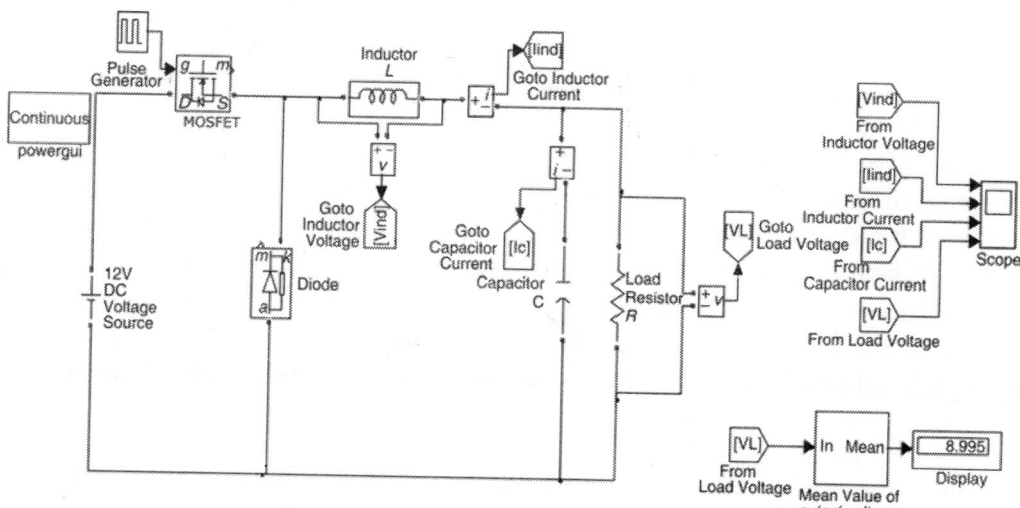

Fig. C14.2: Buck regulator circuit model with measurement blocks

Step 5: Set the parameters of the individual blocks. Components and specifications used in the buck regulator circuit model are:

1. DC voltage source   12 V
2. Load resistor, R 100 Ω
3. Inductor   20 mH
4. Capacitor, C 330 μF
5. Thyristor
6. Pulse generator
7. Diode
8. Powergui block
9. Goto and from blocks
10. Voltage measurement blocks
11. Current measurement blocks
12. Scope
13. Mean value block
14. Display

The parameters of pulse generator used for MOSFET are:

1. Period (secs): 1e-3 (Select, period = 1 ms)
2. Pulse Width (% of period): 75 (Select, duty ratio = 0.75)
3. Phase delay (secs): 0

Note: Switching period = 1/switching frequency = 1/1 kHz = 1 milliseconds.

Step 6: Since the circuit provides DC to DC conversion, the average value of the load voltage can be calculated. For that *mean value block* is used and corresponding value is displayed in the *display block*. Double click on *mean value block* and enter *averaging period* as 0.001 seconds since switching period is selected as 1 msec.

Step 7: Save the Simulink model. Run the simulation by selecting **simulation > start** in the model window or use **simulation button** in the model window toolbar.

The run time can be entered and modified in the space provided in the model window tool bar. For example, run time = 0.02 seconds.

Step 8: Observe and note down the following waveforms by opening the scope and save the result.

(a) Voltage across inductor
(b) Current through inductor
(c) Current through capacitor
(d) Load voltage

The waveforms are shown in the Fig. C14.3.

Also note down the average value of the load voltage from the *display block*.

Step 9: Now adjust the duty ratio of the MOSFET and verify that the load voltage can only be varied from 0 V to supply voltage, 12 V. Also observe corresponding changes in the circuit waveforms.

Note 1: It can be observed that the output voltage waveform has ripples. By increasing the value of filter capacitor to a large value, output voltage can be kept to a constant steady value.

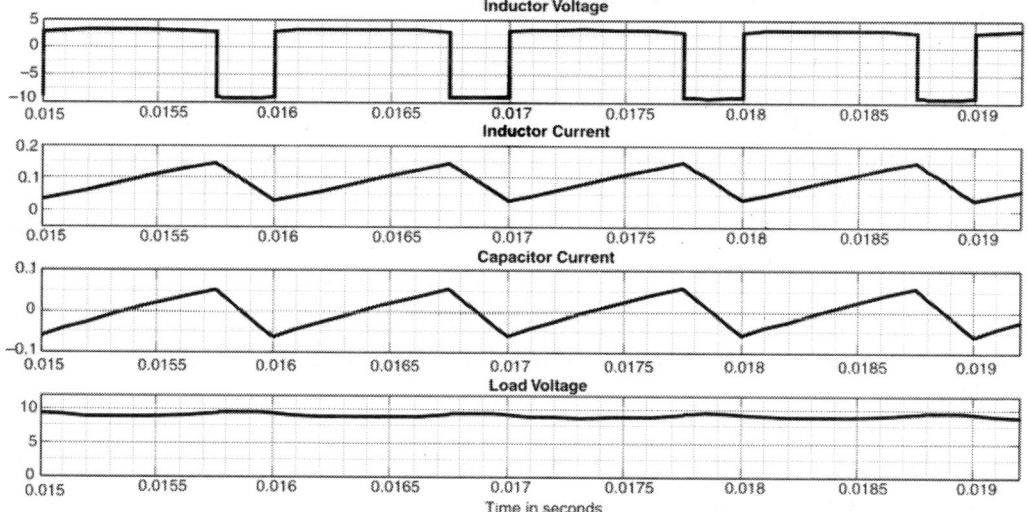

Fig. C14.3: Buck regulator waveforms (a) Voltage across inductor (b) Current through the inductor (c) Current through the capacitor (d) Load voltage

The AC component of the inductor current is passed through the capacitor and DC component is passed through the load resistor.

Note 2: By adjusting the value of inductance, the continuous and discontinuous operation of buck regulator can be observed.

C14.3 SIMULATION RESULT

A buck regulator circuit was modelled and simulated using MATLAB/Simulink and analyzed circuit waveforms.

Experiment C15

Simulation of a Boost Regulator

C15.1 OBJECTIVE

To model and simulate a boost regulator using MATLAB/Simulink and observe the following waveforms

(a) Voltage across inductor

(b) Current through inductor

(c) Current through capacitor

(d) Load voltage

C15.2 MODELLING AND SIMULATION OF BOOST REGULATOR CIRCUIT USING MATLAB/SIMULINK

Theory and design of a boost regulator circuit is discussed in section B16. A basic circuit for the simulation of boost regulator circuit using MATLAB/Simulink is shown in Fig. C15.1.

Following steps are to be followed for analyzing and modelling the boost regulator circuit.

Step 1: Enter "simulink" in the MATLAB command window to open the simulink library browser or click simulink button in the MATLAB toolbar.

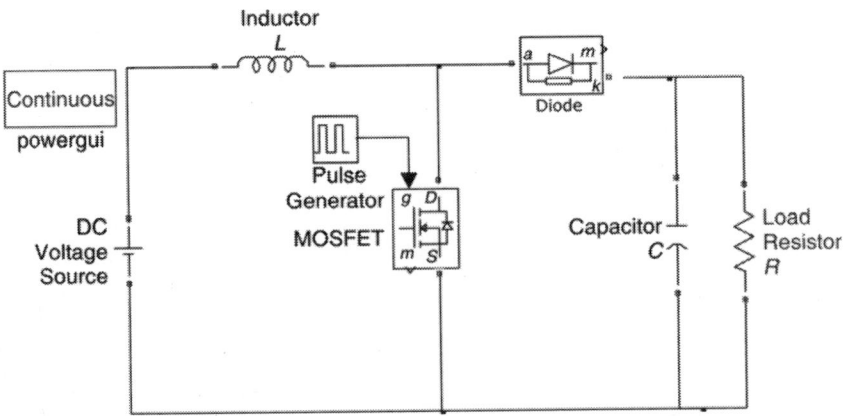

Fig. C15.1: Boost regulator circuit model using MATLAB/Simulink

Step 2: Select **File > New > Model** in the simulink library browser to create a new model or press Ctrl+N. An empty model window is opened and it can be saved by selecting **File** > **Save** in the model window.

Step 3: Drag and place the following building blocks from *Simulink library* and *Sim Power Systems library* to the simulink model file according to the circuit diagram.

   (a) Powergui block [Sim Power Systems]
   (b) DC voltage source [Sim Power Systems > Electrical Sources]
   (c) Series RLC branch (configure it to R) [Sim Power Systems > Elements]
   (d) Series RLC branch (configure it to L) [Sim Power Systems > Elements]
   (e) Series RLC branch (configure it to C) [Sim Power Systems > Elements]
   (f) Diode [Sim Power Systems > Power Electronics]
   (g) Mosfet [Sim Power Systems > Power Electronics]
   (h) Pulse generator [Simulink > Sources]
   (i) Voltage measurement [Sim Power Systems > Measurements]
   (j) Current measurement [Sim Power Systems > Measurements]
   (k) Scope [Simulink > Sinks]
   (l) Mean value [Sim Power Systems > Extra library > Measurements]
  (m) Display [Simulink > Sinks]
   (n) From [Simulink > Signal Routing]
   (o) Goto [Simulink > Signal Routing]

Step 4: Connect the blocks according to the circuit diagram of the boost regulator circuit. Also connect required number of voltage and current measurement blocks to the *scope*, as shown in the Fig. C15.2, to observe the necessary waveforms. The *Goto* and *From* blocks need not be necessarily used, but it will reduce the number of connections in the circuit.

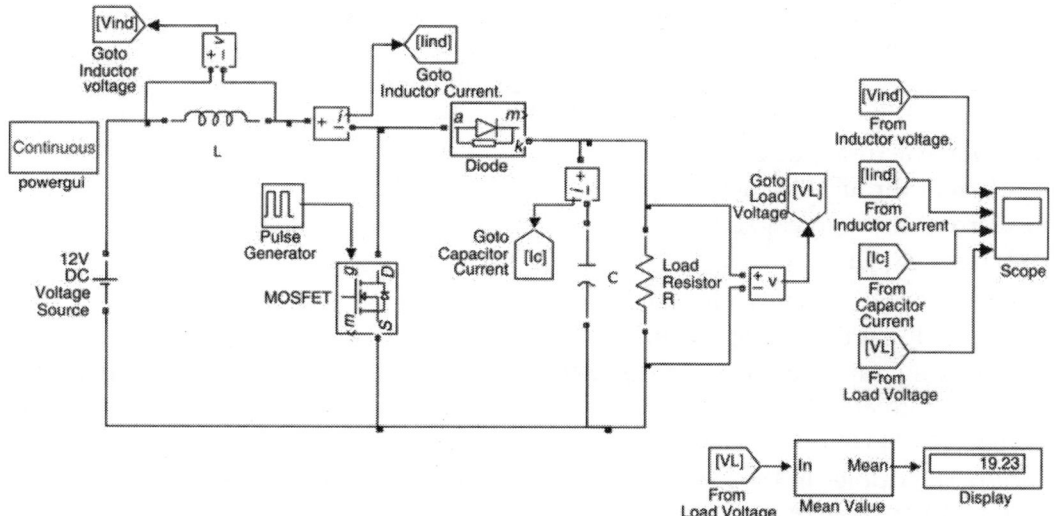

Fig. C15.2: Boost regulator circuit model with measurement blocks

Step 5: Set the parameters of the individual blocks. Components and specifications used in the boost regulator circuit model are:

1. DC voltage source   12 V
2. Load resistor, R   100 Ω
3. Inductor   12 mH
4. Capacitor, C   470 µF
5. MOSFET
6. Pulse generator
7. Diode
8. Powergui block
9. Goto and from blocks
10. Voltage measurement blocks
11. Current measurement blocks
12. Scope
13. Mean value block
14. Display

The parameters of pulse generator used for MOSFET are:
1. Period (secs): 1e-3 (Select period = 1ms)
2. Pulse Width (% of period): 40 (Select duty ratio = 0.4)
3. Phase delay (secs): 0

Note: Switching period = 1/switching frequency =1/1 kHz = 1 milliseconds

Step 6: Since the circuit provides DC to DC conversion, the average value of the load voltage can be calculated. For that *mean value block* is used and corresponding value is displayed in the *display block*. Double click on *mean value block* and enter *averaging period* as 0.001 seconds since switching period is selected as 1 msec.

Step 7: Save the Simulink model. Run the simulation by selecting **simulation > start** in the model window or use **simulation button** in the model window toolbar.

The run time can be entered and modified in the space provided in the model window tool bar. For example, run time = 0.02 seconds

Step 8: Observe and note down the following waveforms by opening the scope and save the result.

(a) Voltage across inductor
(b) Current through inductor
(c) Current through capacitor
(d) Load Voltage

The waveforms are shown in the Fig. C15.3.

Also note down the average value of the load voltage from the *display block*.

Step 9: Now adjust the duty ratio of the MOSFET and verify that the load voltage obtained is greater than or equal to supply voltage, 12 V. Also observe corresponding changes in the circuit waveforms.

Note 1: The ripples in the load voltage waveform can be reduced by choosing a large value of filter capacitor.

The AC components of the diode current are passed through the capacitor and DC component is passed through the load resistor.

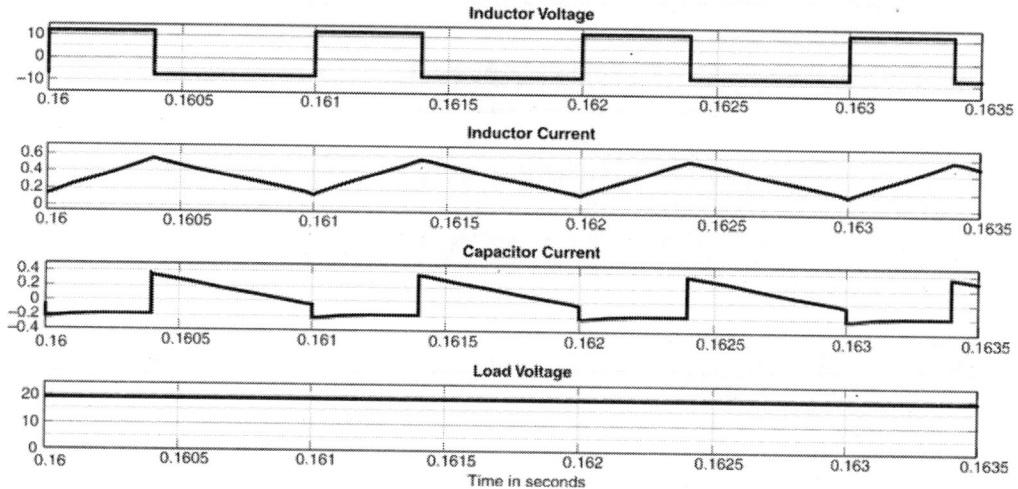

Fig. C15.3: Boost regulator waveforms (a) Voltage across inductor (b) Current through inductor (c) Current through capacitor (d) Load Voltage

Note 2: By adjusting the value of inductance, the continuous and discontinuous operation of boost regulator can be observed.

Note 3: It can be observed, from the capacitor current waveform, that during ON time of the switch, diode does not conduct and capacitor current is negative, i.e. the capacitor discharges its energy to the load during ON time of the switch.

During the OFF time of the switch, the diode starts to conduct and the capacitor gets charged, i.e. during this period, the AC component of the diode current is passed through the capacitor and DC component is passed through the load resistor.

During the ON time, only DC components are passed through the load since capacitor discharges its energy.

C15.3 SIMULATION RESULT

A boost regulator circuit was modelled and simulated using MATLAB/Simulink and analysed circuit waveforms.

Experiment C16

Simulation of a Single Phase Inverter

C16.1 OBJECTIVE

To analyse and simulate a single phase inverter using MATLAB/ Simulink.

C16.2 INTRODUCTION

An inverter convertes a DC voltage into an AC voltage of variable magnitude and variable frequency. Power semicondcutor devices used in inverters are MOSFETs, IGBTs, Thyristors and BJTs. The forced commutation circuits are required to turn off thyristors, if they are used. A very high switching frequency is possible with MOSFET and IGBT and they are fully controlled devices.

Inverters are used in uninterruptible power supplies (UPS), induction heating, high voltage (HV) transmission, variable frequency drives, electroshock weapons, air conditioning, etc.

C16.3 CLASSIFICATION OF INVERTERS

Generally inverters are classified based on number of phases (single phase or three phase) and type of sources (voltage source or current source). Commonly used types of inverters are single and three phase voltage source inverters (VSI). The circuit configurations used in inverters are half bridge and full bridge. Normally full bridge configuration is employed in inverters.

C16.4 PERFORMANCE PARAMETERS

1. **Total harmonic distortion (THD):** It is the ratio of the RMS value of the voltage (or current) excluding the fundamental voltage (or current) to the fundamental voltage (or current).

$$THD = \frac{\sqrt{(V_3^2 + V_5^2 + V_7^2 + \cdots V_n^2)}}{V_1} = \frac{\sqrt{V_{rms}^2 - V_1^2}}{V_1}$$

Where V_{rms} = RMS value of voltage waveform

V_1 = RMS value of the fundamental component of the output voltage

V_n = RMS value of the nth harmonic component of the output voltage

2. **Harmonic factor:** It is ratio of the RMS value of the nth harmonic component of the output voltage to the RMS value of the fundamental component.

$$HF_n = \frac{V_n}{V_1}$$

3. **Distortion factor:** It indicates the amount of harmonic distortion that is associated with a particular waveform after the harmonics of that waveform have been subjected to a *second order attenuation* or divided by n^2.

$$DF = \frac{\sqrt{\left(\frac{V_3}{3^2}\right)^2 + \left(\frac{V_5}{5^2}\right)^2 + \left(\frac{V_7}{7^2}\right)^2 + \cdots + \left(\frac{V_n}{n^2}\right)^2}}{V_1}$$

The distortion factor of an nth harmonic component voltage is

$$DF_n = \frac{V_n}{n^2 V_1}$$

C16.5 SINGLE PHASE FULL BRIDGE VOLTAGE SOURCE INVERTER (VSI)

A full bridge inverter requires only a two wire DC supply whereas half bridge inverter needs a 3 wire DC supply. It can be observed that in a full bridge inverter four switches are required but a half bridge inverter needs only two.

A single phase full bridge inverter consists of two legs and each leg has two controlled swithes. A full bridge inverter feeding a resistive load is shown in Fig. C16.1. If the load is RL, a diode is connected across each switch in anti parallel. These diodes are requied to provide a return path for the energy stored in the inductor to be fed back to the DC source. MOSFETs are having inherent body diodes.

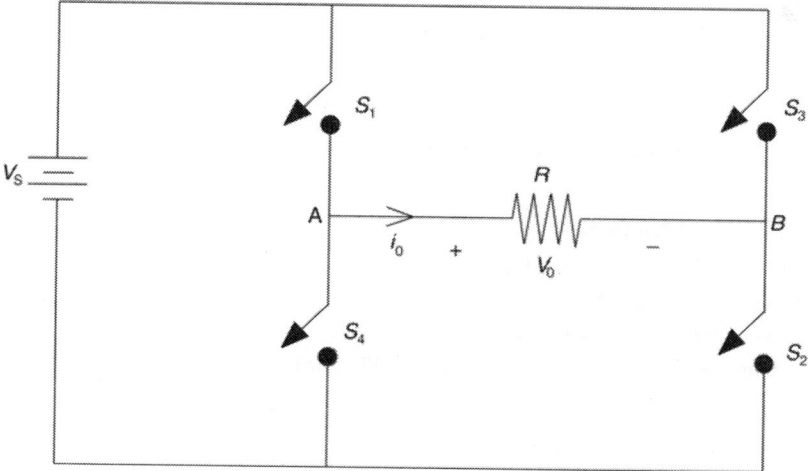

Fig. C16.1: Single phase full bridge inverter with R load

The Table C16.1, given below, shows the operation of single phase square wave inverter circuit.

Table C16.1: Switching Sequence of single phase square wave inverter

S_1 and S_2	S_3 and S_4	Load voltage	Direction of Load current	Time interval
ON	OFF	$+V_S$	Positive	$0 < t < T_S/2$
OFF	ON	$-V_S$	Negative	$T_S/2 < t < T_S$

The output voltage waveform of the inverter is a square wave with amplitude V_S as shown in the Fig. C16.2. Hence this circuit is called as square wave inverter circuit. The wave shape of load current is also a square wave since load is purely resistive. The amplitude of load current is V_S/R.

The voltage appear across non conducting switches is supply voltage V_S. By adjusting the switching frequency of the switches the output voltage frequency can be varied. It is to be noted that the switches S_1 and S_4 or S_2 and S_3 should not remain ON simultaneously; if this happens the DC source is short circuited. This is sometimes called a *shoot-through* fault. The average value of the square waveform is zero. The RMS value of the output voltage waveform is

$$V_{\text{rms}} = \sqrt{\frac{1}{T_S} \int_0^{T_S} V_S^2 \, dt} = \sqrt{\frac{1}{T_S} V_S^2 T_S} = V_S$$

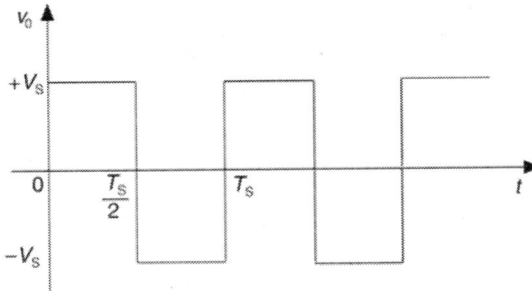

Fig. C16.2: The output voltage waveform of single phase full bridge inverter with *R* load

C16.6 HARMONIC ANALYSIS

The inverter output voltage waveform is a square wave, as shown in Fig. C16.2. It has both *half-wave symmetry* and *odd symmetry*. (Refer to Appendix- Fourier analysis). Hence it contains only sine terms and will not contain cosine terms. Generally a voltage waveform can be expressed in trigonometric Fourier series as

$$v(t) = a_O + \sum_{n=1}^{\infty} \{a_n \cos(n\omega_S t) + b_n \sin(n\omega_S t)\}$$

From the inverter output waveform, it can be seen that the average value is zero. Hence it does not have a DC term, i.e. $a_0 = 0$. Since the waveform has odd symmetry, $a_n = 0$.

The switching frequency in rad/sec, $\omega_S = 2\pi/T_S$. Therefore, the output voltage can be expressed in Fourier series as,

$$v(t) = \sum_{n=1,3,5,\ldots}^{\infty} b_n \sin(n\omega_S t); \quad \omega_S = \frac{2\pi}{T_S}$$

The Fourier coefficient,

$$b_n = \frac{2}{T_S} \int_0^{T_S} v(t)\sin(n\omega_S t)dt; \quad n = 1, 3, 5\ldots$$

$$= \frac{2}{T_S}\left(\int_0^{\frac{T_S}{2}} V_S \sin(n\omega_S t)dt + \int_{\frac{T_S}{2}}^{T_S} (-V_S)\sin(n\omega_S t)dt \right)$$

$$= \frac{2V_S}{T_S}\left[\left(-\frac{\cos n\omega_S t}{n\omega_S} \right)_0^{\frac{T_S}{2}} - \left(-\frac{\cos n\omega_S t}{n\omega_S} \right)_{\frac{T_S}{2}}^{T_S} \right]$$

$$= \frac{2V_S}{n\omega_S T_S}\left[-\cos\left(n\frac{2\pi}{T_S}\frac{T_S}{2} \right) - (-1) + \cos\left(n\frac{2\pi}{T_S}T_S \right) - \cos\left(n\frac{2\pi}{T_S}\frac{T_S}{2} \right) \right]$$

$$= \frac{2V_S}{n\omega_S T_S}\left[-\cos(n\pi) + 1 + \cos(n2\pi) - \cos(n\pi) \right]; \ n \ odd$$

$$= \frac{2V_S}{n\frac{2\pi}{T_S}T_S}[1+1+1+1] = \frac{4V_S}{n\pi}$$

Hence

$$b_n = \frac{4V_S}{n\pi}$$

$$v(t) = \sum_{n=1,3,5,\ldots}^{\infty} \frac{4V_S}{n\pi}\sin(n\omega_S t); \quad \omega_S = \frac{2\pi}{T_S}$$

The load current is

$$i_0(t) = \sum_{n=1,3,5,\ldots}^{\infty} \frac{4V_S}{n\pi R}\sin(n\omega_S t); \quad \omega_S = \frac{2\pi}{T_S}$$

The RMS value of fundamental voltage is

$$i_0(t) = \frac{\left(\dfrac{4V_S}{1\times\pi} \right)}{\sqrt{2}} = \frac{4V_S}{\sqrt{2}\,\pi} = 0.9V_S$$

The RMS value of fundamental current is

$$I_1 = \frac{0.9V_S}{R}$$

Average power delivered to the load

$$P = \sum_{n=1,3,5,7...}^{\infty} P_n = \sum_{n=1,3,5,7}^{\infty} \left(I_{rms}\right)^2 R = \sum_{n=1,3,5,7}^{\infty} \left(\frac{\left(\dfrac{4V_S}{n\pi R}\right)}{\sqrt{2}}\right)^2 R$$

C16.7 MODELLING AND SIMULATION OF A SINGLE PHASE INVERTER CIRCUIT USING MATLAB/SIMULINK

A basic circuit for the simulation of single phase inverter circuit using MATLAB/Simulink is shown in Fig. C16.3.

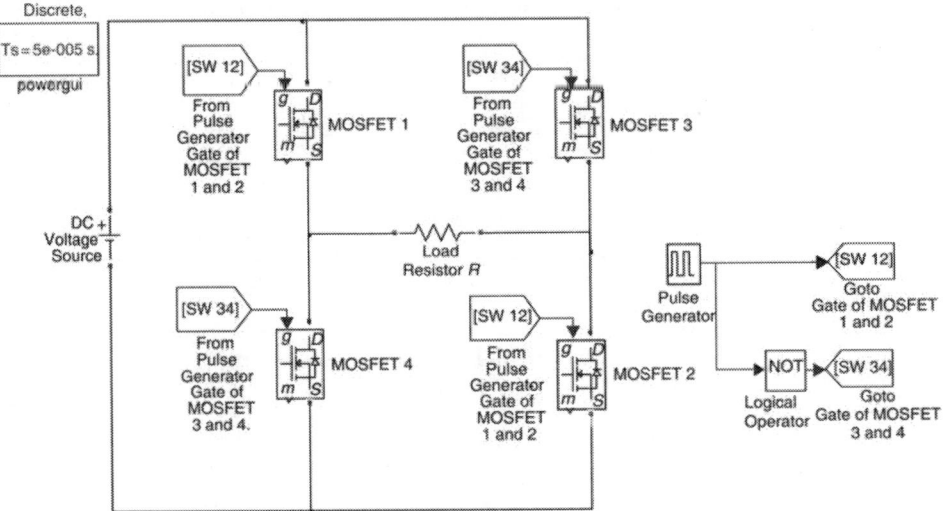

Fig. C16.3: Single phase inverter circuit model using MATLAB/Simulink

Following steps are to be followed for analyzing and modelling the single phase inverter circuit.

Step 1: Enter "simulink" in the MATLAB command window to open the simulink library browser or click simulink button in the MATLAB toolbar.

Step 2: Select **File > New > Model** in the simulink library browser to create a new model or press Ctrl+N. An empty model window is opened and it can be saved by selecting **File > Save** in the model window.

Step 3: Drag and place the following building blocks from *Simulink library* and *Sim Power Systems library* to the simulink model file according to the circuit diagram.

(a) Powergui block [Sim Power Systems]

(b) DC voltage source [Sim Power Systems > Electrical Sources]

(c) Series RLC branch (configure it to R) [SimPowerSystems> Elements]

(d) Mosfet [Sim Power Systems > Power Electronics]

(e) Pulse generator [Simulink > Sources]

(f) Voltage measurement [Sim Power Systems> Measurements]

(g) From [Simulink > Signal Routing]

(h) Goto [Simulink > Signal Routing]

(i) Logical Operator (configure it to NOT gate) [Simulink > Logic and Bit Operations]

(j) Scope [Simulink > Sinks]

Step 4: Connect the blocks according to the circuit diagram of the single phase inverter circuit. Also connect required number of voltage measurement blocks to the *scope*, as shown in the Fig. C16.4, to observe the necessary waveforms. The *Goto* and *From* blocks need not be necessarily used, but it will reduce the number of connections in the circuit.

Step 5: Set the parameters of the individual blocks. Components and specifications used in the single phase inverter are:

1. DC voltage source   100 V
2. Load resistor $100 \, \Omega$
3. Logical operator   NOT
4. MOSFET
5. Pulse generator
6. Goto and From blocks
7. Voltage measurement block
8. Scope
9. Powergui block

The parameters used for pulse generator are:

1. Period (secs): 0.02 sec (Select switching period = 20 ms)
2. Pulse Width (% of period): 50 (Select duty ratio = 0.5)
3. Phase delay (secs): 0

Step 6: Pulses to switches (MOSFETs) 1 and 2 are given directly from pulse generator and pulses to switches 3 and 4 are given through NOT gate as shown in the Fig. C16.4. Pulses are given to switches from pulse generator using *Goto* and *From* blocks for making the connections simpler.

Step 7: Save the Simulink model. Run the simulation by selecting **simulation > start** in the model window or use **simulation button** in the model window toolbar.

The run time can be entered and modified in the space provided in the model window tool bar. For example, run time = 0.06 seconds

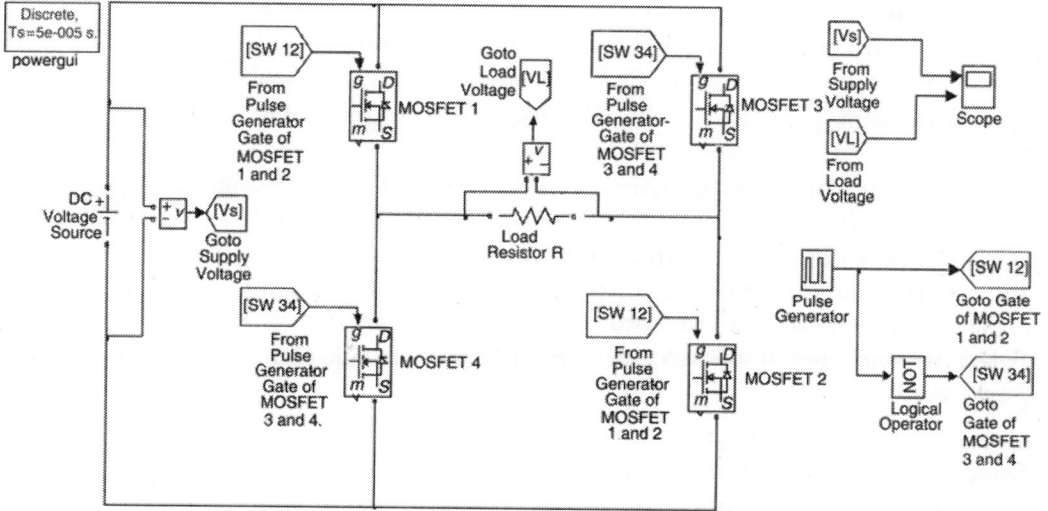

Fig. C16.4: Single phase inverter circuitmodel with measurement blocks

Step 8: Observe and note down the following waveforms by opening the scope and save the result.

(a) Source voltage

(b) Output voltage or voltage across the load

The waveforms are shown in Fig. C16.5. The RMS value of the load voltage can also be measured using *RMS block*.

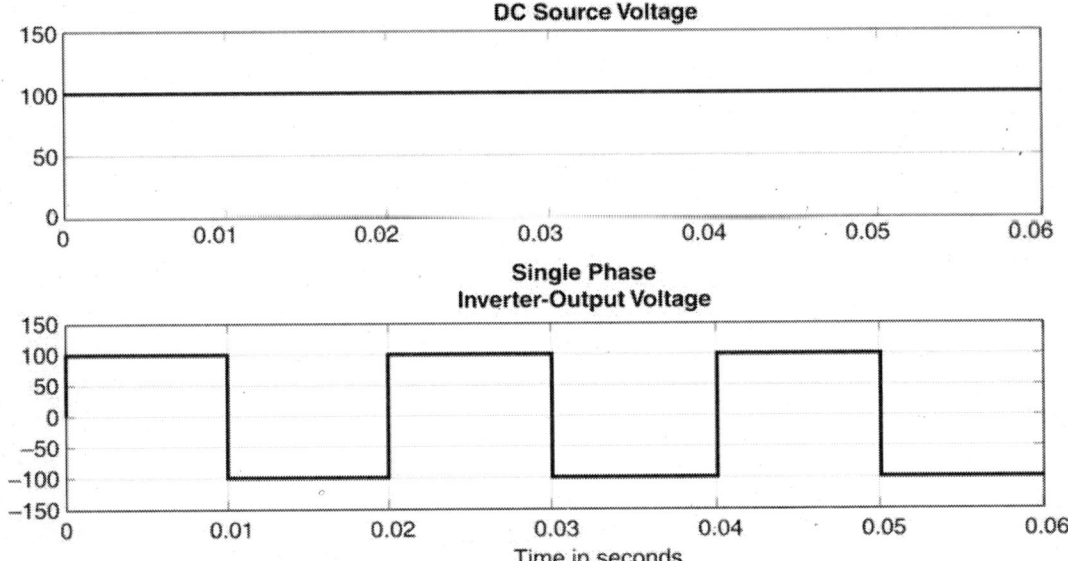

Fig. C16.5: Simulation waveforms of Single phase inverter (a) Source Voltage (b) Load voltage

Step 9: The harmonic analysis of a single phase inverter can be done using MATLAB/ Simulink.

In *Appendix - FFT analysis using MATLAB/Simulink*, the steps that are to be followed to perform harmonic analysis of a waveform are given with an example.

Double click **powergui** block in the model window and click **FFT analysis** tab from the **Analysis tools**. A "Powergui FFT analysis Tool" window will pop up.

Step 10: The settings for FFT analysis in the "Powergui FFT Analysis Tool" window are:

- Select a signal which is to be analysed from "input". The selected signal, load voltage waveform, will be displayed.
- It is to be noted that the variable name which was entered in the *scope parameters* of the scope is same in the "Powergui FFT Analysis Tool". Have a look at *available signal > structure* in the "Powergui FFT Analysis Tool".
- Select the start time of the waveform and number of cycles for which FFT analysis is performed.

Example: Start time = 0.02 and number of cycles = 1.

- Select fundamental frequency as 50 Hz.
- Display style of FFT analysis can be bar type or list type. Select as list type

Step 11: Select "list (relative to fundamental)" and display. The following values can be seen as shown in the Fig. C16.6.

- Fundamental = 127.1 peak (89.85 rms)
- Total harmonic distortion THD = 48.34%
- DC component is negligibly small.
- % of magnitude of frequency components (Refer to *Appendix-Fourier Analysis*) are:
  - 50 Hz components = 100 % (fundamental)
  - 150 Hz components = 33.34%
  - 250 Hz components = 20.00%
  - 350 Hz components = 14.29%
  - And so on.
- All even order harmonics ($n = 2, 4, 6, 8...$) are zero and only odd order harmonics ($n = 3, 5, 7, 9...$) are present.

Step 12: Now change the supply voltage to a new value and observe the waveforms and do FFT analysis.

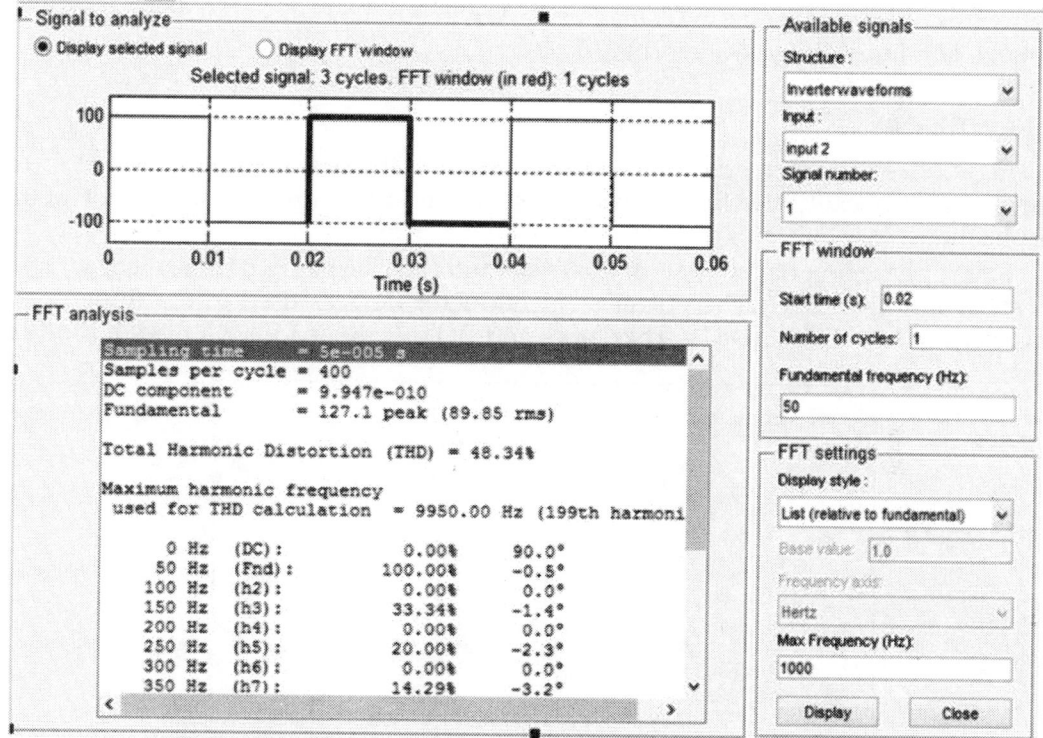

Fig. C16.6: The powergui FFT analysis tool window–single phase inverter

C16.8 SIMULATION RESULT

The single phase inverter circuit was simulated and analysed using MATLAB/ Simulink. The FFT analysis of output voltage waveform is also done. It is observed that the output voltage contains only odd harmonics.

Experiment C17

Simulation of a Three Phase Inverter

C17.1 OBJECTIVE

To analyse and simulate the three phase inverter in 180° and 120° conduction Mode with three phase star connected R load using MATLAB/Simulink and obtain the phase voltage and line voltage waveforms.

C17.2 THEORY

A single phase inverter produces a square wave voltage. It consists of lower order harmonics especially the third, fifth and seventh which are undesirable for many applications.

A three phase inverter consists of three legs and each leg has two controlled switches as shown in Fig. C17.1. The power semiconductor devices used in the three phase inverter are MOSFETs, IGBTs, Thyristors and BJTs. If the load is RL load, a diode is

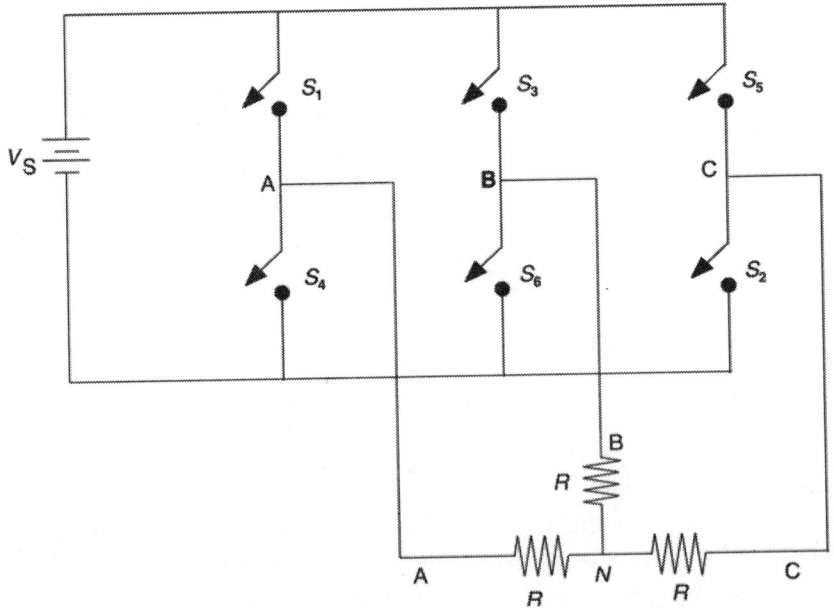

Fig. C17.1: A three phase full bridge inverter with star connected R load

connected across each switch in anti parallel. These diodes are required to provide a return path for the energy stored in the inductor to be fed back to the DC source. MOSFETs have inherent body diodes.

The switches are numbered in accordance with the sequence in which they are triggered in a cyclic manner with a phase difference of 60°. The upper leg switches numbered as 1, 3, 5 and lower leg switches 4, 6, 2. The load terminals are labelled as A, B and C and the load neutral point as N.

The switches are triggered in a proper sequence to obtain the required three phase output voltage waveform which has six step voltages of 60° each in one time period of 360°. So three phase inverter is known as a six step inverter.

The switches in the same leg should not be in ON position at a time, if this happens, a short circuit, sometimes called a *shoot-through* fault, will appear across the DC source.

A three phase inverter has two modes of operation, namely, 180° and 120° which are based on the duration of conduction of the switches. The load connected can be either star or delta with R or RL load. A three phase bridge circuit delivering power to a balanced star connected resistive load is discussed here.

C17.2.1 180° Conduction Mode

In 180° conduction mode, each switch conducts for a period of 180° during a period of 360°. Three switches conduct at a time during each sub interval of 60°. The switching sequence for 180° mode to obtain output voltage are given in the table shown below. The operation of three phase inverter with 180° conduction mode consists of six different steps. The steps are 0°–60°, 60°–120°, 120°–180°, 180°–240°, 240°–300° and 300°–360°.

Duration	0°–60°	60°–120°	120°–180°	180°–240°	240°–300°	300°–360°
Conducting Devices	S_5, S_6, S_1	S_6, S_1, S_2	S_1, S_2, S_3	S_2, S_3, S_4	S_3, S_4, S_5	S_4, S_5, S_6
v_{AN}	$\dfrac{V_S}{3}$	$\dfrac{2V_S}{3}$	$\dfrac{V_S}{3}$	$\dfrac{-V_S}{3}$	$\dfrac{-2V_S}{3}$	$\dfrac{V_S}{3}$
v_{BN}	$\dfrac{-2V_S}{3}$	$\dfrac{-V_S}{3}$	$\dfrac{V_S}{3}$	$\dfrac{2V_S}{3}$	$\dfrac{V_S}{3}$	$\dfrac{-V_S}{3}$
v_{CN}	$\dfrac{V_S}{3}$	$\dfrac{-V_S}{3}$	$\dfrac{-2V_S}{3}$	$\dfrac{-V_S}{3}$	$\dfrac{V_S}{3}$	$\dfrac{2V_S}{3}$

The first step, i.e. 0°–60°, is discussed here. In this step the switches 5, 6, 1 are closed and 2, 3, 4 are opened as shown in the Fig. C17.2(a). The equivalent circuit for this step is shown in Fig. C17.2(b). The terminals A and C are connected to positive of the DC source whereas B is connected to negative of the DC source.

The equations of phase voltages during the interval 0°–60° are,

$$V_{AN} = V_{CN} = V_S \frac{(R/2)}{R+(R/2)} = \frac{V_S}{3}$$

$$V_{BN} = -V_S \frac{R}{R+(R/2)} = \frac{-2V_S}{3}$$

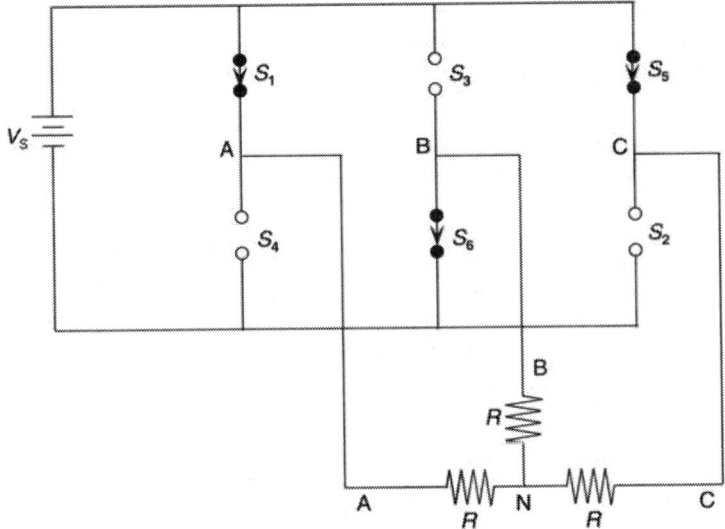

Fig. C17.2: (a) 180° conduction mode with switches S_5, S_6 and S_1 closed 0°–60° interval

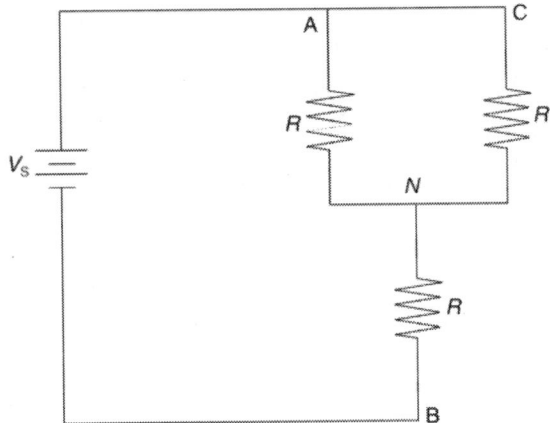

Fig. C17.2: (b) The equivalent circuit for 0°–60° interval–Mode 180°

The equations of line voltages during the interval 0°–60° are,

$$V_{AB} = V_{AN} - V_{BN} = V_S$$
$$V_{BC} = V_{BN} - V_{CN} = -V_S$$
$$V_{CA} = V_{CN} - V_{AN} = 0$$

Similarly the magnitude of phase voltages and line voltages of each sub interval can be found out from corresponding equivalent circuits.

From the equations of the phase voltages and line voltages, the waveforms can be drawn and is shown in Fig. C17.3. The waveforms of phase voltages are stepped whereas the line voltages are seen to be quasi square waves.

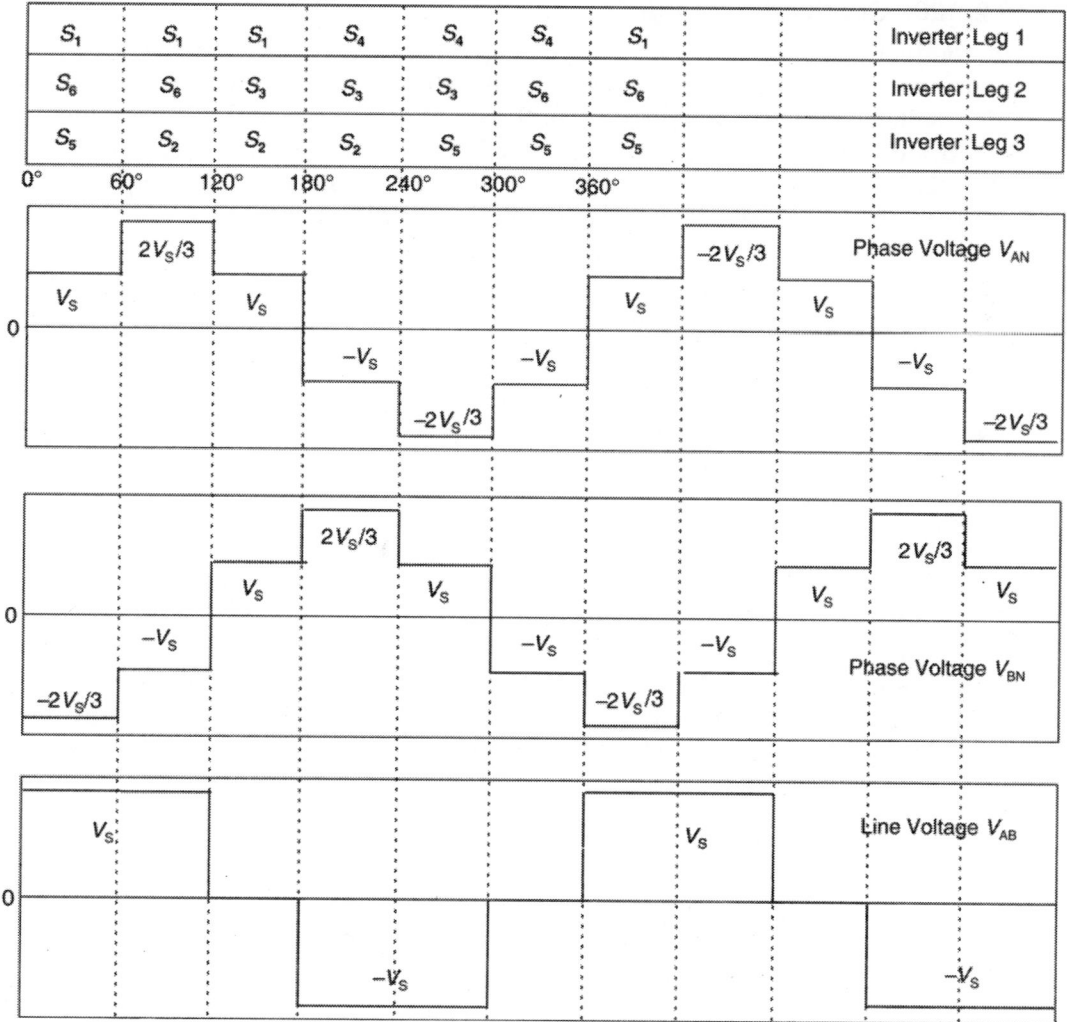

Fig. C17.3: Waveforms of phase voltages V_{AN} and V_{BN} and line voltage V_{AB} 180° mode

The RMS value of phase voltage is

$$V_{prms} = \sqrt{\frac{1}{2\pi} 2 \times \int_0^\pi v_0^2 \, d\omega t}$$

$$= \sqrt{\frac{1}{\pi}\left[\left(\frac{V_S}{3}\right)^2 \frac{\pi}{3} + \left(\frac{2V_S}{3}\right)^2 \frac{\pi}{3} + \left(\frac{V_S}{3}\right)^2 \frac{\pi}{3}\right]} = \frac{\sqrt{2}}{3} V_S$$

The RMS value of line voltage is

$$V_{Lrms} = \sqrt{\frac{1}{2\pi} 2 \times \int_0^\pi v_0^2 \, d\omega t} = \sqrt{\frac{1}{\pi}\left[(V_S)^2 \frac{\pi}{3} + (V_S)^2 \frac{\pi}{3} + 0\right]} = \sqrt{\frac{2}{3}} V_S$$

C17.2.2 120° Conduction Mode

In 120° conduction mode only two switches are conducting at a time. The ON time duration of each switch is 120° in one period. At a time one upper leg switch and one lower leg switch conduct. The operation of three phase inverter with 120° conduction mode is divided in to six sub intervals of 60° each. The sub intervals are 0°–60°, 60°–120°, 120°–180°, 180°–240°, 240°–300° and 300°–360°. The switching sequence for 120⁰ mode to obtain output voltage is given in the table shown below.

Duration	0°–60°	60°–120°	120°–180°	180°–240°	240°–300°	300°–360°
Conducting Devices	S_6, S_1	S_1, S_2	S_2, S_3	S_3, S_4	S_4, S_5	S_5, S_6
v_{AN}	$\dfrac{V_S}{2}$	$\dfrac{V_S}{2}$	0	$\dfrac{-V_S}{2}$	$\dfrac{-V_S}{2}$	0
v_{BN}	$\dfrac{-V_S}{2}$	0	$\dfrac{V_S}{2}$	$\dfrac{V_S}{2}$	0	$\dfrac{-V_S}{2}$
V_{CN}	0	$\dfrac{-V_S}{2}$	$\dfrac{-V_S}{2}$	0	$\dfrac{V_S}{2}$	$\dfrac{V_S}{2}$

The first sub interval, i.e. 0°–60°, is discussed here. In this interval the switches S_6 and S_1 conduct with all other switches open as shown in Fig. C17.4 (a). The equivalent circuit for this interval is shown in Fig. C17.4(b). The terminal A is connected to positive of the DC source whereas B is connected to negative of the DC source.

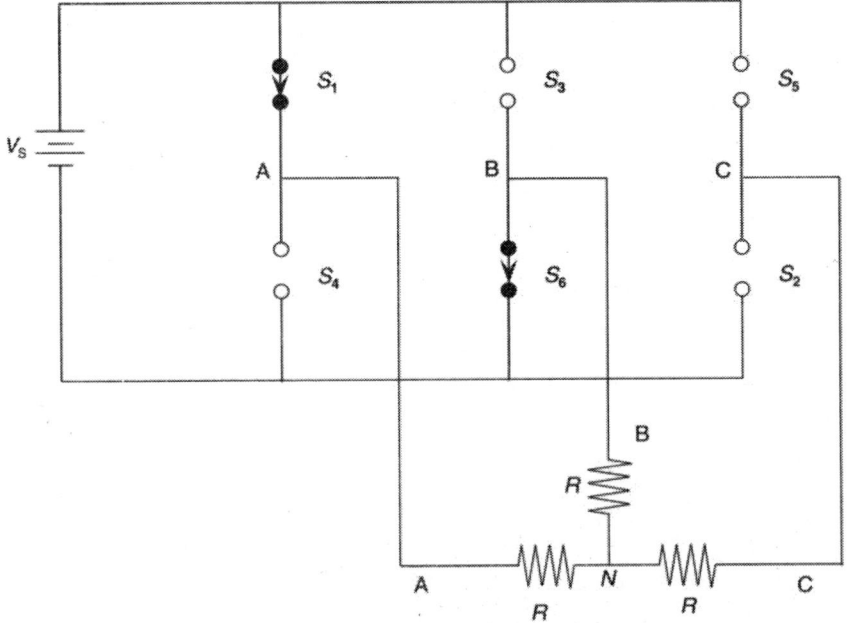

Fig. C17.4 (a): 120° mode 0°–60° interval switches – S_6 and S_1 closed

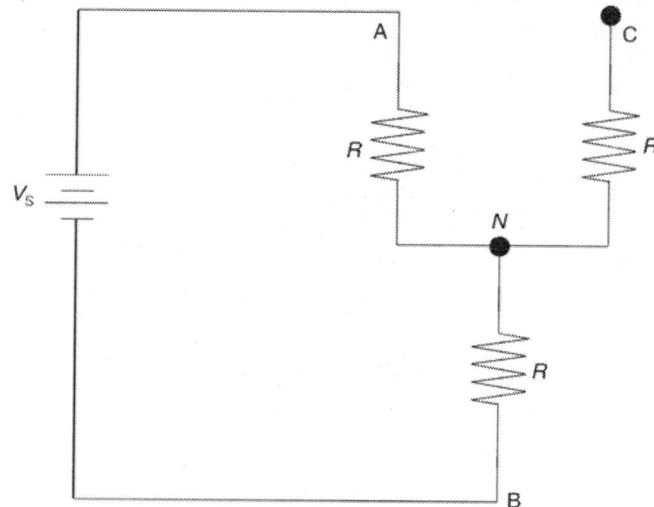

Fig. C17.4 (b): The equivalent circuit for 0°–60° interval – Mode 120°

The equations of phase voltages during the interval 0°–60° are,

$$V_{AN} = V_S \frac{R}{R+R} = \frac{V_S}{2}$$

$$V_{BN} = -V_S \frac{R}{R+R} = \frac{-V_S}{2}$$

$$V_{CN} = 0$$

The equations of line voltages during the interval 0°–60° are,

$$V_{AB} = V_{AN} - V_{BN} = \frac{V_S}{2} + \frac{V_S}{2} = V_S$$

$$V_{BC} = V_{BN} - V_{CN} = -\frac{V_S}{2} - 0 = -\frac{V_S}{2}$$

$$V_{CA} = V_{CN} - V_{AN} = 0 - \frac{V_S}{2} = -\frac{V_S}{2}$$

Similarly the magnitude of phase voltages and line voltages of each sub interval can be found out from corresponding equivalent circuits. These values are shown in the above table. From the equations of the phase voltages and line voltages, the waveforms can be drawn and is shown in Fig. C17.5. The waveforms of phase voltages are quasi square wave whereas the line voltages are seen to be stepped wave.

The RMS values of phase voltages and line voltages can be determined from waveforms as shown in Fig. C17.5

The RMS value of phase voltage

$$V_{Prms} = \sqrt{\frac{1}{2\pi} 2 \times \int_0^\pi v_0^2 \, d\omega t} = \sqrt{\frac{1}{\pi}\left[\left(\frac{V_S}{2}\right)^2 \frac{\pi}{3} + \left(\frac{V_S}{2}\right)^2 \frac{\pi}{3} + 0\right]} = \frac{V_S}{\sqrt{6}}$$

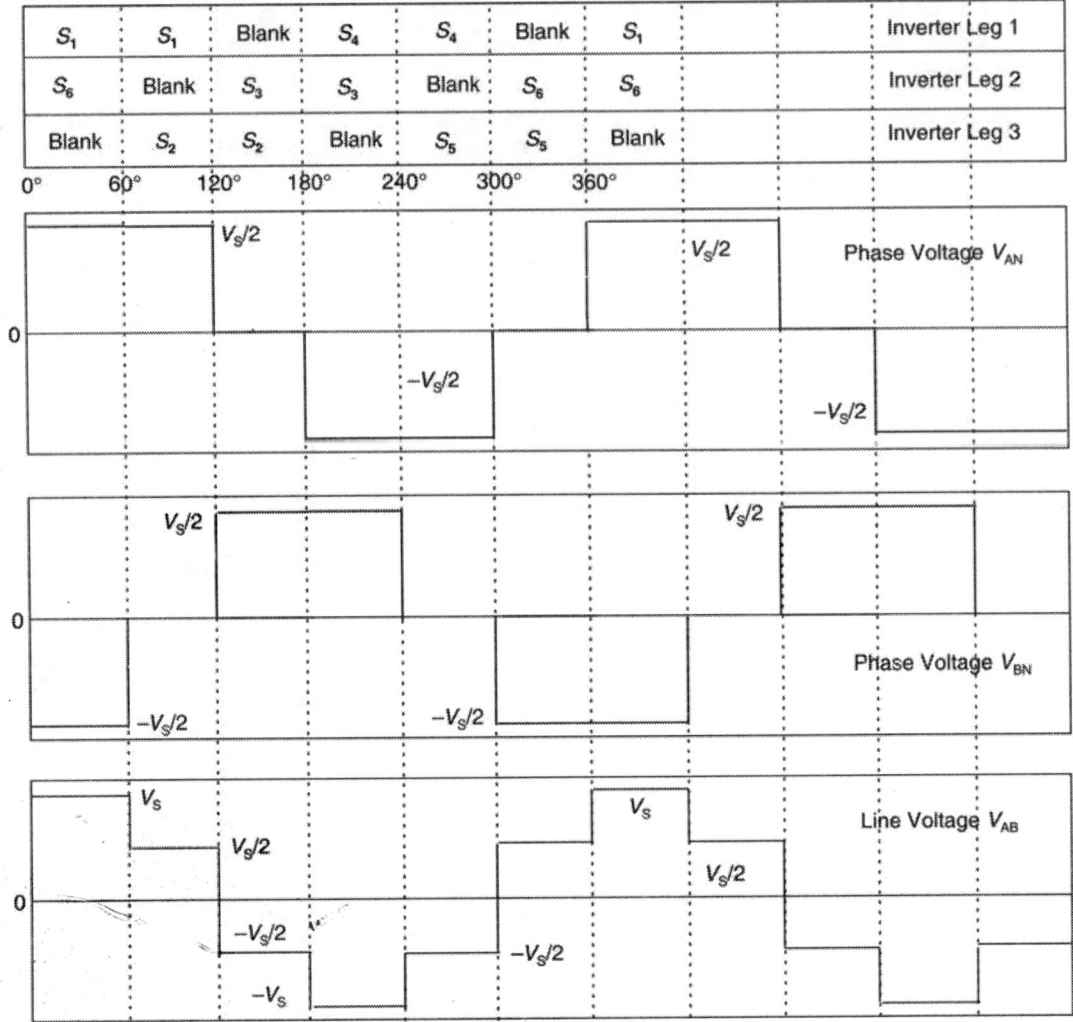

S_1	S_1	Blank	S_4	S_4	Blank	S_1			Inverter Leg 1
S_6	Blank	S_3	S_3	Blank	S_6	S_6			Inverter Leg 2
Blank	S_2	S_2	Blank	S_5	S_5	Blank			Inverter Leg 3

Fig. C17.5: Waveforms of phase voltages V_{AN} and V_{BN} and line voltage V_{AB} −120° mode

The RMS value of line voltage

$$V_{Lrms} = \sqrt{\frac{1}{2\pi} 2 \times \int_0^\pi v_0^2 \, d\omega t} = \sqrt{\frac{1}{\pi}\left[(V_S)^2 \frac{\pi}{3} + \left(\frac{V_S}{2}\right)^2 \frac{\pi}{3} + \left(-\frac{V_S}{2}\right)^2 \frac{\pi}{3}\right]} = \frac{V_S}{\sqrt{2}}$$

C17.3 MODELLING AND SIMULATION OF A THREE PHASE INVERTER CIRCUIT USING MATLAB/SIMULINK

A basic circuit for the simulation of three phase inverter circuit using MATLAB/Simulink is shown in Fig. C17.6.

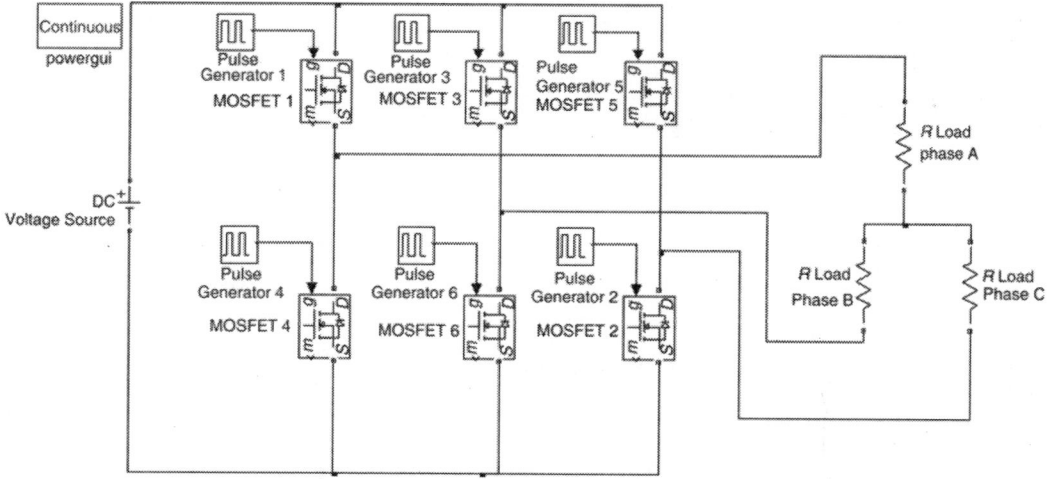

Fig. C17.6: Three phase inverter circuit model using MATLAB/Simulink

Following steps are to be followed for analyzing and modelling the single phase inverter circuit.

Step 1: Enter "simulink" in the MATLAB command window to open the simulink library browser or click simulink button in the MATLAB toolbar.

Step 2: Select **File > New > Model** in the simulink library browser to create a new model or press Ctrl+N. An empty model window is opened and it can be saved by selecting **File > Save** in the model window.

Step 3: Drag and place the following building blocks from *Simulink library* and *Sim Power Systems library* to the simulink model file according to the circuit diagram.

(a) Powergui block [Sim Power Systems]

(b) DC voltage source [Sim Power Systems > Electrical Sources]

(c) Series RLC branch (configure it to R) [Sim Power Systems > Elements]

(d) Mosfet [Sim Power Systems > Power Electronics]

(e) Pulse Generator [Simulink > Sources]

(f) Voltage Measurement [Sim Power Systems > Measurements]

(g) Scope [Simulink > Sinks]

(h) Neutral blocks [SimPowerSystems > Elements > Neutral]

Step 4: Connect the blocks according to the circuit diagram of the three phase inverter circuit. Also connect required number of voltage measurement blocks to the *scope*, as shown in the Fig. C17.7, to observe the necessary waveforms.

A *Neutral* block can be used to interconnect two points without drawing a connection line. The *Neutral* block implements a node with a specific node number as shown in Fig. C17.7. Select required numbers of *Neutral* blocks and double click on it and enter corresponding node numbers, as shown in the Fig. C17.7. The *Neutral* blocks need not be necessarily used, but it will reduce the number of connections in the circuit.

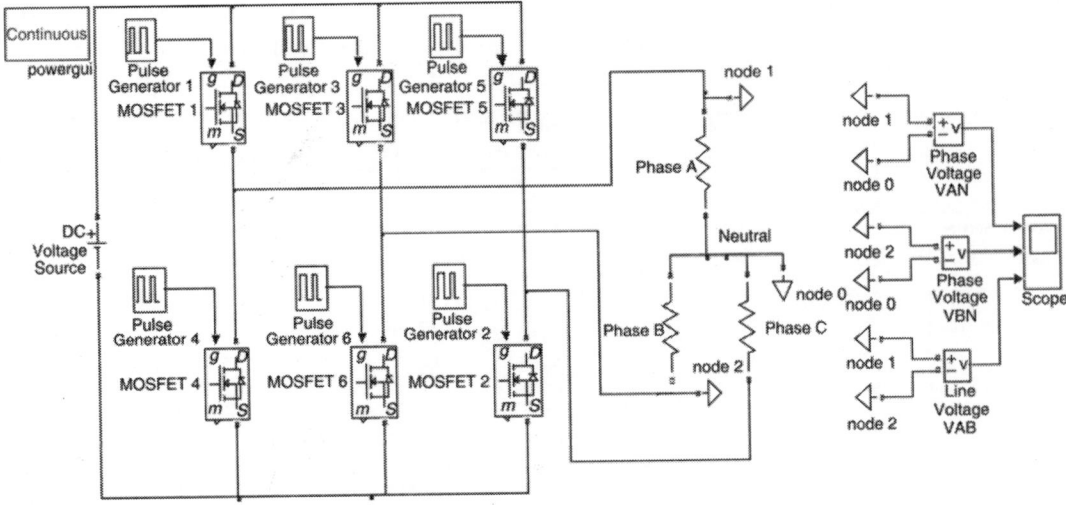

Fig. C17.7: Three phase inverter circuit model using measuremet blocks

Simulation of 180° Mode Operation

Step 5: Set the parameters of the individual blocks. Components and specifications used in the three phase inverter are:

1. DC voltage source   100 V
2. Load resistor   100 Ω
3. MOSFET
4. Pulse generator
5. Neutral blocks
6. Voltage measurement block
7. Scope
8. Powergui block

The parameters used for pulse generators are:

1. Period (secs): 0.02 sec
2. Pulse width (% of period): 100/2 (one half of full period), for all the switches.
3. Phase delay (secs):

    i. Pulse generator 1 (to switch S_1): 0 ms

    ii. Pulse generator 2 (to switch S_2): $\dfrac{60}{180} \times 10\,\text{ms} = \dfrac{10}{3}\,\text{ms}$

    iii. Pulse generator 3 (to switch S_3): $\dfrac{60}{180} \times 20\,\text{ms} = \dfrac{20}{3}\,\text{ms}$

    iv. Pulse generator 4 (to switch S_4): $\dfrac{60}{180} \times 30\,\text{ms} = 10\,\text{ms}$

    v. Pulse generator 5 (to switch S_5): $\dfrac{60}{180} \times 40\,\text{ms} = \dfrac{40}{3}\,\text{ms}$

    vi. Pulse generator 6 (to switch S_6): $\dfrac{60}{180} \times 50\,\text{ms} = \dfrac{50}{3}\,\text{ms}$

Note: The numbering of switches must be made in accordance with the sequence of switching.

Step 6: Save the simulink model. Run the simulation by selecting **simulation > start** in the model window or use **simulation button** in the model window toolbar.

The run time can be entered and modified in the space provided in the model window tool bar. For example, run time = 0.08 seconds.

Step 7: Observe and note down the following waveforms by opening the scope and save the result.

(a) Phase voltages V_{AN}, V_{BN}

(b) Line voltage, V_{AB}

The waveforms are shown in Fig. C17.8 for 180°. The RMS value of the line and phase voltages can also be measured using *RMS block*.

Note that the expected waveforms (as shown in Fig. C17.3 and in Fig. C17.5) are seen from 0.02 sec only in the simulated waveforms since all pulses to the switches from pulse generators are activated only after the first one cycle.

Step 8: The waveforms of phase voltages V_{AN}, V_{BN}, V_{CN} and line voltages V_{AB}, V_{BC}, V_{CA} can also be checked. The harmonic analysis on three phase inverter waveforms can also be done using MATLAB/Simulink (Refer to section C16 and *Appendix - harmonic analysis*).

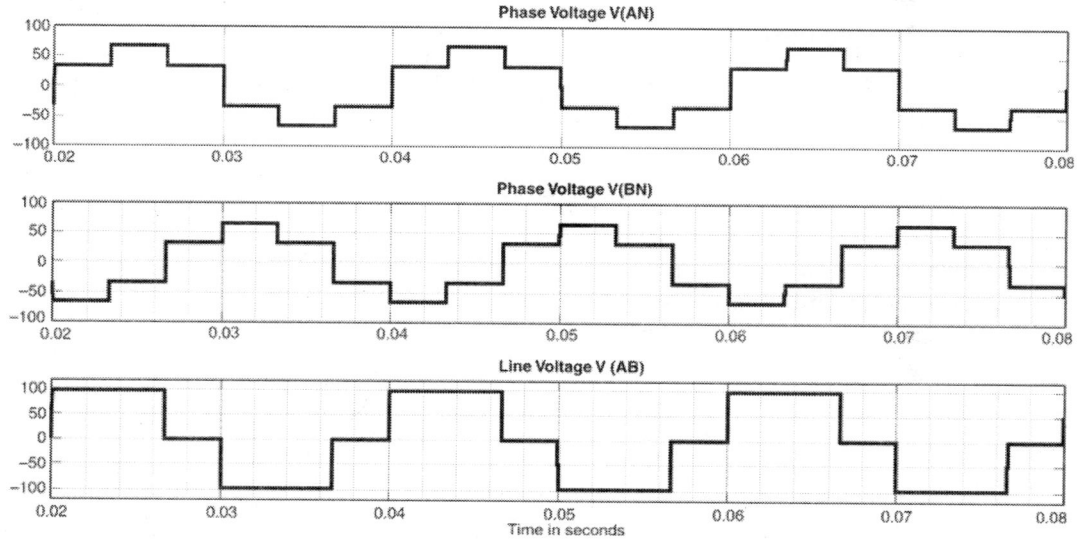

Fig. C17.8: Simulation waveforms of three phase inverter −180° conduction mode
(a) Phase Voltage V_{AN} (b) Phase Voltage V_{BN} (c) Line Voltage V_{AB}

Simulation of 120° Mode Operation

For simulating the 120° mode operation, the only change that is to be made is in the pulse width.

The parameters used for pulse generators are:

Pulse width (% of period): 100/3 (one third of full period), for all the switches.

Repeat steps 7 to 9 to get the waveforms. The waveforms are shown in Fig. C17.9 for 120°.

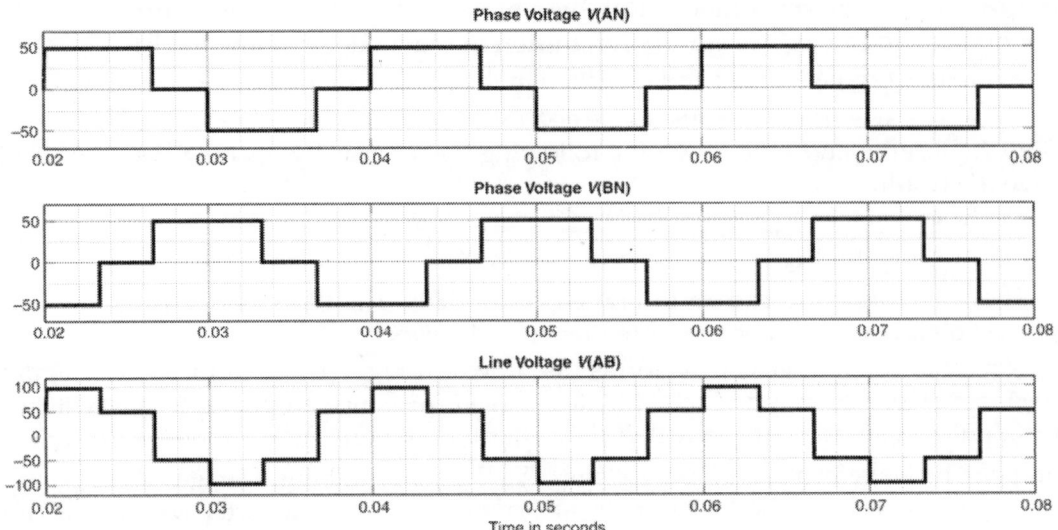

Fig. C17.9: Simulation waveforms of three phase inverter −120° conduction mode (a) Phase Voltage V_{AN} (b) Phase Voltage V_{BN} (c) Line Voltage V_{AB}

C17.4 SIMULATION RESULT

The three phase inverter circuit was simulated using MATLAB/Simulink.

It is observed that the phase voltages of

(a) 180° conduction mode are stepped waves

(b) 120° conduction mode are quasi square waves

Again it is observed that the line voltages of

(a) 180° conduction mode are quasi square waves

(b) 120° conduction mode are stepped waves.

Experiment C18

Power Quality Analysis of Power Electronic Circuits—A Simulation Study

C18.1 OBJECTIVE

i. To study power quality issues caused by nonlinear loads in power sytem.
ii. To simulate and analyze the power quality issues caused by a single phase diode bridge rectifier circuit using MATLAB/Simulink.

C18.2 LOADS THAT CAUSE POWER QUALITY PROBLEMS

Most of the electrical loads have nonlinear behaviour at the AC mains. Nonlinear loads draw harmonic current, reactive power component of current, fluctuating current and unbalanced current from the AC mains. So these loads are known as nonlinear loads.

At present, power electronics based converters are used in large numbers because of their energy conservation, reduced size, reduced overall cost, etc. Even though it possesses a number of benefits, nonlinear loads draw non sinusoidal and increased current from the AC mains in addition to the fundamental active power component of current. Some of the nonlinear loads are:

  a. Fans with electronic voltage regulators
  b. Adjustable speed drives
  c. Microwave ovens and induction heating devices
  d. Computers, printer, scanner, fax machines and television sets
  e. High frequency welding machines, Arc furnaces
  f. Switch mode power supplies (SMPS)
  g. Photocopying machines (Xerox machines) and medical equipments
  h. Battery chargers
  i. Electric traction
  j. HVDC transmission systems

C18.2.1 Classification of Nonlinear Loads

The nonlinear loads can be generally classified based on
  i. Non-solid state and solid state device
  ii. Converter
  iii. Supply system

C18.3 NONSOLID STATE DEVICE TYPES OF NONLINEAR LOADS

Non solid state device means it does not have any power electronic switches. Many electrical loads such as induction motors, transformers, etc. do not have any solid state devices.

Most of the electrical machines comes under the category of nonlinear loads. This is because,

    i. Saturation in magnetic material of the electrical machines
    ii. Saturation in the electromagnetic devices
   iii. Skin and proximity effects in conductors
    iv. Non uniform air gap in rotating machines
    v. Effect of teeth and slotting, etc.

Because of above reasons, these nonlinear loads produce harmonic currents under steady and transient conditions in the AC mains.

Practical examples of these types of nonlinear loads are:

a. Magnetic ballast of fluorescent lamps
b. Transformers operating at no load or light load conditions and
c. Single phase induction motors (since they are usually designed with high level of no load current to reduce the size and cost of motors).

These nonlinear loads draw reactive power component of current and harmonic currents along with active power component of current and also cause excessive neutral current in the three phase four wire supply system.

C18.4 SOLID STATE DEVICE TYPES OF NONLINEAR LOADS

Nowadays, most of the loads are based on power converter circuits which are made of solid state devices such as power diodes, thyristors, MOSFTEs, IGBTs, etc. and they draw non-sinusoidal currents from AC mains and behave as nonlinear loads. This non-sinusoidal current consists of active power component of current, reactive power component of current and harmonic current.

Examples include both domestic and commercial equipments such as microwave oven, television sets, electronic ballasts based lighting systems, inverters, AC voltage controller based fans, adjustable speed drives, computers, scanners, photocopying machines, fax machines, etc.

C18.5 CONVERTER BASED NONLINEAR LOADS

Converter based nonlinear loads mainly consists of AC–DC converters, AC voltage controllers, cycloconverters or a combination of all.

For many applications, such as computers, SMPS, fax machines, battery chargers, HVDC transmission systems, electric traction, adjustable speed drives etc, AC to DC conversion is necessary at the front end. This can be achieved using single phase and three phase, uncontrolled, half controlled and fully controlled converters depending upon the application. Filters are used to smoothen the rectified DC voltage. Depending on the filters used, AC–DC converter behaviour varies a lot at the AC mains. The AC to DC converter draw excessive (or moderate) harmonic current with high (or low) crest factor. They exhibit poor power factor at the AC mains generally due to harmonics as well as reactive power.

C18.6 SUPPLY SYSTEM BASED NONLINEAR LOADS

Nonlinear loads can be classified based on the AC supply system as
 i. Two wire (single-phase) system
 ii. Three wire (Three phase without neutral conductor) system
 iii. Four wire (Three phase with neutral conductor) system

C18.7 TWO WIRE (SINGLE PHASE) NONLINEAR LOADS

Most of the domestic appliances and fairly a large number of commercial equipments belongs to this category. Examples of such nonlinear loads are television sets, computers, power supplies, electronic fan regulators, electronic ballast based vapour lighting system and so on. These types of nonlinear loads draw harmonic currents and sometimes reactive power from the AC mains along with active power.

C18.8 THREE PHASE THREE WIRE NONLINEAR LOADS

Adjustable speed drives, HVDC transmission system and wind power conversion are the typical examples of three wire nonlinear loads. These types of nonlinear loads inject harmonic current and sometimes they draw reactive power along with active power from the AC mains. Three wire nonlinear loads may also have unbalanced currents.

C18.9 THREE PHASE FOUR WIRE NONLINEAR LOADS

Most of the single phase (two wires) nonlinear loads are supplied from the three phase four wire supply system. Four wire nonlinear loads causes harmonic currents, unbalanced current and reactive power. They also cause excessive neutral current due to harmonic currents and unbalancing of single phase loads on the 3 phases.

 Computers and electronic ballasts based vapour lighting systems are the typical examples. Four wire nonlinear loads also creates voltage distortion and voltage imbalance at the point of common coupling (PCC) and some potential at the neutral terminal.

C18.10 POWER QUALITY PROBLEMS CAUSED BY NONLINEAR LOADS

The nonlinear loads cause a large number of power quality problems in the distribution system. The nonlinear loads inject harmonic currents into the supply mains which increases the rms value of the supply current, increases losses, cause poor utilization and heating of components of the distribution system, and also cause distortion and notching in voltage waveforms at the point of common coupling due to voltage drop in the source impedance.

A few effects of harmonic currents are as follows:
1. Increased rms value of the supply current
2. Increased losses
3. Poor power factor
4. Poor utilization of distribution system
5. Heating of distribution system components
6. Derating of the distribution system
7. Distortion of the voltage waveform at the point of common coupling

8. Disturbance to the nearby consumers
9. Malfunctioning of protection systems such as relays
10. Interference in communication system
11. Capacitor bank failure due to overload, resonance, harmonic amplification
12. Excessive neutral current
13. Harmonic voltage at the neutral point

As already discussed, many nonlinear loads, along with harmonics, also require reactive power which creates unbalancing. This cause addition problems, along with above mentioned issues, such as

1. Voltage regulation and voltage fluctuations
2. Imbalance in three phase voltages
3. Derating of cables and feeders

C18.11 PERFORMANCE PARAMETERS FOR POWER QUALITY ANALYSIS

1. RMS value of source current

RMS value of the periodic source current with fundamental frequency component and its integral multiples is defined as

$$I_{SRMS} = \sqrt{\left((I_{DC})^2 + (I_{S1RMS})^2 + (I_{S2RMS})^2 + (I_{S3RMS})^2 + (I_{S4RMS})^2\right) + \cdots}$$

where, I_{DC} = DC component of source current

I_{S1RMS} = Fundamental or 50 Hz component of current waveform

$I_{S2RMS}, I_{S3RMS}, I_{S4RMS} \cdots$ are the RMS value of harmonic components of the source current corresponds to 100 Hz (50 × 2), 150 Hz (50 × 3), 200 Hz (50 × 4) and so on.

2. Active power

It is the power delivered by a source or the power absorbed by the circuit elements. Hence active power,

$$P = P_1 = V_{S1RMS} I_{S1RMS} \cos \phi_1 = V_{SRMS} I_{S1RMS} \cos \phi_1$$

where, V_{S1RMS} = RMS value of fundamental component of source voltage.

V_{S1RMS} = V_{SRMS} if source voltage is purely sinusoidal.

I_{S1RMS} = RMS value of fundamental component of source current.

ϕ_1 = phase angle between fundamental component of source current and source voltage.

Active power is produced by the fundamental component of voltage and current and its unit is Watts (W).

3. Reactive power

Reactive power is defined as

$$Q = Q_1 = V_{S1RMS} I_{S1RMS} \sin \phi_1 = V_{SRMS} I_{S1RMS} \sin \phi_1$$

Reactive power is produced by the current at the same frequency as that of the supply voltage and its unit is Volt-Ampere-Reactive (VAR).

4. Apparent power

It is defined as,

$$S = V_{SRMS} S_{RMS}$$

where, V_{SRMS} = RMS value of source voltage.

V_{SRMS} = V_{S1RMS} if source voltage is purely sinusoidal.

I_{SRMS} = RMS value of source current.

Unit of apparent power is Volt-Ampere (VA).

5. Distortion power

The apparent power is

$$S = V_{SRMS}\, I_{SRMS}$$

$$= V_{S1RMS}\, I_{SRMS} \text{ Since supply voltage is purely sinusoidal}$$

$$= \sqrt{V_{S1RMS}^2 \left(I_{DC}^2 + I_{S1RMS}^2 + I_{S2RMS}^2 + I_{S3RMS}^2 + I_{S4RMS}^2 + \cdots\right)}$$

$$= \sqrt{I_{DC}^2 V_{S1RMS}^2 + I_{S1RMS}^2 V_{S1RMS}^2 + I_{S2RMS}^2 V_{S1RMS}^2 + I_{S3RMS}^2 V_{S1RMS}^2 + I_{S4RMS}^2 V_{S1RMS}^2 + \cdots}$$

$$= \sqrt{\left[I_{S1RMS}^2 V_{S1RMS}^2\right] + I_{DC}^2 V_{S1RMS}^2 + I_{S2RMS}^2 V_{S1RMS}^2 + I_{S3RMS}^2 V_{S1RMS}^2 + I_{S4RMS}^2 V_{S1RMS}^2 + \cdots}$$

$$= \sqrt{\left[V_{S1RMS}^2 I_{S1RMS}^2\right] + V_{S1RMS}^2 \left(I_{DC}^2 + I_{S2RMS}^2 + I_{S3RMS}^2 + I_{S4RMS}^2 + \cdots\right)}$$

$$= \sqrt{S_1^2 + V_{S1RMS}^2 \left(I_{DC}^2 + I_{S2RMS}^2 + I_{S3RMS}^2 + I_{S4RMS}^2 + \cdots\right)}$$

$$= \sqrt{S_1^2 + D^2}$$

where, $$D^2 = V_{S1RMS}^2 \left(I_{DC}^2 + I_{S2RMS}^2 + I_{S3RMS}^2 + I_{S4RMS}^2 + \cdots\right)$$

$$D^2 = V_{S1RMS}^2 \left(I_{SRMS}^2 - I_{S1RMS}^2\right)$$

$$D = V_{S1RMS} \sqrt{\left(I_{SRMS}^2 - I_{S1RMS}^2\right)}$$

D is called distortion power and its unit is Volt-Ampere-Distortion (VAD). A purely sinusoidal voltage having harmonic currents provide an additional volt-ampere called distortion power, D.

$$S = \sqrt{S_1^2 + D^2}$$

$$S = \sqrt{P_1^2 + Q_1^2 + D^2}$$

In general, apparent power,

$$S = \sqrt{P^2 + Q^2 + D^2}$$

Where, S = Apparent power in VA

P = Active power in W

Q = Reactive power in VAR

D = Distortion power in VAD

6. Displacement power factor (*DPF*)

It is defined as the cosine of the phase angle between fundamental component of source current and voltage.

$$DPF = \cos \phi_1$$

Where ϕ_1 = phase angle between fundamental component of source current and (fundamental component of) source voltage.

The phase angle is called displacement angle.

7. Distortion factor (*DF*)

It is the ratio of rms value of fundamental component of the source current to the rms value of the source current.

$$DF = \frac{I_{S1RMS}}{I_{SRMS}}$$

8. Power factor (*PF*)

It is defined as the ratio of active power to the apparent power.

$$
\begin{aligned}
PF &= \frac{\text{Active power}}{\text{Apparent power}} \\
&= \frac{V_{S1RMS} I_{S1RMS} \cos\phi_1}{V_{SRMS} I_{SRMS}} \\
&= \frac{V_{SRMS} I_{S1RMS} \cos\phi_1}{V_{SRMS} I_{SRMS}} \quad \left(\begin{array}{c} V_{S1RMS} = V_{SRMS}, \text{because the source voltage} \\ \text{is purely sinusoidal} \end{array} \right) \\
&= \frac{I_{S1RMS}}{I_{SRMS}} \cos\phi_1 \\
PF &= DF \times DPF
\end{aligned}
$$

So power factor can also be defined as the product of distortion factor and displacement power factor.

9. Total harmonic distortion (*THD*) or harmonic factor (*HF*)

THD of source current is the ratio of the rms value of the non-fundamental frequency currents to the rms value of the fundamental frequency current.

$$THD = \sqrt{\frac{\left(I^2_{SRMS} - I^2_{S1RMS} \right)}{I^2_{S1RMS}}}$$

The harmonic factor or *THD* is a measure of the distortion of the source current or source voltage. Therefore harmonic factor is often referred to as total harmonic distortion.

10. Crest factor (*CF*)

$$CF = \frac{I_{\text{Peak-source}}}{I_{SRMS}}$$

It is the ratio of peak value of source current to rms value of source current.

C18.12 SIMULATION FOR POWER QUALITY ANALYSIS OF SINGLE PHASE DIODE BRIDGE RECTIFIER CIRCUIT USING MATLAB/SIMULINK

At the front end of most of the commercial equipments and domestic appliances, AC to DC conversion is inevitable. Generally AC to DC conversion with a large DC capacitor at the DC side is provided to get a DC voltage source for the remaining conversion process and this causes to draw peaky current from the supply mains with high crest factor. Power factor of AC to DC conversion is generally poor at the AC mains due to harmonics as well as reactive power.

AC to DC conversion can be achieved with single phase or three phase half-wave, semi, full-wave controlled or uncontrolled rectifier circuits.

A basic setup of the single phase diode bridge rectifier with filter capacitor for low power application is shown in the Fig. C18.1. A step down transformer is used to reduce the input supply voltage 230 V to 24 V.

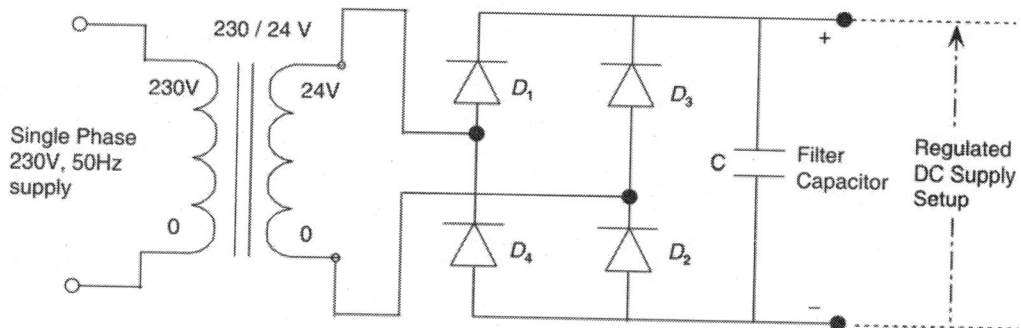

Fig. C18.1: Basic setup of regulated DC supply with single phase diode bridge rectifier having filter capacitor

Let us model, simulate and analyse the power quality of the supply system having single phase diode bridge (uncontrolled) rectifier which is the most commonly used AC to DC conversion circuit.

Consider a single phase uncontrolled bridge converter with sinusoidal input supply of 24 V rms, 50 Hz.

Following steps are to be followed for analyzing and modelling the single phase diode bridge rectifier circuit.

Step 1: Enter "simulink" in the MATLAB command window to open the simulink library browser or click simulink button in the MATLAB toolbar.

Step 2: Select **File > New > Model** in the simulink library browser to create a new model or press Ctrl+N. An empty model window is opened and it can be saved by selecting **File > Save** in the model window.

Step 3: Drag and place the following building blocks from *Simulink library* and *Sim Power Systems library* to the simulink model file according to the circuit diagram.

(a) Powergui block [Sim Power Systems]
(b) AC voltage source [Sim Power Systems > Electrical Sources]
(c) Series RLC branch (configure it to R) [Sim Power Systems > Elements]
(d) Series RLC branch (configure it to R) [Sim Power Systems > Elements]
(e) Diode [Sim Power Systems > Power Electronics]
(f) Voltage measurement [Sim Power Systems > Measurements]
(g) Current measurement [Sim Power Systems > Measurements]
(h) Scope [Simulink > Sinks]
(i) RMS [Sim Power Systems > Extra library > Measurements]
(j) Fourier [Sim Power Systems > Extra library > Measurements]
(k) Active and reactive power [Sim Power Systems > Extra library > Measurements]
(l) Total harmonic distortion [Sim Power Systems > Extra library > Measurements]
(m) Display [Simulink > Sinks]
(n) From [Simulink > Signal Routing]
(o) Goto [Simulink > Signal Routing]

Step 4: Connect the blocks according to the circuit diagram of the single phase diode bridge rectifier circuit. Also connect required number of current and voltage measurement blocks to the *scope*, as shown in the Fig. C18.2, to observe the necessary waveforms. The "From" and "Goto" blocks need not be necessarily used, but it will reduce the number of connections in the circuit.

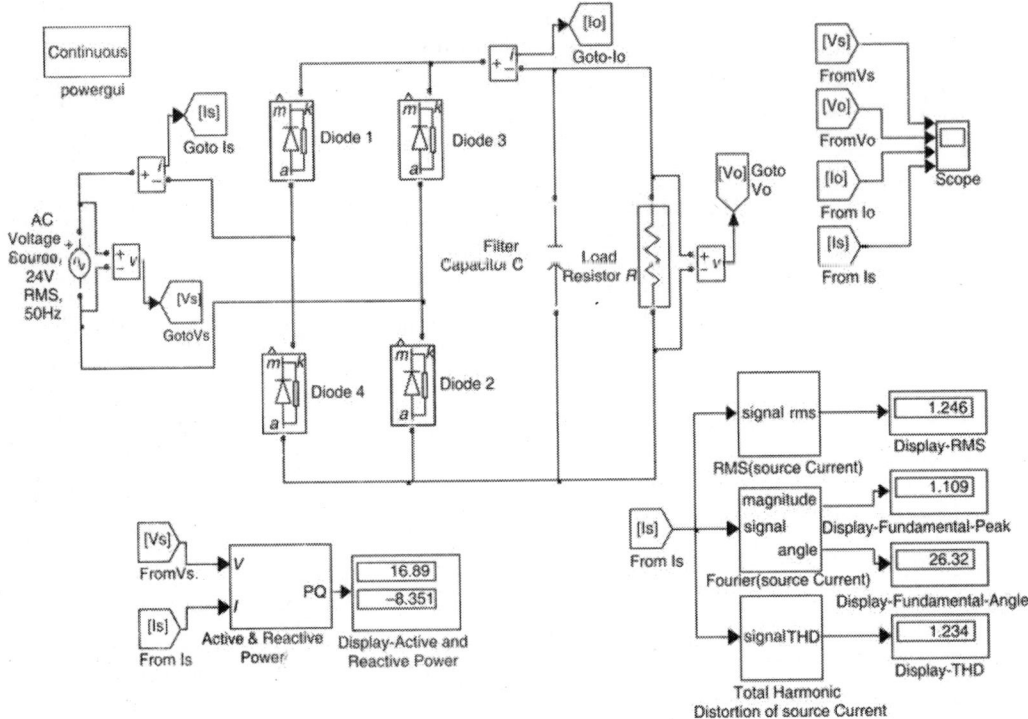

Fig. C18.2: The single phase diode bridge rectifier circuit model for power quality analysis with measurement blocks using MATLAB/Simulink

Step 5: Set the parameters of the individual blocks. Components and specifications used in single phase diode bridge rectifier circuit are:

1. AC voltage source	24 V (RMS), 50 Hz
2. Load resistor	50 Ω
3. Diodes	
4. Filter capacitor C	470 µF
5. Powergui block	
6. Current measurement block	
7. Voltage measurement block	
8. Scope	
9. RMS block	
10. Fourier block	
11. Total garmonic distortion block	
12. Active and reactive power block	
13. Display	
14. From	
15. Goto	

Step 6: RMS and the total harmonic distortion of source current is to be calculated. Magnitude and phase angle of fundamental component of source current can be calculated using *Fourier block*. Active power and reactive power burden of the supply can be measured using *Active and Reactive Power block*.

Now, double click on *RMS block*, *Total Harmonic Distortion block* and *Active and Reactive Power block* and enter fundamental frequency as 50 Hz since supply voltage frequency is 50 Hz. Then double click on *Fourier block* and enter fundamental frequency as 50 Hz and *Harmonic n* as 1 since we need fundamental component. Here *Fourier block* gives maximum (peak) value of fundamental component of supply current and corresponding phase angle.

Step 7: Save the simulink model. Run the simulation by selecting **simulation > start** in the model window or use **simulation button** in the model window toolbar.

The run time can be entered and modified in the space provided in the model window tool bar. For example, run time = 0.1 seconds.

Step 8: Observe and note down the following waveforms by opening the scope and save the result.

(a) Source voltage

(b) Load voltage [Output voltage or voltage across the load resistor R or C]

(c) Load current, [i.e. capacitor current + load resistor current]

(d) Source current

The waveforms are shown in the Fig. C18.3. From the *display blocks*, note down the respective values.

The measured values are as follows:

(a) RMS of the source current, I_{1RMS} = 1.246 A

(b) Maximum value of the fundamental component of source current = 1.109 A.

Therefore, RMS value of fundamental component of source current,

$$I_{1RMS} = 1.109/\sqrt{2} = 0.7842 \text{ A.}$$

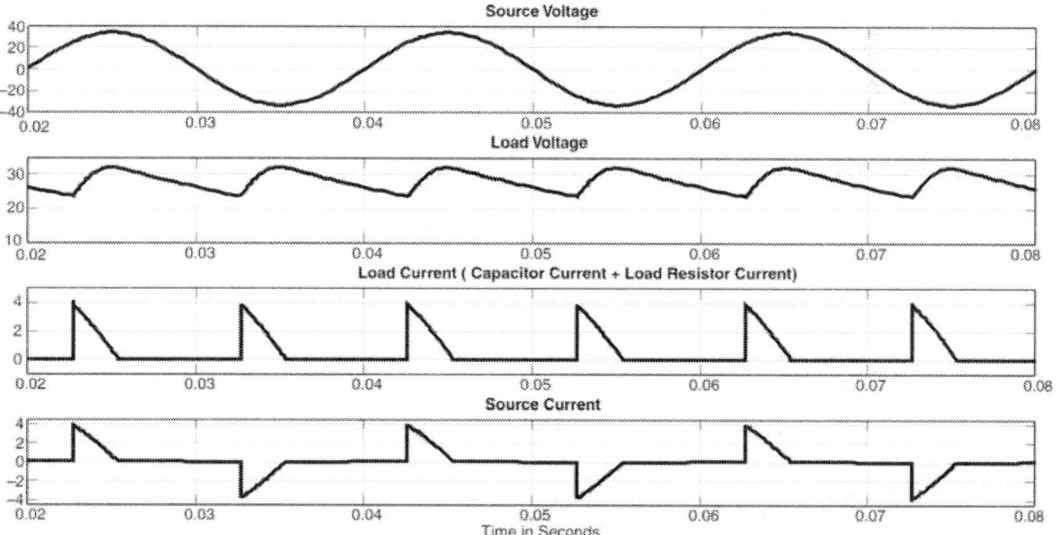

Fig. C18.3: Waveforms of single phase diode bridge rectifier (a) Source voltage (b) Load voltage (c) Load current (d) source current

(c) Phase angle of fundamental component of source current, $\phi_1 = 26.32°$.

(d) Total harmonic distortion (*THD*) of source current = 1.234 = 123.4%

(e) Active or real power to be supplied from the source, $P = 16.89$ W

(f) Reactive power burden of the source, $Q = 8.351\ VAR$

Step 9: From the waveform as shown in the Fig. C18.3 (d), it can be observed that source current is highly distorted and contains a lot of harmonics. That means source is heavily polluted with current harmonics.

Now let us do power quality analysis of single phase diode bridge rectifier by analysing above simulated data.

RMS value of source voltage (Here this value is the same as the RMS value of fundamental component of source voltage since source voltage waveform is not distorted).

$$V_{SRMS} = V_{S1RMS} = 24 \text{ V}$$

RMS value source current,

$$I_{SRMS} = 1.246 \text{ A}$$

RMS value of fundamental component of source current,

Phase angle between fundamental component of source current and source voltage, $\phi = 26.32°$.

1. Apparent power, $S = V_{SRMS}\ I_{SRMS} = 24 \times 1.246 = 29.904$ VA

2. Active power, $P = V_{S1RMS}\ I_{S1RMS} \times \cos(\phi_1) = 24 \times 0.7842 \times \cos(26.32) = 16.869$ W
   Note that calculated value of active power is same as simulated value.

3. Reactive power $= V_{S1RMS}\ I_{S1RMS} \times \sin(\phi_1) = 24 \times 0.78 \times \sin(26.32) = 8.3448\ VAR$

4. Distortion power,

$$D = \sqrt{(S^2 - P^2 - Q^2)}$$

$$= \sqrt{(29.904^2 - 16.8696^2 - 8.3448^2)} = 23.238\ VAD$$

Distortion power can also be calculated using following equation,

$$D = V_{S1RMS}\sqrt{(I_{SRMS})^2 - (I_{S1RMS})^2}$$

$$D = 24\sqrt{(1.246)^2 - (0.7842)^2} = 23.238\ VAD$$

5. Displacement power factor,
$$DPF = \cos(\phi_1) = \cos(26.32) = 0.8963$$

6. Distortion factor,

$$DF = \frac{I_{S1RMS}}{I_{SRMS}}$$

7. Power factor, $PF = DF \times DPF$
$$= 0.694 \times 0.8963 = 0.564$$

So power factor is poor

8. Total harmonic distortion,

$$THD = \frac{\sqrt{(I_{SRMS})^2 - (I_{S1RMS})^2}}{I_{S1RMS}}$$

$$= \sqrt{\frac{(1.246)^2 - (0.7842)^2}{0.7842}} = 1.234 = 123.4\%$$

Note that calculated value of THD is same as simulated value. THD value shows that source is highly distorted with current harmonics. According to IEEE 519–1992 standard, the maximum allowable limit of current THD is 5%.

9. Crest factor,

$$CF = \frac{I_{\text{Peak-source}}}{I_{\text{SRMS}}}$$

Peak value of supply (source) current can obtained by the direct inspection from the source current waveform, as shown in the Fig. C18.3(d), and which is equal to 4A.

$$\therefore \quad CF = \frac{4}{1.246} = 3.21$$

Step 10: Harmonic analysis and total harmonic distortion can also be found out by using *FFT analysis* tab in the *powergui* block.

Reader are requested to go through *Appendix B: FFT analysis using MATLAB/Simulink* for better understanding of harmonic analysis using simulink.

Figure C18.4 shows the harmonic analysis of source current waveform using *powergui FFT analysis tool* window.

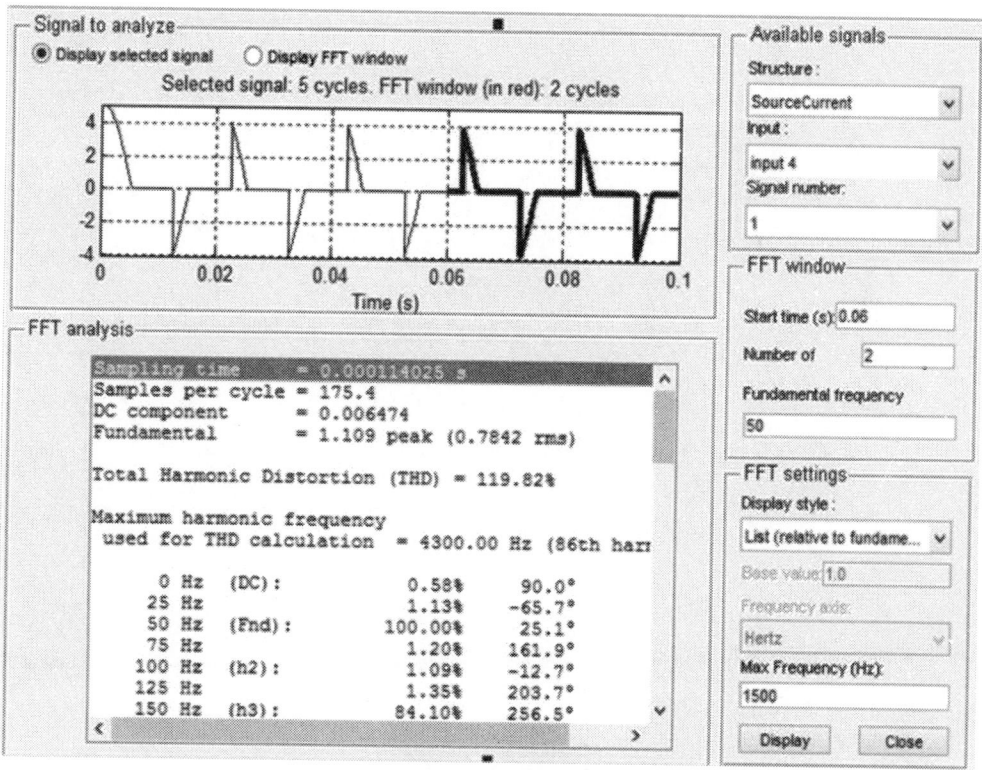

Fig. C18.4: Harmonic analysis of source current waveform using *the powergui FFT analysis tool* window (display style: list)

Harmonic analysis of source current is as follows:

(a) RMS value of fundamental component of source current = 0.7842 A

(b) Total harmonic distortion, *THD* = 119.82%

(c) DC component of source current is negligible

(d) The 3rd harmonic component is 84.10%, 5th harmonic component is 58.51%, 7th harmonic component is 33.56%, 9th harmonic component is 21.3%.

C18.13 SIMULATION RESULT

Power quality problems caused by single phase diode bridge rectifier with filter capacitor circuit was simulated and analyzed using MATLAB/Simulink. By the simulation and calculation of performance parameters, it is shown that the power factor is poor and current THD limit is well above the IEEE 519–1992 standard limit.

Section D

Circuit Simulation Using Matlab Program

Experiment D1

Simulation of R Triggering Circuit for a Thyristor Using MATLAB Program

D1.1 OBJECTIVE

To simulate a resistance triggering circuit for a thyristor using MATLAB program

D1.2 MATLAB CODE

```
% MATLAB Code for R triggering Circuit
clc;
clear all;
    % Generation of source or supply voltage
f = 50;                    % frequency of the supply voltage
T=1/f;                     % Time period of the supply voltage
tt = 0:T/200:2*T;          % set time axis limits from 0 to 2T in
                           % steps of T/200 to plot 2 cycles of Vs
Vrms=input('Enter R.M.S value of the Supply Voltage in Volts, Vrms=');
Vm=Vrms*sqrt(2);           % Peak or Maximum value of supply voltage
Vs = Vm*sin(2*pi*50*tt);   % Instantaneous values of supply voltage, Vs
% Code for R triggering
Rload=input('Enter Load Resistance value in Ohm (Range 1 to 100 ohm)=');
R1=220;
Rvar=input('Enter variable Resistor value in Ohm, R2 (Range 1 to 3.38 k)=');
Rgk=100;   % Resistance between Gate and Cathode Terminals of Thyristor
for t1=0:T/200:T
    v=Vm*sin(2*pi*50*t1);
    i=v/(Rload+R1+Rvar+Rgk);
    vgk=Rgk*i;
    if(vgk>=0.8)
        break;
    end
end
Trig_ang_in_sec=t1;        % firing angle in seconds
```

257

```
alpha_deg=t1*2*pi*50*180/pi;     % firing angle in degrees
                                 % [wt in radians & wt*180/pi  in degree]
alpha_rad=t1*2*pi*50;            % firing angle in radians
td1 = pi/256;                    % increment time step in radians

t0 = [0:td1:alpha_rad];          % calculation of time divisions for two cycles
t1 = [alpha_rad:td1:pi];
t2 = [pi:td1:2*pi];
t3 = [2*pi:td1:alpha_rad+2*pi];
t4 = [2*pi+alpha_rad:td1:3*pi];
t5 = [3*pi:td1:4*pi];

t_rad=[t0 t1 t2 t3 t4 t5];       % time divisions in radian for two cycles
t=t_rad/(2*pi*f);                % time divisions in seconds for two cycles
             % t= t_rad *(180/pi); for time division in degrees

vo = [0*sin(t0) Vm*sin(t1) 0*sin(t2) 0*sin(t3) Vm*sin(t4) 0*sin(t5)];
                                 % Instantaneous values of output voltage
io=vo/Rload;                     % Instantaneous values of output current

Vth=[Vm*sin(t0) 0*Vm*sin(t1) Vm*sin(t2) Vm*sin(t3) 0*Vm*sin(t4) Vm*sin(t5)];
                                 % Instantaneous values of Thyristor voltage

% To Display the Results in Command Window
     disp(' ****** Results ******')
 Triggering_Angle_in_Seconds=Trig_ang_in_sec
 Triggering_Angle_in_Degree=alpha_deg

% TO plot the waveforms in a single Window
            % subplot can be used to plot the waveforms in a single window
subplot(4,1,1)
plot(tt,Vs)
grid on;
title('Supply Voltage')

subplot(4,1,2)
plot(t,vo)
grid on;
title('Load Voltage')

subplot(4,1,3)
plot(t,io)
grid on;
title('Load Current')

subplot(4,1,4)
plot(t,Vth)
grid on;
xlabel('Time in Seconds');
title('Voltage Across Thyristor')
```

% *To plot a single waveform*
% Example: MATLAB code to plot supply voltage waveform
   % figure(2);
   % plot(tt,Vs);
   % grid on;
   % xlabel('Time in Seconds');
   % ylabel('Amplitude (V)');
   % title('Supply Voltage')

D1.3 EXECUTION OF MATLAB CODE

Click on the *play* button to compile the program. If any error messages are displayed in the command window, make necessary corrections and compile the program again. Enter the relevant input parameters in the *command window* while running the program.

The sample input parameters are given below and the corresponding waveforms are shown in the following figure.

Command Window

Enter the R.M.S value of the supply voltage in volts, Vrms = 24
Enter load resistance value in Ohm (Range 1 to 100 ohm) = 100
Enter a value of variable resistor in Ohm, R2 (Range 1 to 3.38 k) = 2e3

****** Results ******
Triggering_Angle_in_Seconds = 0.0020
Triggering_Angle_in_Degree = 36

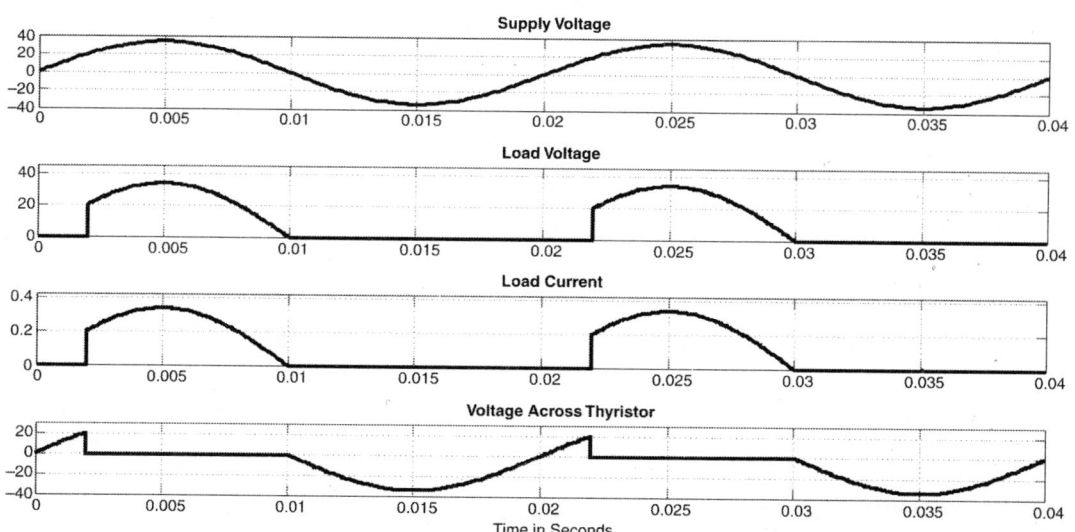

Waveforms: Resistance triggering using MATLAB program

Experiment D2

Simulation of an RC Triggering Circuit for a Thyristor Using MATLAB Program

D2.1 OBJECTIVE

To simulate a half-wave resistance-capacitance triggering circuit for a thyristor using MATLAB program.

D2.2 MATLAB CODE

```
%  MATLAB Code for Half-wave RC triggering Circuit
clc;
clear all;
        % Generation of source or supply voltage
f = 50;                         % frequency of the supply voltage
T=1/f;                          % Time period of the supply voltage
tt = 0:T/200:2*T;               % set time axis limits from 0 to 2T in
                                % steps of T/200 to plot 2 cycles of Vs
Vrms=input( ' Enter the R.M.S value of the Supply Voltage in Volts, Vrms=');
Vm=Vrms*sqrt(2);                % Peak or Maximum value of supply voltage
Vs = Vm*sin(2*pi*50*tt);        % Instantaneous values of supply voltage, Vs

   % Code for Half-wave RC triggering circuit

Rload=input('Enter Load Resistance value in Ohm (Range 1ohm to 100ohm) =');
R1=220;         % R1 is kept to limit the charging current when R2=0
Rvar=input('Enter Variable Resistor value in Ohm, R2(Range 5k to 900k)=');
C=0.22e-6;      % Value of the Capacitor
VcPre=input('Enter Initial Value of Capacitor Voltage in volts =');
phi= atan(1/(2*pi*50*(Rvar+R1+Rload)*C));

for t1 = 0:T/200:2*T

i=Vm*sin(2*pi*50*t1+phi+(atan(1/(2*pi*50*(Rvar+R1+Rload)*C))))/(sqrt(((Rvar+R1+Rload)^2)+(1/(((2*pi*50)^2*C^2)))));
                        % i=Vm/sqrt(R^2+1/w^2*C^2) * sin(wt+tan-1(1/wRC)
    Vc=i*1/(2*pi*50*C); % Capacitor voltage=Current*Capacitive Reactance
    Vc1=Vc+VcPre;            % Present value of Capacitor voltage
```

```
     VcPre=Vc1;
     if(Vc1>=0.8)
        break;      % Capacitor voltage reaches Gate trigger voltage of 0.8V
     end
  end
  Trig_ang_in_sec=t1;                % firing angle in seconds
  alpha_deg=t1*2*pi*50*180/pi;       % firing angle in degrees
                                     % wt in radians & wt*180/pi in degree
  alpha_rad=t1*2*pi*50;              % firing angle in radians
  td1 = pi/256;                      % increment time step in radians
  t0 = [0:td1:alpha_rad];            % calculation of time divisions for two cycles
  t1 = [alpha_rad:td1:pi];
  t2 = [pi:td1:2*pi];
  t3 = [2*pi:td1:alpha_rad+2*pi];
  t4 = [2*pi+alpha_rad:td1:3*pi];
  t5 = [3*pi:td1:4*pi];

  t_rad=[t0 t1 t2 t3 t4 t5];         % time divisions in radian for two cycles
  t=t_rad/(2*pi*f);                  % time divisions in seconds for two cycles
                                     % t= t_rad *(180/pi); for time division in degrees
  vo = [0*sin(t0) Vm*sin(t1) 0*sin(t2) 0*sin(t3) Vm*sin(t4) 0*sin(t5)];
                                     % Instantaneous values of output voltage
  io=vo/Rload;                       % Instantaneous values of output current
  Vth=[Vm*sin(t0) 0*Vm*sin(t1) Vm*sin(t2) Vm*sin(t3) 0*Vm*sin(t4) Vm*sin(t5)];
                                     % Instantaneous values of Thyristor voltage
  % To display the results in command window
       disp(' ****** Results ******')
  Triggering_Angle_in_Seconds=Trig_ang_in_sec
  Triggering_Angle_in_Degree=alpha_deg
       % To plot the waveforms in a single Window
  subplot(4,1,1)
  plot(tt,Vs)
  grid on;
  title('Supply Voltage')

  subplot(4,1,2)
  plot(t,vo)
  grid on;
  title('Load Voltage')

  subplot(4,1,3)
  plot(t,io)
  grid on;
  title('Load Current')
```

```
subplot(4,1,4)
plot(t,Vth)
grid on;
xlabel('Time in Seconds');
title('Voltage Across Thyristor')
```

D2.3 EXECUTION OF MATLAB CODE

Click on the *play* button to compile the program. If any error messages are displayed in the command window, make necessary corrections and compile the program again. Enter the relevant input parameters in the *command window* while running the program.

The sample input parameters are given below and the corresponding waveforms are shown in the following figure.

Command Window

Enter the R.M.S value of the supply voltage in volts, Vrms = 24
Enter load resistance value in Ohm (Range 1 ohm to 100 ohm) = 100
Enter variable resistor value in Ohm, R2 (Range 5k to 900k) = 700e3
Enter initial value of capacitor Voltage in volts = –30

****** Results ******
Triggering_Angle_in_Seconds = 0.0061
Triggering_Angle_in_Degree = 109.8000

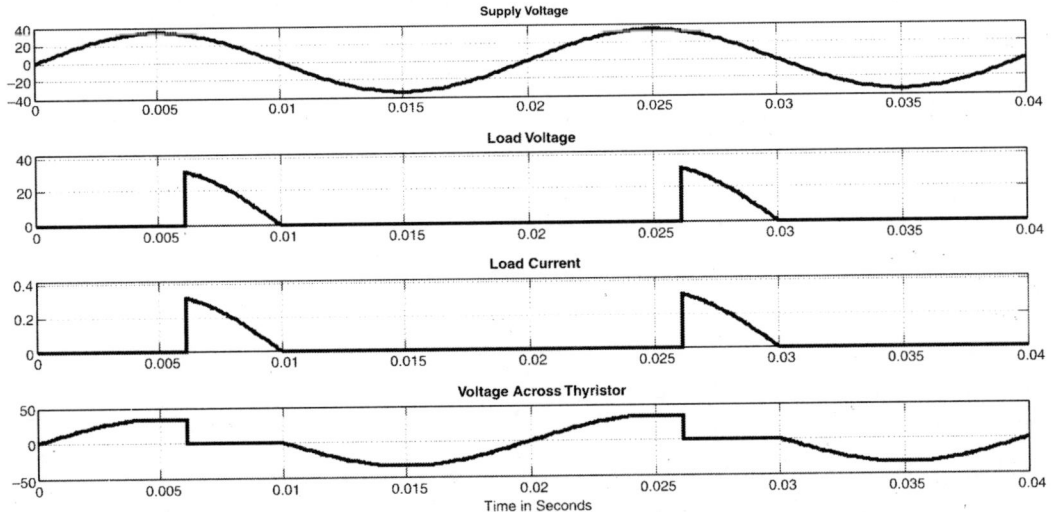

Waveforms: Half-wave RC triggering using MATLAB program

Experiment D3

Simulation of UJT Triggering Circuit for a Thyristor Using MATLAB Program

D3.1 OBJECTIVE

To simulate a half-wave UJT Triggering circuit for a thyristor using MATLAB program.

D3.2 MATLAB CODE

```
% MATLAB code for half-wave UJT triggering circuit for a thyristor
clc;
clear all;

    % Generation of source or supply voltage
f = 50;                         % frequency of the supply voltage
T=1/f;                          % Time period of the supply voltage
tt = 0:T/200:2*T;               % set time axis limits from 0 to 2T in
                                % steps of T/200 to plot 2 cycles of Vs
Vrms=input( ' Enter the R.M.S value of the Supply Voltage in Volts, Vrms=');
Vm=Vrms*sqrt(2);                % Peak or Maximum value of supply voltage
Vs = Vm*sin(2*pi*50*tt);        % Instantaneous values of supply voltage, Vs
% Code for UJT Triggering
Rload=100;                              % Load Resistor
vd = 0.7;                               % diode voltage drop
eta=input ('Enter intrinsic standoff ratio of UJT (between 0 and 1) =');
Vbb = 12;                       % consider VBB (base to base) of the UJT is 12 V
R=input('Enter Variable Resistor value in ohm, R2(Range 1k to 50k)=');
C = 0.22e-6;                            % capacitance value is 0.22uF
Vp = eta*Vbb + vd;                      % peak voltage of the UJT
tc = R*C*log(1/(1-eta));
                    % charging time of the capacitor tc=RCln[1/(1-eta)]
                    % 'log' in matlab represent natural logarithm .i.e.,
                    % log to the base e (not log to the base 10)
Charging_Time_of_Capacitor_to_Reach_Vp =tc;
tt1 = 0:T/400:tc;
```

```
Vc = Vbb*(1-exp((-tt1/(R*C))));
        % Instantaneous Capacitor Voltage during Charging, Vc=VBB[1-e(-t/RC)]
        % To plot the waveform of Voltage across Capacitor during charging
figure(1);
plot(tt1,Vc);
grid on;
xlabel('Charging Time of Capacitor (Seconds) ');
ylabel('Voltage across Capacitor during Charging (Volts)');
title (' Waveform - Charging of Capacitor (Approximate) ');
% Code for Half-wave Rectifier
alpha=tc*2*pi*50*180/pi;        % firing angle in degrees
                                % [wt in radians & wt*180/pi  in degree]
alpha_rad=tc*2*pi*50;           % firing angle in radians
td1 = pi/256;                   % increment time step in radians

t0 = [0:td1:alpha_rad];     % calculation of time divisions for two cycles
t1 = [alpha_rad:td1:pi];
t2 = [pi:td1:2*pi];
t3 = [2*pi:td1:alpha_rad+2*pi];
t4 = [2*pi+alpha_rad:td1:3*pi];
t5 = [3*pi:td1:4*pi];

t_rad=[t0 t1 t2 t3 t4 t5];          % time divisions in radian for two cycles
t=t_rad/(2*pi*f);                   % time divisions in seconds for two cycles
                    % t= t_rad *(180/pi); for time division in degrees

vo = [0*sin(t0) Vm*sin(t1) 0*sin(t2) 0*sin(t3) Vm*sin(t4) 0*sin(t5)];
                        % Instantaneous values of output voltage
io=vo/Rload;            % Instantaneous values of output current

Vthy=[Vm*sin(t0) 0*Vm*sin(t1) Vm*sin(t2) Vm*sin(t3) 0*Vm*sin(t4) Vm*sin(t5)];
                    % Instantaneous values of Thyristor voltage

% To Display the results in command window
    disp(' ****** Results ******')
  Peak_Voltage_Vp_of_UJT_in_Volts=Vp
  Charging_Time_of_Capacitor_in_Seconds_to_Reach_Vp =tc
  Firing_Angle_in_Degree=alpha

    % TO plot the waveforms in a single Window
figure(2);
subplot(4,1,1)
plot(tt,Vs)
grid on;
title('Supply Voltage')

subplot(4,1,2)
plot(t,vo)
```

```
grid on;
title('Load Voltage')
subplot(4,1,3)
plot(t,io)
grid on;
title('Load Current')
subplot(4,1,4)
plot(t,Vthy)
grid on;
xlabel('Time in Seconds');
title('Voltage Across Thyristor')
```

D3.3 EXECUTION OF MATLAB CODE

Click on the *play* button to compile the program. If any error messages are displayed in the command window, make necessary corrections and compile the program again. Enter the relevant input parameters in the *command window* while running the program.

The sample input parameters are given below and the corresponding waveforms are shown in the following figure.

Command Window

Enter R.M.S value of the supply voltage in volts, Vrms = 24
Enter intrinsic standoff ratio of UJT (between 0 and 1) = .6
Enter variable resistor value in ohm, R2 (Range 1k to 50k) = 40e3

****** Results ******

Peak_Voltage_Vp_of_UJT_in_Volts = 7.9000
Charging_Time_of_Capacitor_in_Seconds_to_Reach_Vp = 0.0081
Firing_Angle_in_Degree = 145.1405

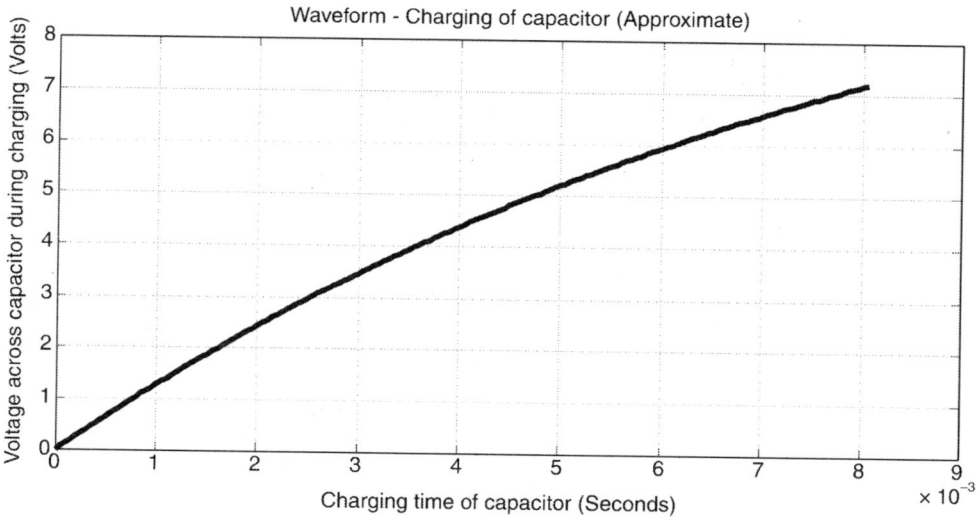

Waveform: Capacitor voltage during charging - UJT triggering circuit

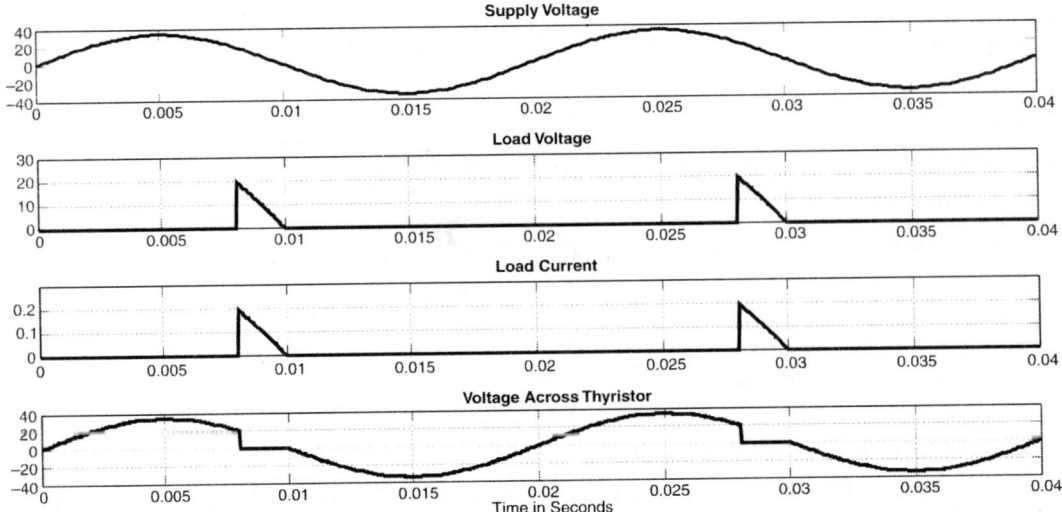

Waveform: Half-wave UJT triggering circuit using MATLAB program

Experiment D4

Simulation of Single Phase Half-wave Controlled Rectifier Using MATLAB Program

D4.1 OBJECTIVE

To simulate a single phase half-wave controlled rectifier with resistive load using MATLAB program.

D4.2 MATLAB CODE

```
% MATLAB code for single phase half-wave controlled rectifier with R load
clc;
clear all;
% Generation of source or supply voltage
f = 50;                     % frequency of the supply voltage
T=1/f;                      % Time period of the supply voltage
tt = 0:T/200:2*T;           % set time axis limits from 0 to 2T in
                            % steps of T/200 to plot 2 cycles of Vs
Vrms=input( 'Enter R.M.S value of the Supply Voltage in Volts, Vrms=');
Vm=Vrms*sqrt(2);            % Peak or Maximum value of supply voltage
Vs = Vm*sin(2*pi*50*tt);    % Instantaneous values of supply voltage, Vs

    % Code for single phase half-wave rectifier
Rload=input('Enter Load Resistance value in Ohm =');
alpha = input('Enter Firing Angle of the Thyristor in degree = ');
alpha_rad=alpha*pi/180;     % firing angle in radians
td1 = pi/256;               % increment time step in radians

t0 = [0:td1:alpha_rad];     % calculation of time divisions for two cycles
t1 = [alpha_rad:td1:pi];
t2 = [pi:td1:2*pi];
t3 = [2*pi:td1:2*pi+alpha_rad];
t4 = [2*pi+alpha_rad:td1:3*pi];
t5 = [3*pi:td1:4*pi];
```

```
t_rad=[t0 t1 t2 t3 t4 t5];        % time divisions in radian for two cycles
t=t_rad/(2*pi*f);                 % time divisions in seconds for two cycles
            % t= t_rad *(180/pi); for time division in degrees
vo = [0*sin(t0) Vm*sin(t1) 0*sin(t2) 0*sin(t3) Vm*sin(t4) 0*sin(t5)];
                                  % Instantaneous values of output voltage
io=vo/Rload;                      % Instantaneous values of output current
Vthy=[Vm*sin(t0) 0*Vm*sin(t1) Vm*sin(t2) Vm*sin(t3) 0*Vm*sin(t4) Vm*sin(t5)];
                    % Instantaneous values of Thyristor voltage
    % Calculation of Average value, RMS value , Real power and Power Factor
Vdc = (Vm/(2*pi))*(1 + cos(alpha_rad))  ;
            % Average value of output voltage=Vm/2pi*(1+ cos(theta))
            % In matlab use theta in redians
Idc=Vdc/Rload;                    % Average value of Load Current
Vorms = sqrt(mean(vo.*vo));       % RMS value of output voltage
Iorms=Vorms/Rload;                % RMS value of Load Current
P=Iorms^2*Rload;                  % Power Delivered to Load=Real power
PF=P/(Vrms*Iorms);
                    % Input Power Factor=Real power/Apparent power
                    % Apprent power=Vsrms*Isrms=Vrms*Iorms
                    % In Single phase Half-wave rectifier circuit,
                    % Source current, Isrms=Load current,Iorms
% To display the results in command window
disp(' ****** Results ******')
Average_Value_of_Output_Voltage_in_Volts=Vdc
Average_Value_of_Load_Current_in_Amperes=Idc
RMS_Value_of_Output_Voltage_in_Volts=Vorms
RMS_Value_of_Load_Current_in_Amperes=Iorms
Power_delivered_to_the_Load_in_Watts=P
Input_Power_Factor=PF
        % TO plot the waveforms in a single Window
subplot(4,1,1)
plot(tt,Vs)
grid on;
title('Supply Voltage')

subplot(4,1,2)
plot(t,vo)
grid on;
title('Load Voltage')

subplot(4,1,3)
plot(t,io)
grid on;
title('Load Current')
```

```
subplot(4,1,4)
plot(t,Vthy)
grid on;
xlabel('Time in Seconds');
title('Voltage Across Thyristor')
```

D4.3 EXECUTION OF MATLAB CODE

Click on the *play* button to compile the program. If any error messages are displayed in the command window, make necessary corrections and compile the program again. Enter the relevant input parameters in the *command window* while running the program.

The sample input parameters are given below and the corresponding waveforms are shown in the following figure.

Command Window

Enter R.M.S value of the Supply voltage in Volts, Vrms = 24
Enter Load Resistance value in Ohm = 100
Enter Firing Angle of the Thyristor in degree = 60
****** Results ******

Average_Value_of_Output_Voltage_in_Volts = 8.1028
Average_Value_of_Load_Current_in_Amperes = 0.0810
RMS_Value_of_Output_Voltage_in_Volts = 15.2195
RMS_Value_of_Load_Current_in_Amperes = 0.1522
Power_delivered_to_the_Load_in_Watts = 2.3163
Input_Power_Factor = 0.6341

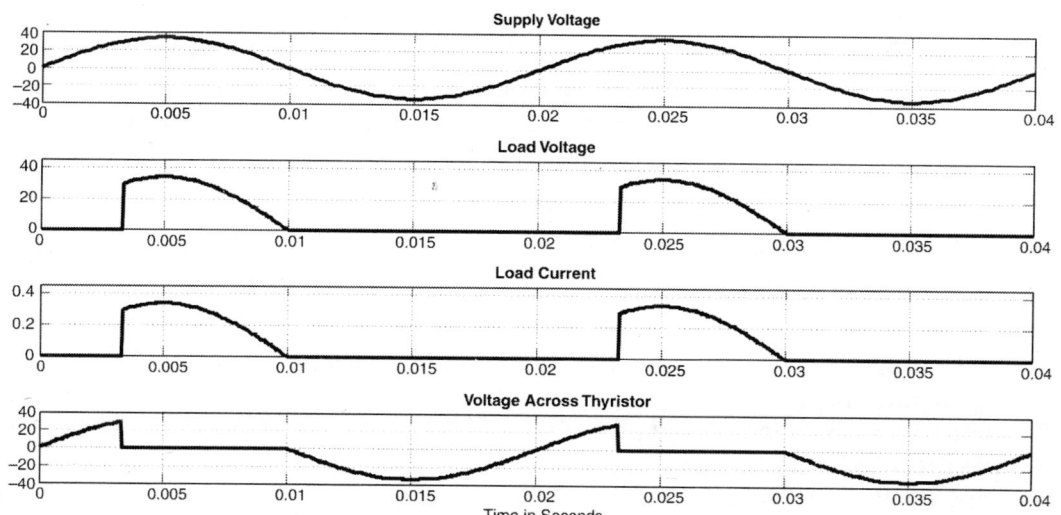

Waveforms: Single phase Half-wave controlled rectifier with R load using MATLAB program

Experiment D5

Simulation of Single Phase Full-wave Fully Controlled Bridge Rectifier Using MATLAB Program

D5.1 OBJECTIVE

To simulate a single phase full-wave fully controlled bridge rectifier with resistive load using MATLAB program.

D5.2 MATLAB CODE

```
% MATLAB code for single phase full-wave fully controlled bridge
                % Rectifier with R Load
clc;
clear all;
% Generation of source or supply voltage
f = 50;                              % frequency of the supply voltage
T=1/f;                               % Time period of the supply voltage
tt = 0:T/200:2*T;                    % set time axis limits from 0 to 2T in
                                     % steps of T/200 to plot 2 cycles of Vs
Vrms=input( 'Enter R.M.S value of the Supply Voltage in Volts, Vrms=');
Vm=Vrms*sqrt(2);                     % Peak or Maximum value of supply voltage
Vs = Vm*sin(2*pi*50*tt);             % Instantaneous values of supply voltage, Vs
    % Code for single phase Full-wave Bridge Rectifier
Rload=input('Enter Load Resistance value in Ohm =');
alpha = input('Enter Firing Angle of the Thyristor in degree = ');
alpha_rad=alpha*pi/180;              % firing angle in radians
td1 = pi/256;                        % increment time step in radians
t0 = 0:td1:alpha_rad;                % calculation of time divisions for two cycles
t1 = alpha_rad:td1:pi;
t2 = pi:td1:pi+alpha_rad;
t3 = pi+alpha_rad:td1:2*pi;
t4 = 2*pi:td1:2*pi+alpha_rad;
t5 = 2*pi+alpha_rad:td1:3*pi;
t6 = 3*pi:td1:3*pi+alpha_rad;
t7 = 3*pi+alpha_rad:td1:4*pi;
```

```
t_rad=[t0 t1 t2 t3 t4 t5 t6 t7];    % time divisions in radian for two cycles
t=t_rad/(2*pi*f);                   % time divisions in seconds for two cycles
            % t= t_rad *(180/pi); for time division in degrees
vo = [0*sin(t0) Vm*sin(t1) 0*sin(t2) -Vm*sin(t3) 0*sin(t4) Vm*sin(t5)
    0*sin(t6) -Vm*sin(t7)];
                            % Instantaneous values of output voltage
io = vo/Rload;              % Instantaneous values of output current
Vthy = [(Vm/2)*sin(t0) 0*Vm*sin(t1) (Vm/2)*sin(t2) Vm*sin(t3) (Vm/2)*sin(t4)
    0*sin(t5) (Vm/2)*sin(t6) Vm*sin(t7) ];
                            % Instantaneous values of Thyristor voltage T1 or T2
    % Calculation of Average value, RMS value, Real power and Power Factor
Vdc =  (Vm/pi)*(1 + cos(alpha_rad));
                    % Average value of Output or Load voltage=Vm/pi*(1+cos(theta))
                    % In matlab use theta in redians
Idc=Vdc/Rload;              % Average value of Load Current
Vorms = sqrt(mean(vo.*vo));  % RMS value of Load voltage
Iorms=Vorms/Rload;         % RMS value of Load Current
P=Iorms^2*Rload;           % Power Delivered to Load=Real power
PF=P/(Vrms*Iorms);         % Input Power Factor=Real power/Apparent power
        % To Display the Results in Command Window
disp(' ****** Results ******')

Average_Value_of_Load_Voltage_in_Volts=Vdc
Average_Value_of_Load_Current_in_Amperes=Idc
Average_Value_of_Thyristor_Current_in_Amperes=Idc/2
RMS_Value_of_Load_Voltage_in_Volts=Vorms
RMS_Value_of_Load_Current_in_Amperes=Iorms
Power_delivered_to_the_Load_in_Watts=P
Input_Power_Factor=PF

        % TO plot the waveforms in a single Window
subplot(4,1,1)
plot(tt,Vs)
grid on;
title('Supply Voltage')

subplot(4,1,2)
plot(t,vo)
grid on;
title('Load Voltage')

subplot(4,1,3)
plot(t,io)
grid on;
title('Load Current')
```

subplot(4,1,4)
plot(t,Vthy)
grid on;
xlabel('Time in Seconds');
title('Voltage Across Thyristor T1 or T2')

D5.3 EXECUTION OF MATLAB CODE

Click on the *play* button to compile the program. If any error messages are displayed in the command window, make necessary corrections and compile the program again. Enter the relevant input parameters in the *command window* while running the program.

The sample input parameters are given below and the corresponding waveforms are shown in the following figure.

Command Window

Enter R.M.S value of the Supply voltage, Vrms = 24
Enter Load Resistance value = 100
Enter Firing Angle of the Thyristor in degree = 60
****** Results ******
Average_Value_of_Load_Voltage_in_Volts = 16.2057
Average_Value_of_Load_Current_in_Amperes = 0.1621
Average_Value_of_Thyristor_Current_in_Amperes = 0.0810
RMS_Value_of_Load_Voltage_in_Volts = 21.5236
RMS_Value_of_Load_Current_in_Amperes = 0.2152
Power_delivered_to_the_Load_in_Watts = 4.6327
Input_Power_Factor = 0.8968.

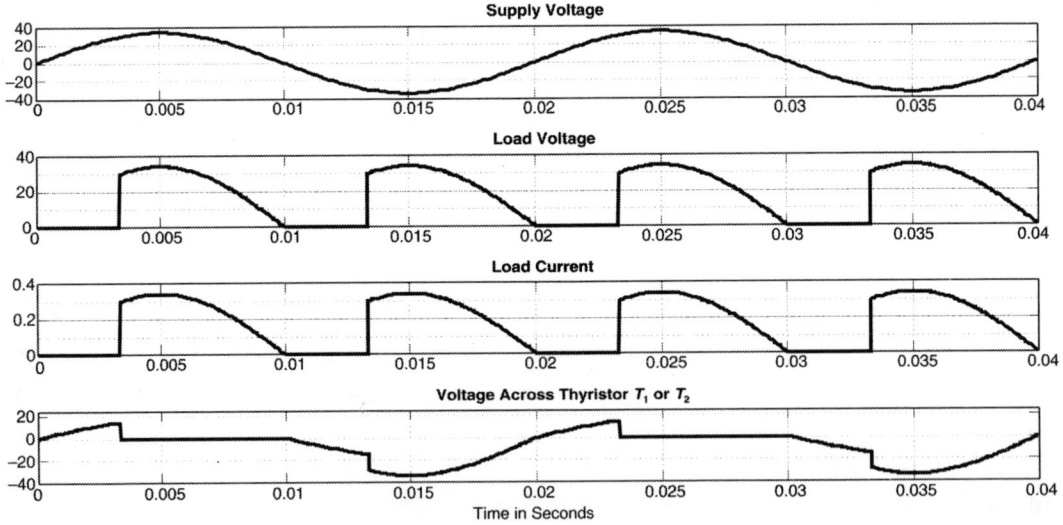

Waveforms: Single phase full-wave fully controlled bridge rectifier with R load using MATLAB program

Experiment D6

Simulation of a Basic Step Down DC–DC Chopper Using MATLAB Program

D6.1 OBJECTIVE

To simulate a basic DC–DC step down converter using MATLAB program.

D6.2 MATLAB CODE

```
% MATLAB code for a Basic DC–DC step down chopper
clc;
clear all;
Vs=input('Enter the value of DC Supply Voltage in Volts, Vs=');
                                        % Input DC supply voltage, Vs
R=input('Enter Load Resistance value in Ohm =');
Fs=input('Enter switching frequency of the step down chopper in Hz =');
Ts = 1/Fs;          % Switching or chopping Period,Ts = 1/Fs
D = input ('Enter the value of duty ratio D (Range 0 to 1) = ');
                                        % Duty Ratio,D=ton/Ts
ton = D*Ts;                  % ton= ON Time of the switch
t1 = 0:Ts/400:ton;           % calculation of time divisions for three cycles
t2 = ton:Ts/400:Ts;
t3 = Ts:Ts/400:Ts+ton;
t4 = Ts+ton:Ts/400:2*Ts;
t5 = 2*Ts:Ts/400:2*Ts+ton;
t6 = 2*Ts+ton:Ts/400:3*Ts;
t = [t1 t2 t3 t4 t5 t6]; % set time divisions in seconds for three cycles
vo = [Vs*ones(size(t1)) zeros(size(t2)) Vs*ones(size(t3)) zeros(size(t4)) Vs*ones(size(t5))
    zeros(size(t6))];
                    % instantaneous values of output voltage for three cycles
io=vo/R;            % Instantaneous values of load current
Vsw=Vs-vo;          % Instantaneous values of voltage across switch
Vdc = D*Vs;     % Average value of output or load voltage of step down chopper
Idc = Vdc/R;    % Average value of load current of step down chopper
```

273

```
% To Display the Results in Command Window
disp(' ****** Results ******')

Average_Value_of_Load_Voltage=Vdc
Average_value_of_Load_Current=Idc

    % TO plot the waveforms in a single Window
subplot(5,1,1)
Vs_plot = [Vs*ones(size(t))];        % To Plot source voltage
plot(t,Vs_plot)
grid on;
title('Source voltage')

subplot(5,1,2)
Vdc_plot = [Vdc*ones(size(t))];     % To Plot Load voltage
plot(t,Vdc_plot)
grid on;
title('Average Load voltage')

subplot(5,1,3)
plot(t,vo)
grid on;
title('Instantaneous Load Voltage')

subplot(5,1,4)
plot(t,io)
grid on;
title('Instantaneous Load Current')

subplot(5,1,5)
plot(t,Vsw)
grid on;
xlabel('Time in Seconds');
title('Instantaneous Switch Voltage')
```

D6.3 EXECUTION OF MATLAB CODE

Click on the play button to compile the program. If any error messages are displayed in the command window, make necessary corrections and compile the program again. Enter the relevant input parameters in the command window while running the program.

The sample input parameters are given below and the corresponding waveforms are shown in the following figure.

Command Window

Enter the value of DC supply voltage in volts, Vs = 12
Enter load resistance value in Ohm = 100

Enter switching frequency of the step down chopper in Hz =1e3
Enter the value of duty ratio D (Range 0 to 1) = .4

****** Results ******

Average_Value_of_Load_Voltage = 4.8000
Average_value_of_Load_Current = 0.0480

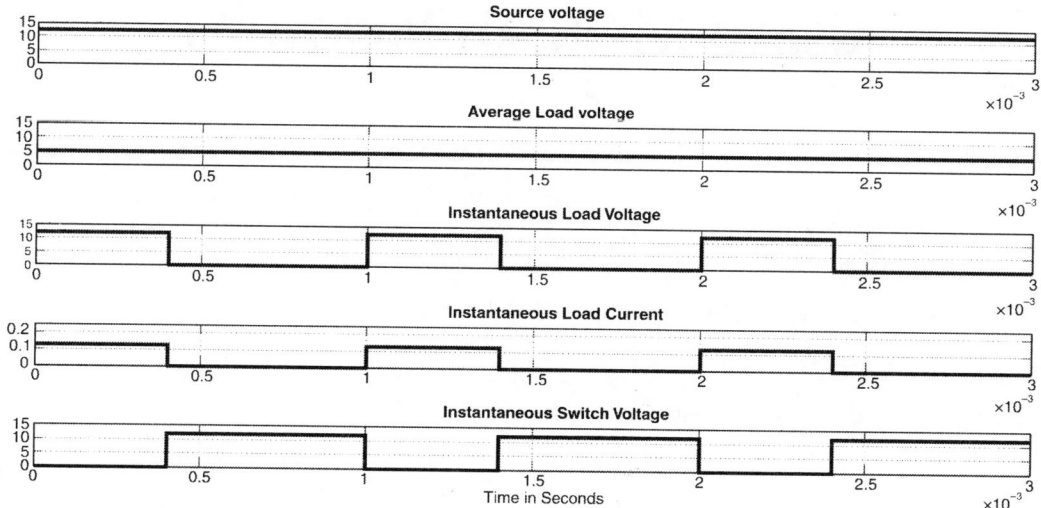

Waveforms: A Basic DC–DC step down chopper using MATLAB program

Experiment D7

Simulation of a Buck Regulator Using MATLAB Program

D7.1 OBJECTIVE

To simulate a buck regulator using MATLAB program.

D7.2 MATLAB CODE

```
% MATLAB code for buck regulator or converter
clc;
clear all;
Vs=input('Enter the value of DC Source Voltage in Volts, Vs=');
Vdc=input('Enter required value of DC output Voltage in Volts, Vdc=');
R=input('Enter Load Resistance value in Ohm =');
Fs=input('Enter switching frequency of the Buck regulator in Hz =');
Ts = 1/Fs;          % Switching or chopping Period, Ts = 1/Fs

D=Vdc/Vs;           % Duty Ratio,D= Vdc/Vs)
                    % Average value of Output or Load voltage, Vdc = Vs*D
Vdc = D*Vs;                  % Average value of Output or Load voltage
Io = Vdc/R;                  % Average value of Load current
ton = D*Ts;                  % ON time of the Switch (Duty Ratio,D= ton/Ts)

Lmin=R*(1-D)/(2*Fs);         % Min value of Inductor
L=Lmin+.25*Lmin;             % Inductance value for Contionus Load Current
                             % Select L, 25% greater than Lmin

dI=Vdc*(1-D)/(L*Fs);         % Change in inductor current (Peak to Peak Value)
I2=Io+(dI/2);                % Maximum value of inductor current
                             % In Buck converter, average value of inductor
                             % current, IL= Average value of load current,Io

I1=Io-(dI/2);                % Minimum value of inductor current
dVdc=.01*Vdc;                % delta Vdc, dVdc = Peak to Peak ripple voltage or
                             % output voltage ripple
                             % Assuming dVdc= 1% of output voltage
C=Vdc*(1-D)/(8*L*Fs*Fs*dVdc);        % Value of Filter capacitance C
         % Selection of C based on standard capacitor value table
```

```
t1 = 0:Ts/400:ton;                % calculation of time divisions for two cycles
t2 = ton:Ts/400:Ts;
t3 = Ts:Ts/400:Ts+ton;
t4 = Ts+ton:Ts/400:2*Ts;

t = [t1 t2 t3 t4];                % set time divisions in seconds for two cycles
       % To plot Inductor current waveform
d=Ts/400;         % Minimum Time step for the change in inductor current
iL1=I1;                  % Minimum value of inductor current
S1=(Vs-Vdc)/L;           % Slope of Inductor Current during ON time (di/dt)
i=1;

for cycle = 1:2    % "For Loop" to plot two cycles of output waveforms
   i1=0;
   for D1=0:Ts/400:ton         % For loop -To Plot Inductor current during Ton
     d1=d*i1;                   % time step for integration
     iL(i)=iL1+(S1*d1);         % Instantaneous inductor current, during ON time
                                %, stored in matrix iL

     i1=i1+1;
     i=i+1;
   end

     iL2=iL(i-1);               % Maximum value of Inductor current
     S2=Vdc/L;                  % Slope of inductor current during OFF time
     i1=0;

   for D1=ton:Ts/400:Ts        % For loop -To Plot Inductor current during Toff
     d1=d*i1;                   % time step for integration
     iL(i)=iL2-(S2*d1);         % Instantaneous inductor current, during OFF time
                                %, stored in matrix,iL

     i=i+1;
     i1=i1+1;
   end

end

Ic=iL-Io;   % Instantaneous Capacitor Current=Inductor Current-Load Current
VL = [(Vs-Vdc)*ones(size(t1)) -Vdc*ones(size(t2)) (Vs-Vdc)*ones(size(t3)) -
      Vdc*ones(size(t4))];
                                        % Instantaneous Inductor Voltage
Vload = [Vdc*ones(size(t))];        % Instantaneous Load voltage
       % To Display the Results in Command Window
   disp(' ****** Results ******')
Duty_Ratio_of_the_Buck_Regulator= D
Average_value_of_Load_Current_in_Amperes=Io
Value_of_Inducance_for_contionus_Load_Current_in_Henry=L
```

```
Peak_to_Peak_Inductor_Current_in_Amperes=dI
Maximum_value_of_Inductor_Current_in_Amperes=I2
Minimum_value_of_Inductor_Current_in_Amperes=I1
Value_of_Filter_Capacitor_in_Farad=C

        %To Plot the waveforms in a single Window
subplot(4,1,1)
plot(t,VL)
grid on;
title('Voltage Across Inductor')

subplot(4,1,2)
plot(t,iL)
grid on;
title('Inductor  Current')

subplot(4,1,3)
plot(t,Ic)
grid on;
title('Capacitor  Current')

subplot(4,1,4)
plot(t,Vload)
grid on;
xlabel('Time in Seconds');
title('Load voltage')
```

D7.3 EXECUTION OF MATLAB CODE

Click on the *play* button to compile the program. If any error messages are displayed in the command window, make necessary corrections and compile the program again. Enter the relevant input parameters in the *command window* while running the program.

The sample input parameters are given below and the corresponding waveforms are shown in the following figure.

Command Window

Enter the value of DC Source voltage in volts, Vs = 12
Enter required value of DC output voltage in volts, Vdc = 9
Enter Load Resistance value in Ohm =100
Enter switching frequency of the Buck regulator in Hz = 1e3

****** Results ******

Duty_Ratio_of_the_Buck_Regulator = 0.7500
Average_value_of_Load_Current_in_Amperes = 0.0900
Value_of_Inducance_for_contionus_Load_Current_in_Henry = 0.0156
Peak_to_Peak_Inductor_Current_in_Amperes = 0.1440
Maximum_value_of_Inductor_Current_in_Amperes = 0.1620
Minimum_value_of_Inductor_Current_in_Amperes = 0.0180
Value_of_Filter_Capacitor_in_Farad = 2.0000e-004

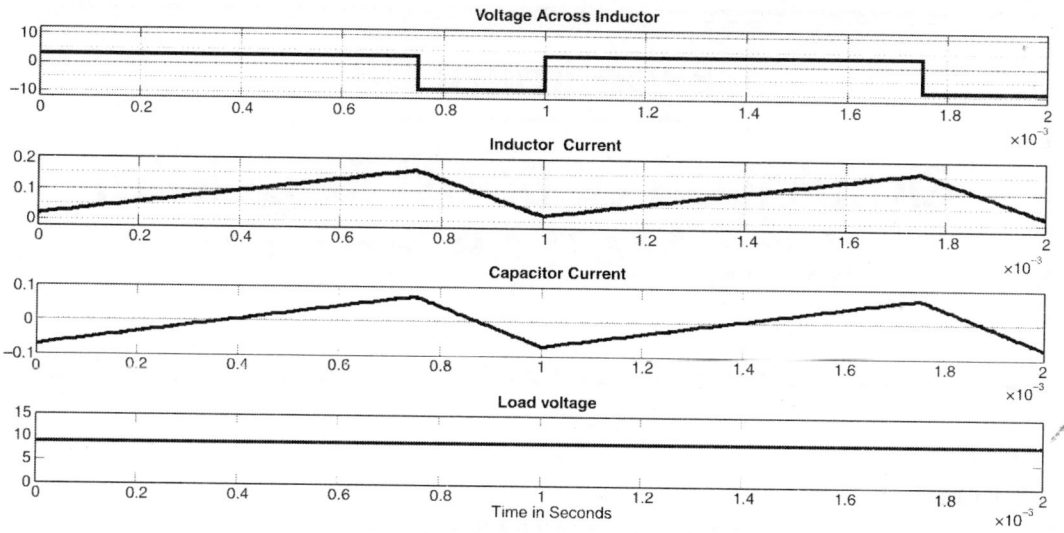

Waveforms: Buck regulator using MATLAB program

Experiment D8

Simulation of Boost Regulator Using MATLAB Program

D8.1 OBJECTIVE

To simulate a boost regulator using MATLAB program.

D8.2 MATLAB CODE

```
% MATLAB code for boost regulator or converter
clc;
clear all;
Vs=input('Enter the value of DC Source Voltage in Volts, Vs=');
Vdc=input('Enter required value of DC output Voltage in Volts, Vdc=');
R=input('Enter Load Resistance value in Ohm =');
Fs=input('Enter switching frequency of the Boost regulator in Hz =');
Ts = 1/Fs;                  % Switching or chopping Period,Ts = 1/Fs
D=1-(Vs/Vdc);        % Duty Ratio, D= 1-(Vs/Vdc)
                     % Average value of Output or Load voltage, Vdc = Vs/(1-D)
Io = Vdc/R;                 % Average value of Load current
ton = D*Ts;                 % ON time of the Switch (Duty Ratio, D= ton/Ts)
Lmin=R*D*((1-D)^2)/(2*Fs);  % Minimum value of Inductor
L=Lmin+.25*Lmin;            % Inductance value for Contionus Inductor Current
                            % Select L, 25% greater than Lmin
dI=Vdc*D*(1-D)/(L*Fs);      % Change in inductor current (Peak to Peak Value)
IL=Io/(1-D);                % Average value of inductor current
I2=IL+(dI/2);               % Maximum value of inductor current
I1=IL-(dI/2);               % Minimum value of inductor current
dVdc=.01*Vdc;               % delta Vdc, dVdc = Peak to Peak ripple voltage or
                            % output voltage ripple
                            % Assuming dVdc= 1% of output voltage
C=Io*D/(dVdc*Fs);           % Value of Filter capacitance C. Actual Selection
                            % of C based on Standard Capacitor value Table
```

```
t1 = 0:Ts/400:ton;               % calculation of time divisions for two cycles
t2 = ton:Ts/400:Ts;
t3 = Ts:Ts/400:Ts+ton;
t4 = Ts+ton:Ts/400:2*Ts;
t = [t1 t2 t3 t4];               % set time divisions in seconds for two cycles
               % To plot Inductor, Diode and Capacitor current waveforms
d=Ts/400;                % Minimum Time step for the change in inductor current
iL1=I1;                  % Minimum value of inductor current
S1=Vs/L;                 % Slope of Inductor Current during ON time (.i.e., di/dt)
i=1;

for cycle = 1:2          % "For Loop" to plot two cycles of waveforms
   i1=0;
    for D1=0:Ts/400:ton  % For loop- To Plot Inductor current during Ton
      d1=d*i1;           % Time step for integration
      iL(i)=iL1+(S1*d1); % Instantaneous inductor current, during ON time
                         %, Stored in matrix iL
        iD(i)=0;              % To plot Instantaneous diode current,
                             %  iD, during ON time
      i1=i1+1;
      i=i+1;
    end
      iL2=iL(i-1);           % Maximum value of Inductor current
      S2=-(Vs-Vdc)/L;        % Slope of inductor current during OFF time
      i1=0;
    for D1=ton:Ts/400:Ts     % For loop- To Plot Inductor current during Toff
      d1=d*i1;               % Time step for integration
      iL(i)=iL2-(S2*d1);     % Instantaneous inductor current,during OFF time
                             %, Stored in matrix, iL
        iD(i)=iL(i);               % To plot Diode Current iD during OFF time
      i=i+1;
      i1=i1+1;
    end
  end
end
iC=iD-Io;              % To plot Instantaneous Capacitor Current iC
VL = [Vs*ones(size(t1)) (Vs-Vdc)*ones(size(t2)) Vs*ones(size(t3))
     (Vs-Vdc)*ones(size(t4))];
                                 % Instantaneous Inductor Voltage
Vload = [Vdc*ones(size(t))];     % Instantaneous Load voltage
       % To Display the Results in Command Window
  disp(' ****** Results ******')
Duty_Ratio_of_the_Boost_Regulator= D
Average_value_of_Load_Current_in_Amperes=Io
```

Value_of_Inducance_for_contionus_inductor_Current_in_Henry=L
Average_value_of_Inductor_Current_in_Amperes=IL
Peak_to_Peak_Inductor_Current_in_Amperes=dI
Maximum_value_of_Inductor_Current_in_Amperes=I2
Minimum_value_of_Inductor_Current_in_Amperes=I1
Value_of_Filter_Capacitor_in_Farad=C

```
        %To Plot the waveforms in a single Window
subplot(5,1,1)
plot(t,VL)
grid on;
title('Voltage Across Inductor')

subplot(5,1,2)
plot(t,iL)
grid on;
title('Inductor  Current')

subplot(5,1,3)
plot(t,iD)
grid on;
title('Diode Current')

subplot(5,1,4)
plot(t,iC)
grid on;
title('Capacitor Current')

subplot(5,1,5)
plot(t,Vload)
grid on;
xlabel('Time in Seconds');
title('Load Voltage')
```

D8.3 EXECUTION OF MATLAB CODE

Click on the *play* button to compile the program. If any error messages are displayed in the command window, make necessary corrections and compile the program again. Enter the relevant input parameters in the *command window* while running the program.

The sample input parameters are given below and the corresponding waveforms are shown in the following figure.

Command Window

Enter the value of DC source voltage in volts, Vs = 12
Enter required value of DC output Voltage in volts, Vdc = 20
Enter Load Resistance value in Ohm =100
Enter switching frequency of the Boost regulator in Hz =1e3

****** Results ******

Duty_Ratio_of_the_Boost_Regulator = 0.4000

Average_value_of_Load_Current_in_Amperes = 0.2000

Value_of_Inducance_for_contionus_inductor_Current_in_Henry = 0.0090

Average_value_of_Inductor_Current_in_Amperes = 0.3333

Peak_to_Peak_Inductor_Current_in_Amperes = 0.5333

Maximum_value_of_Inductor_Current_in_Amperes = 0.6000

Minimum_value_of_Inductor_Current_in_Amperes = 0.0667

Value_of_Filter_Capacitor_in_Farad = 4.0000e-004

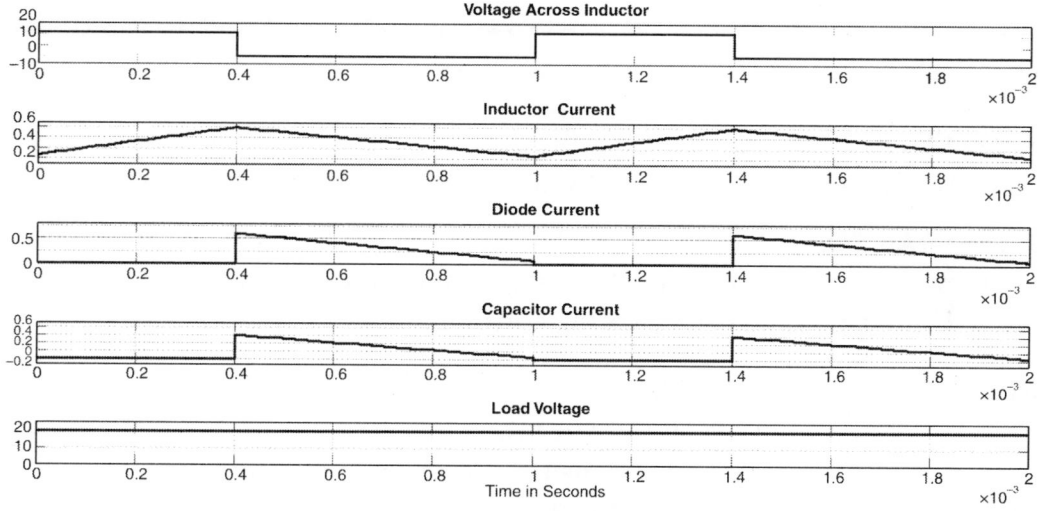

Waveforms: Boost regulator using MATLAB program

Experiment D9

Simulation of an AC Voltage Regulator Using MATLAB Program

D9.1 OBJECTIVE

To simulate an AC voltage regulator feeding resistive load using MATLAB program.

D9.2 MATLAB CODE

```
% MATLAB code for single phase AC voltage regulator with two
        % anti-parallel thyristors feeding R load
clc;
clear all;
    % Generation of source or supply voltage
f = 50;                         % frequency of the supply voltage
T=1/f;                          % Time period of the supply voltage
tt = 0:T/200:2*T;               % set time axis limits from 0 to 2T in
                                % steps of T/200 to plot 2 cycles of Vs
Vrms=input( 'Enter R.M.S value of the Supply Voltage in Volts, Vrms=');
Vm=Vrms*sqrt(2);                % Peak or Maximum value of supply voltage
Vs = Vm*sin(2*pi*50*tt);        % Instantaneous values of supply voltage, Vs
    % Code for single phase AC voltage regualtor with R load
Rload=input('Enter Load Resistance value in Ohm =');
alpha = input('Enter Firing Angle of the Thyristor T1 in degree =');
                    % Firing angle of the Thyristor T2, AlphaT2= 180+ AlphaT1
alpha_rad=alpha*pi/180;         % firing angle in radians
td1 = pi/256;                   % increment time step in radians
t0 = 0:td1:alpha_rad;           % calculation of time divisions for two cycles
t1 = alpha_rad:td1:pi;
t2 = pi:td1:pi+alpha_rad;
t3 = pi+alpha_rad:td1:2*pi;
t4 = 2*pi:td1:2*pi+alpha_rad;
t5 = 2*pi+alpha_rad:td1:3*pi;
t6 = 3*pi:td1:3*pi+alpha_rad;
t7 = 3*pi+alpha_rad:td1:4*pi;
```

```
t_rad=[t0 t1 t2 t3 t4 t5 t6 t7];     % time divisions in radian for two cycles
t=t_rad/(2*pi*f);                    % time divisions in seconds for two cycles
                                     % t= t_rad *(180/pi); for time division in degrees
vo = [0*sin(t0) Vm*sin(t1) 0*sin(t2) Vm*sin(t3) 0*sin(t4) Vm*sin(t5) 0*sin(t6) Vm*sin(t7)];
                                     % Instantaneous values of output voltage
io = vo/Rload;                       % Instantaneous values of output current
Vthy =  [Vm*sin(t0) 0*Vm*sin(t1) Vm*sin(t2) 0*Vm*sin(t3) Vm*sin(t4) 0*Vm*sin(t5)
         Vm*sin(t6) 0*Vm*sin(t7)];
                                     % Instantaneous values of voltage across Thyristor T1
         % Calculation of  RMS value , Real power and Power Factor
Vorms = sqrt(mean(vo.*vo));     % RMS value of Load or output voltage
                                % Average Value of Load voltage=0
Iorms=Vorms/Rload;              % RMS value of Load Current
P=Iorms^2*Rload;                % Power Delivered to Load=Real power
PF=P/(Vrms*Iorms);              % Input Power Factor= Real power/Apparent power
IT_dc = (Vm/(2*pi*Rload))*(1 + cos(alpha_rad));
         % Average value of Thyristor current=(Vm/(2*p*iR))*(1+ cos(alpha))
         % Thyristor Current, IT1 = Thyristor Current, IT2
IT_rms = Iorms/sqrt(2);
         %RMS value of thyristor current = RMS value of load current/sqrt(2)
     % To Display the Results in Command Window
  disp(' ****** Results ******')
RMS_Value_of_Load_Voltage_in_Volts=Vorms
RMS_Value_of_Load_Current_in_Amperes=Iorms
Power_delivered_to_the_Load_in_Watts=P
Input_Power_Factor=PF
Average_value_of_Thyristor_Current_in_Amperes=IT_dc
RMS_value_of_Thyristor_Current_in_Amperes=IT_rms
     % To plot the waveforms in a single window
subplot(4,1,1)
plot(tt,Vs)
grid on;
title('Supply Voltage')
subplot(4,1,2)
plot(t,vo)
grid on;
title('Load Voltage')
subplot(4,1,3)
plot(t,io)
grid on;
title('Load Current')
```

subplot(4,1,4)
plot(t,Vthy)
grid on;
xlabel('Time in Seconds');
title('Voltage Across Thyristor T1')

D9.3 EXECUTION OF MATLAB CODE

Click on the *play* button to compile the program. If any error messages are displayed in the command window, make necessary corrections and compile the program again. Enter the relevant input parameters in the *command window* while running the program.

The sample input parameters are given below and the corresponding waveforms are shown in the following figure.

Command Window

Enter R.M.S value of the Supply Voltage in Volts, Vrms = 230
Enter Load Resistance value in Ohm = 5
Enter Firing Angle of the Thyristor T_1 in degree = 45

****** Results ******
RMS_Value_of_Load_Voltage_in_Volts = 218.6860
RMS_Value_of_Load_Current_in_Amperes = 43.7372
Power_delivered_to_the_Load_in_Watts = 9.5647e+003
Input_Power_Factor = 0.9508
Average_value_of_Thyristor_Current_in_Amperes = 17.6748
RMS_value_of_Thyristor_Current_in_Amperes = 30.9269

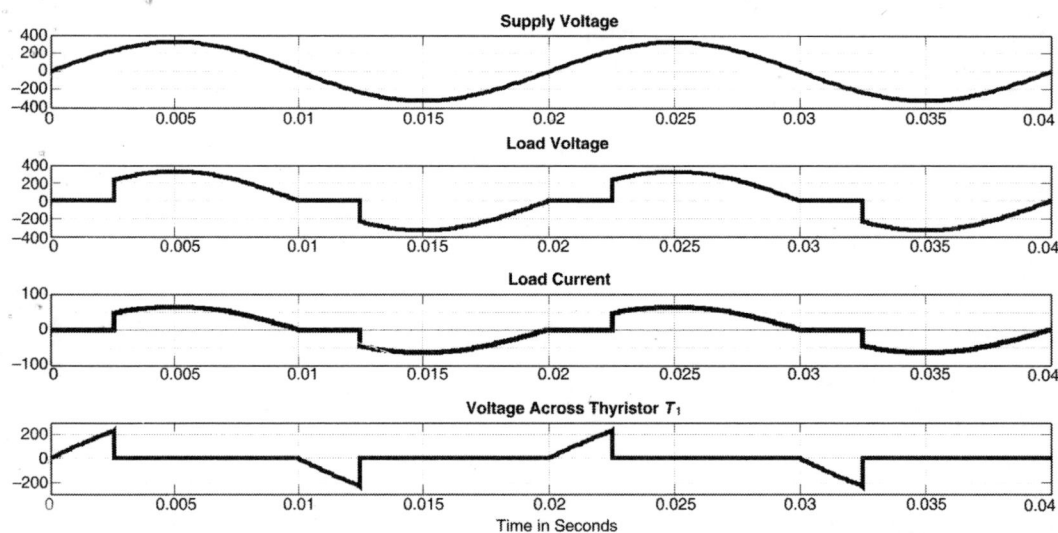

Waveforms: AC voltage regulator with R load using MATLAB program

Simulation of a Single Phase Bridge Inverter Using MATLAB Program

D10.1 OBJECTIVE

To simulate a single phase bridge inverter feeding resistive load using MATLAB program.

D10.2 MATLAB CODE

```
% MATLAB code for single phase bridge inverter feeding R load
        % to produce a square wave AC output voltage waveform
clc;
clear all;
Vs=input('Enter the value of D.C Source Voltage in Volts, Vs=');
R=input('Enter Load Resistance value in Ohm =');
Fs=input('Enter Load Voltage Frequency in Hz =');
        % For single phase square wave Inverter,
        % output or load frequency = switching frequency of the switches
n=input('Enter the order of Harmonics to be analyzed,n =');
Ts = 1/Fs;                      % Switching Period,Ts = 1/Fs
t1 = 0:Ts/400:Ts/2;         % calculation of time divisions for two cycles
t2 = Ts/2:Ts/400:Ts;
t3 = Ts:Ts/400:Ts+Ts/2;
t4 = Ts+Ts/2:Ts/400:2*Ts;
t = [t1 t2 t3 t4];      % set time divisions in seconds for two cycles
V_RMS=Vs;               % RMS value of output or load voltage waveform=supply
                          voltage Vs
V_1RMS=0.9*Vs;          % RMS value of Fundamental component of load voltage
                        % V_1RMS=(2*sqrt(2)*Vs)/pi =0.9*Vs
I_1RMS=0.9*Vs/R;  % RMS value of Fundamental component of load current
P=V_RMS^2/R;      % Output power of inverter
                  % To determine THD, HF and DF of output voltage waveform
THD=(sqrt(V_RMS^2-V_1RMS^2)/V_1RMS)*100;  %THD of output voltage waveform
```

```
HF=1/n;            %Harmonic Factor for nth harmonic voltage,HF= Vn/V1 , where
                   %Vn=RMS value of nth harmonic voltage=(2*sqrt(2)*Vs)/(n*pi)
                   %V1= RMS value of Fundamental Voltage=(2*sqrt(2)*Vs)/(1*pi)
                   % Therfore, HF=1/n
DF=1/(n^2*n);      % Distortion Factor for nth harmonic voltage,DF=Vn/(n^2*V1)
                   % since Vn/V1 is HF, DF=1/(n^2*n)
HF_3rdOrder=1/3; % Harmonic Factor for Lowest Order (3rd) Harmonic Voltage
V_3RMS=0.9*Vs/3; % RMS value of Lowest Order(3rd)Harmonics of load Voltage
% To plot Source Voltage, Load Voltage, Load Current and
% Fundamental component of Load Voltage waveforms
V_Load = [Vs*ones(size(t1)) -Vs*ones(size(t2)) Vs*ones(size(t3)) -Vs*ones(size(t4))];
                   % Instantaneous Load Voltage
I_Load=V_Load/R;   % Instantaneous Load Current
V_Source = [Vs*ones(size(t))]; % Instantaneous source voltage
Vm_1=V_1RMS*sqrt(2); % Maximum value of Fundamental output voltage
Vs1 = Vm_1*sin(2*pi*Fs*t); % Instantaneous values of Fundamental voltage
          % To Display the Results in Command Window
disp(' ****** Results ******')
RMS_Value_of_Load_Voltage_in_Volts=V_RMS
RMS_Value_of_Fundamental_Component_of_Load_Voltage_in_Volts=V_1RMS
RMS_Value_of_Fundamental_Component_of_Load_Current_in_Amperes=I_1RMS
Ouput_Power_in_Watts=P
Order_of_Harmonics_n_for_which_HF_and_DF_to_be_determined=n
Harmonic_Factor_for_nth_Harmonic_Voltage=HF
Distortion_Factor_for_nth_Harmonic_Voltage=DF
Total_Harmonic_Distortion_of_Load_Voltage_in_Percentage=THD
Harmonic_Factor_of_Third_voltage_Harmic=HF_3rdOrder
RMS_value_of_3rd_harmonic_component_of_Output_Voltage_in_Volts=V_3RMS
        %To Plot the waveforms in a single Window
subplot(4,1,1)
plot(t,V_Source)
grid on;
title('Source Voltage')
subplot(4,1,2)
plot(t,V_Load)
grid on;
title('Load Voltage')
subplot(4,1,3)
plot(t,I_Load)
grid on;
title('Load Current')
```

```
subplot(4,1,4)
plot(t,Vs1)
grid on;
xlabel('Time in Seconds');
title('Fundamental Component of Output Voltage')
```

D10.3 EXECUTION OF MATLAB CODE

Click on the *play* button to compile the program. If any error messages are displayed in the command window, make necessary corrections and compile the program again. Enter the relevant input parameters in the *command window* while running the program.

The sample input parameters are given below and the corresponding waveforms are shown in the following figure.

Command Window

Enter the value of DC Source Voltage in Volts, Vs = 100
Enter Load Resistance value in Ohm = 100
Enter Load Voltage Frequency in Hz = 50
Enter the order of Harmonics to be analyzed, n =5

****** Results ******

RMS_Value_of_Load_Voltage_in_Volts = 100
RMS_Value_of_Fundamental_Component_of_Load_Voltage_in_Volts = 90
RMS_Value_of_Fundamental_Component_of_Load_Current_in_Amperes = 0.9000
Ouput_Power_in_Watts = 100
Order_of_Harmonics_n_for_which_HF_and_DF_to_be_determined = 5
Harmonic_Factor_for_nth_Harmonic_Voltage = 0.2000
Distortion_Factor_for_nth_Harmonic_Voltage = 0.0080
Total_Harmonic_Distortion_of_Load_Voltage_in_Percentage = 48.4322
Harmonic_Factor_of_Third_Voltage_Harmic = 0.3333
RMS_value_of_3rd_Harmonic_component_of_Output_Voltage_in_Volts = 30

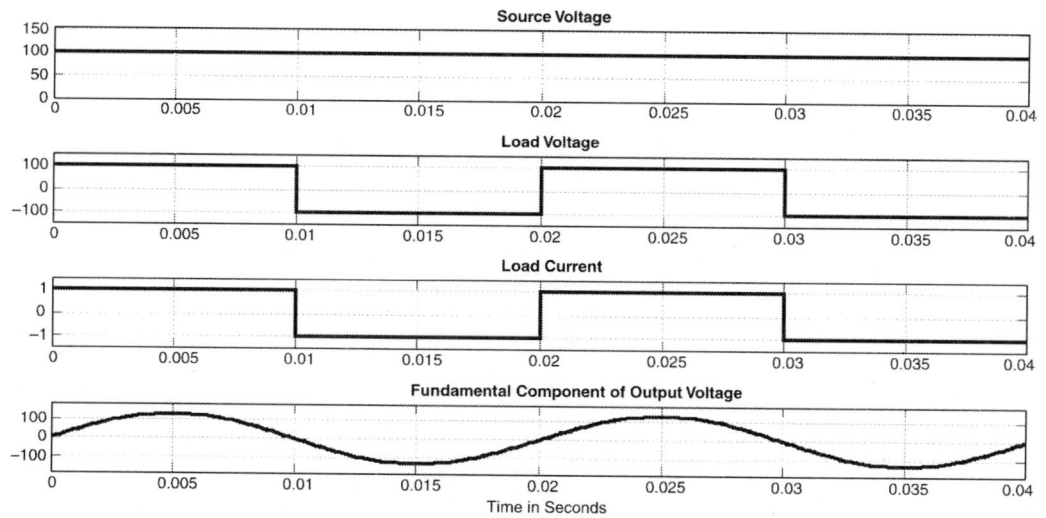

Waveforms: Single phase bridge inverter using MATLAB program

Section   E

Experiment E1

Viva Voce Questions

CHARACTERISTICS OF DEVICES

Thyristor

1. Explain the basic silicon structure of a thyristor.
2. Define latching current of a thyristor.
3. Define holding current of a thyristor.
4. What is meant by forward blocking mode of a thyristor?
5. What is meant by forward conducting mode of a thyristor?
6. What is meant by reverse blocking mode of a thyristor?
7. What is meant by triggering of a thyristor?
8. Explain why the gate triggering signal is inactive after the thyristor is turned into conduction?
9. Why a thyristor is called a latching device?
10. Why a thyristor is called a current controlled device?
11. What is (di/dt) effect in a thyristor?
12. What is (dv/dt) effect in a thyristor?
13. What is a snubber circuit?
14. Define turn OFF time of a thyristor.
15. Why a thyristor is not suitable for high frequency switching applications?
16. Can a thyristor block both forward and reverse voltages?
17. How is a thyristor turned OFF while working in AC circuits?
18. How is a thyristor turned OFF while working in DC circuits?
19. Is thyristor a fully controlled device?
20. Is a thyristor a bidirectional device?
21. How can a thyristor be used as a semiconductor switch?
22. Define the rate of rise of anode voltage. Is it a performance parameter of a thyristor?

Triac

1. Explain the basic silicon structure of a triac.
2. Is the device triac a member of thyristor family?
3. Is it possible to represent electrically a triac using thyristors?
4. Is it a unidirectional or bidirectional device?
5. Can it block voltages in both directions?

6. Can it be triggered using either a positive or negative current at the gate terminal?
7. In which mode of operation the device exhibits forward characteristics?
8. In which mode of operation the device exhibits reverse characteristics?
9. Is the triac a latching device?
10. Can a triac be operated in both positive and negative half cycles of the AC supply voltage?
11. What is the problem of using a triac with large inductive loads?
12. Why a triac is not suitable for high frequency switching applications?
13. Mention any three features of a triac.
14. Mention any three applications of a triac.
15. Compare a triac with a thyristor.

BJT

1. Explain the basic silicon structure of a power BJT.
2. How does a power BJT different from a signal transistor?
3. Why an npn BJT is more commonly used in power electronics circuits?
4. Is a BJT a current controlled device or voltage controlled device?
5. Why silicon based devices are preferred to germanium based devices?
6. Define the current gain of a BJT.
7. What is meant by input characteristics of power BJT?
8. What is meant by output characteristics of power BJT?
9. What is meant by transfer characteristics of power BJT?
10. What is the relation between the ratio of the collector current to emitter current and the ratio of collector current to the base current?
11. Define cut-off region, active region and saturation region in the operation of a BJT.
12. Explain how a BJT can be operated as a power electronic switch.
13. Why it is called a bipolar device?
14. Mention any three features of a power BJT.
15. Mention any three applications of a power BJT.
16. Compare BJT with a thyristor.

MOSFET

1. Explain the basic silicon structure of a power MOSFET.
2. Which are the two different enhancement type MOSFETs?
3. Why an n-channel enhancement MOSFET is more generally used in power applications?
4. Is the MOSFET a unipolar device?
5. Is the MOSFET a current controlled device or a voltage controlled device?
6. For a MOSFET does the gate-source voltage have a threshold value?
7. What is meant by output characteristics of a power MOSFET?
8. What is meant by transfer characteristics of a power MOSFET?
9. How can a MOSFET be operated as an electronic switch?
10. Mention the reason why an n-channel enhancement MOSFET is generally used for low power applications.
11. What is the reason that to trigger a MOSFET, the gate current require is very small?
12. Mention a few features of power MOSFETs.

13. Mention a few applications of a power MOSFET.
14. Compare the operational features of a power MOSFET with that of a thyristor.
15. Compare the operational features of a power MOSFET with that of a triac.

IGBT

1. Explain the basic silicon structure of an IGBT.
2. An IGBT combines the advantages of a BJT and a MOSFET. Justify this statement.
3. What is the phenomenon of latch up in an IGBT?
4. How can an IGBT be used as an electronic switch?
5. What is meant by output characteristics of an IGBT?
6. What is meant by transfer characteristics of an IGBT?
7. Is it a current controlled device or a voltage controlled device?
8. How is an IGBT turned ON?
9. How is an IGBT turned OFF?
10. Compare the switching time of an IGBT with that of a MOSFET.
11. Is the forward voltage blocking capability of an IGBT better than that of a power MOSFET?
12. Is there any threshold value for the gate to emitter voltage?
13. Mention a few features of an IGBT.
14. Is it a latching device? If not is there any possibility of latch up?
15. Why an IGBT has a low switching frequency compared to a MOSFET?
16. Is the IGBT a bipolar device?
17. Name the scientist who discovered the IGBT.

TRIGGERING CIRCUITS

1. Is an AC signal or a DC signal desirable for triggering the gate of a thyristor?
2. What is meant by triggering of a device?
3. What is meant by continuous triggering?
4. What is meant by pulse triggering?
5. Why pulse triggering is preferred to continuous triggering?
6. What is meant by (di/dt) triggering of a thyristor?
7. What is meant by (dv/dt) triggering of a thyristor?
8. What is meant by "light triggering"of a thyristor?
9. What should be the qualities of a triggering current pulse to ensure turn ON a thyristor?
10. Why the gate circuit of a thyristor is to be isolated from the power circuit involving anode and cathode?
11. What is the principle of Resistance triggering?
12. Define the term triggering angle or delay angle for a thyristor.
13. What is the range of delay angle possible in R-triggering method?
14. What is the function of a gate stabilizing resistance for a thyristor?
15. In the R-triggering circuit, why is it necessary to keep a fixed resistance in series with a variable resistance?
16. What is the function of a diode in the gate circuit?
17. Mention the reason why the R-triggering circuit is not generally adopted for triggering a thyristor.

18. What is the principle of RC-triggering?
19. In what way the delay angle can be varied in an RC-triggering circuit?
20. In an RC-triggering circuit, how can the charging time of the capacitor be varied?
21. In an RC-triggering circuit, how can a small delay angle be obtained?
22. In an RC-triggering circuit, how can a large delay angle be obtained?
23. What is the range of delay angle possible in RC-triggering method?
24. How can the RC-triggering circuit, used for triggering a half-wave thyristor rectifier, be modified for using with a full-wave rectifier?
25. What is the approximate range of time constant for which an RC triggering circuit is designed for using with a half-wave thyristor rectifier?
26. What is the approximate range of time constant for which an RC triggering circuit is designed for using with a full-wave thyristor rectifier?
27. What are the advantages of using RC-triggering circuit rather than an R-triggering circuit?
28. Explain the basic silicon structure of a UJT.
29. Define intrinsic stand-off ratio.
30. Define peak voltage.
31. Define valley voltage.
32. Explain the $v - i$ characteristics of a UJT.
33. How does a UJT exhibit negative resistance characteristics?
34. Discuss how triggering pulses are generated by a UJT triggering circuit.
35. What is meant by line synchronized triggering circuit?
36. How is the triggering angle varied in a UJT triggering circuit?
37. In what way the gate circuit can be isolated from the power circuit?
38. What is an opto-coupler?
39. What is an isolation transformer?

RECTIFIERS

1. What is a controlled rectifier?
2. How can a thyristor function as a controlled rectifier?
3. How can the output voltage be varied in thyristor controlled rectifier?
4. In a half-wave rectifier feeding an RL load, how does the thyristor conduct a little more time during the negative half cycle of the AC supply voltage?
5. For a half-wave rectifier feeding a purely resistor, if the delay angle is α what is the conduction time of the thyristor?
6. Define the circuit turn-off time for a half-wave thyristor controlled rectifier.
7. For a half-wave rectifier feeding a purely resistor, what is the expression for the average value of the output voltage?
8. For a half-wave rectifier feeding a purely resistor, define the form factor?
9. For a half-wave rectifier feeding a purely resistor, define the ripple factor?
10. Define the current extinction angle.
11. In a thyristor controlled rectifier, the expression for the instantaneous value of the load current is $i = I_{max} \sin(\omega t - \phi)$, where ϕ is the impedance angle of the load. If the firing angle of the thyristor is $\alpha = \phi$, what is the nature of the load current and what is the conducting time of the thyristor?
12. What are the conditions under which the load current in a thyristor controlled rectifier is continuous?

13. What is a freewheeling diode?
14. Explain the operation of freewheeling diode.
15. What are the advantages of connecting a freewheeling diode in a power electronic circuit?
16. Why a freewheeling diode is called a commutating diode?
17. What are the demerits of a half-wave rectifier?
18. What are the different configurations of a single phase full-wave rectifier?
19. For a single phase full-wave rectifier with the mid-tap configuration, if the supply voltage is $f_S = V_{max} \sin \omega t$, what is the peak inverse voltage that can appear across a thyristor?
20. What is meant by semi-controlled bridge rectifier?
21. What is meant by fully-controlled bridge rectifier?
22. Which are the different configurations of single phase thyristor controlled bridge rectifier?
23. What is the effect of half waving?
24. How can the effect of half waving be prevented?
25. Why the symmetrical connection for a single phase bridge rectifier is unsuitable for high inductive loads?
26. In a symmetrical connection of thyristors for a single phase bridge rectifier, under what conditions the thyristors carry freewheeling current?
27. How does the asymmetrical connection for a single phase bridge rectifier offer an inherent freewheeling action?
28. What are the effects of source inductance in the performance of a thyristor controlled rectifier?
29. What is meant by line commutated converter?
30. What is the performance of a rectifier during the period in which the output voltage is negative and the load current is positive?
31. Why for a high inductive load it customary to connect a freewheeling diode across the load?
32. Why the semi-controlled bridge rectifiers are called single quadrant converters?
33. Why the fully-controlled bridge rectifiers are called two quadrant converters?
34. In a rectifier circuit, a thyristor conducts from $\omega t = \alpha$ to $\omega t = \beta$. What is the circuit turn-off time?
35. What is a two-pulse rectifier?
36. What is meant by commutation overlap?
37. What is the reason for the commutation overlap to occur?
38. What happens during a commutation overlap period?
39. Define the term pulse number in the case of a rectifier.
40. What are the demerits of using mid-tap configuration for a full-wave rectifier?
41. The input power factor of a single phase semi converter is better than that of a full converter. Why?
42. What type of single bridge rectifier is suitable for highly inductive loads?
43. What can be done to make the load current for an RL load?
44. How is the firing angle of one thyristor related to that of the other in a single phase semi-converter?
45. On what all factors the speed of a DC motor depends?
46. What are the different methods of speed control of a DC motor?

47. How is the speed of a Dc motor related to armature voltage and field current?
48. What is meant by field current control?
49. What is meant by armature voltage control?
50. How can a thyristor controlled rectifier be used for the speed control of a DC shunt motor?

AC VOLTAGE CONTROLLERS

1. What is an AC voltage controller?
2. In what way the output voltage in an AC voltage controller is varied?
3. Which are the different methods of output voltage control in an AC voltage regulator?
4. Explain ON-OFF control.
5. What is an integral cycle control?
6. What is phase control?
7. What is meant by a unidirectional voltage controller?
8. What are the demerits of unidirectional controllers?
9. What is a bidirectional voltage controller?
10. What are the advantages of bidirectional voltage controllers?
11. What is the disadvantage of a bidirectional voltage controller?
12. In an AC voltage controller supplying an RL load, firing angle control below the impedance angle of the load is not possible? Explain the reason.
13. What is a cycloconverter?
14. Can a triac be used as an AC voltage controller?
15. Why is difficult to use a triac as an AC voltage controller with high inductive loads?

COMMUTATION CIRCUITS

1. What is meant by commutation of a conducting thyristor?
2. What are the different methods of commutation of a thyristor?
3. What is meant by natural commutation?
4. What is meant by self commutation?
5. What is meant by forced commutation?
6. Under what circumstances a forced commutation circuit is necessary for a thyristor?
7. What is the influence of circuit turn-off time for reliable commutation of a thyristor?
8. What are the conditions to be satisfied for reliable commutation of a thyristor?
9. What are the main functions of a forced commutation circuit?
10. What is the principle of resonance commutation?
11. Under what conditions a reliable commutation takes place in a resonance commutation circuit?
12. In resonance commutation, what must be the relation between the load current and the maximum value of the capacitor current for successful commutation?
13. What is the principle of impulse commutation?
14. What is the difference between the turn-off time of a thyristor and circuit turn-off time?
15. In an impulse commutation circuit, what must be the minimum time for which the main thyristor should conduct?
16. What is complementary commutation?

17. In a complementary commutation circuit, what must be the minimum time for which the main thyristor should conduct?
18. What is meant by current commutation?
19. What is meant by voltage commutation?
20. What is the difference between line-side commutation and load-side commutation?

DC–DC CHOPPERS

1. What is DC–DC chopper?
2. Explain the working of a basic DC–DC chopper, using a power semiconductor switch.
3. What is meant by duty ratio?
4. In a DC–DC chopper how can the output voltage be controlled?
5. Which all power electronic devices can be as switches for a chopper?
6. Explain the principle of a step-down chopper.
7. What is the expression for output voltage in a step-down chopper?
8. Define peak to peak ripple current.
9. How does a large load inductance minimize the ripple current?
10. How does a high switching frequency minimize the ripple current?
11. What is meant by time ratio control (TRC)?
12. Which are the two different types of TRC?
13. Explain the principle of step-up chopper.
14. What is the expression for output voltage in step-up chopper?
15. In a step-down chopper why the input current is discontinuous?
16. What methods are adopted to minimize the ripple current in a chopper?
17. Explain the principle of step-up/down chopper.
18. What is a buck regulator?
19. What is a boost regulator?
20. Mention a few applications of a DC–DC chopper.

INVERTERS

1. What is an inverter?
2. Which are the two different configurations of a single phase bridge inverter?
3. What are the drawbacks of a half bridge inverter?
4. What is meant by a square wave inverter?
5. What are the essential features of a well designed inverter?
6. Distinguish between a voltage source inverter and a current source inverter?
7. Compare the features of a single phase half bridge inverter with that of a full bridge inverter.
8. Which are the two different conduction modes of three phase inverter?
9. Explain the principle of 180° conduction mode of a three phase inverter.
10. Explain the principle of 120° conduction mode of a three phase inverter.
11. Explain the principle of pulse width modulation.
12. Define modulation index.
13. Define frequency modulation ratio.
14. Explain the principle of sinusoidal pulse width modulation.

General

1. What are the advantages of using power electronic devices for switching applications?
2. What are the essential requirements of an ideal switch?
3. What are the characteristics of a power electronic switch that are different from an ideal switch?
4. What is meant by ON-state power loss in a power semiconductor switch?
5. What is meant by OFF-state power loss in a power semiconductor switch?
6. What is meant by switching losses?
7. What are the advantages of power electronic systems?
8. What are the disadvantages of power electronic systems?
9. Explain how harmonics are injected in to the supply systems while using power electronic devices as switches.
10. For a thyristor controlled rectifier, define the term distortion factor?
11. For a thyristor controlled rectifier, define the term displacement factor?
12. Define the input power factor in terms of the distortion factor and the displacement factor.
13. Define the term total harmonic distortion.
14. What is the relation between total harmonic distortion and the distortion factor?
15. Define term form factor.
16. Define term ripple factor.
17. What is the relation between the ripple factor and the form factor?
18. Define the term crest factor.
19. Define the term distortion power.
20. How do the terms apparent power, active power, reactive power and distortion power related?
21. What are the assumptions made in analyzing a power electronic circuit?

CHARACTERISTICS OF DEVICES

A1. Static characteristics of a thyristor

1. Set up an experiment to plot static V-I characteristics of the given device TYN 616. Find out minimum gate voltage and gate current to turn ON the device.
2. What is meant by latching current and holding current of a device? Set up an experiment to obtain these values practically, given the device TYN 616. Also plot the forward characteristics.
3. Design and set up a circuit to plot static V-I characteristics of the given thyristor TYN616. What is the minimum value of gate voltage and gate current to turn ON the device TYN 616?

A2. Static characteristics of a triac

4. Set up an experiment to plot static V-I characteristics of given device BT 136.
5. Design and set up a circuit to plot forward and reverse characteristics of the device BT 136.
6. Set up an experiment to plot static V-I characteristics of the device which can be operated in both first and third quadrant.

A3. Static characteristics of a BJT

7. Set up an experiment to find the input characteristics of the given device 2N3055.
8. Design and set up a circuit to plot input and output characteristics of 2N3055.
9. Design and set up a circuit to plot collector emitter voltage versus collector current for a constant base current for the given device 2N3055.

A4. Static characteristics of a MOSFET

10. Set up an experiment to find threshold voltage of IRF 740. Also plot its transfer characteristics.
11. Design and set up a circuit to plot transfer characteristics and output characteristics of IRF 740.
12. Set up an experiment to plot the transfer and output characteristics of the given MOSFET.

A5. Static characteristics of an IGBT

13. Set up an experiment to find threshold voltage of CT60 and to plot its transfer characteristics.

14. Design and set up a circuit to plot the transfer characteristics and the output characteristics of CT60.

15. Set up an experiment to plot transfer and output characteristics of the given IGBT.

DESIGN AND TESTING

B1. R triggering

16. Design and set up a firing circuit that trigger a thyristor with a maximum possible firing angle of 90°. Derive the RMS and average value of output voltage and pre determine the values for the firing angle of 45°. Obtain and plot the following waveforms at a firing angle α = 45°.
    (a) Source voltage (b) load voltage and (c) voltage across the thyristor. Use an isolation transformer 230/24 V.

17. Design a triggering circuit for a thyristor to obtain the given voltage waveform. Consider the firing angle α = 30°. The triggering circuit should not have any capacitor. Use a 230/24 V transformer for isolation.

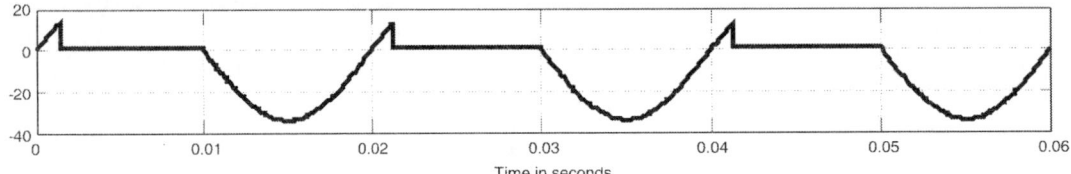

18. Design and set up a half-wave controlled rectifier with a firing circuit having maximum possible firing angle of 90°. Draw the relevant wave forms. Use a 230/24 V transformer for isolation.

B2. RC Triggering

19. Design and set up a triggering circuit for a thyristor to obtain given waveform. The firing circuit has to be a maximum possible firing angle of 180°.

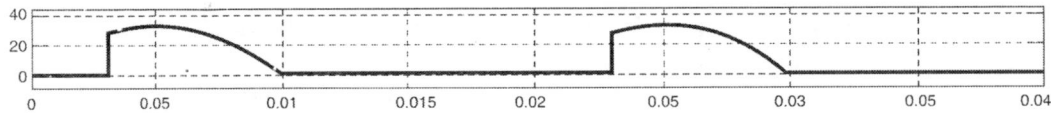

Also obtain relevant wave forms at the power side for α = 120°. Use a 230/24 V transformer for isolation. Predetermine average and rms values of the load voltage.

20. Design and setup a triggering circuit for a thyristor to obtain following waveform. The firing circuit has only passive components in the triggering part. Use a 230/24 V transformer for isolation.

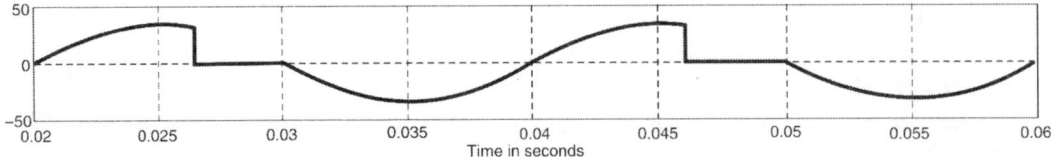

Also find out average and rms values of the load voltage.

21. Design and set up RC full-wave triggering circuit. Obtain relevant wave forms at the power side for α = 30°. Given 230/24 V Transformer.

B3 and B4. UJT Triggering

22. Design and set up a half-wave thyristor controlled rectifier using UJT firing circuit. Draw the relevant waveforms both at power side and control side for $\alpha = 45°$. Predetermine the average and rms values of the load voltage. Use a 230/24 V isolation transformer.

23. Design and set up a full-wave thyristor controlled rectifier using UJT firing circuit. Draw relevant waveforms both at power side and control side for $\alpha = 90°$. Predetermine the average and rms values of the load voltage. Use a 230/24 V isolation transformer.

24. Design and set up a triggering circuit for a thyristor controlled half-wave rectifier with firing angle of $\alpha = 120°$ which has good reliability and satisfactory operation over a wide range of temperature. Use a 230/24 V isolation transformer.

25. Design and set up a triggering circuit for full-wave thyristor rectifier output voltage which has low current requirement for triggering. Use a 230/24 V isolation transformer.

26. Obtain the given wave form for a thyristor controlled rectifier. Use minimum gate power dissipation circuit to fire the thyristor. Use a 230/24 V isolation transformer.

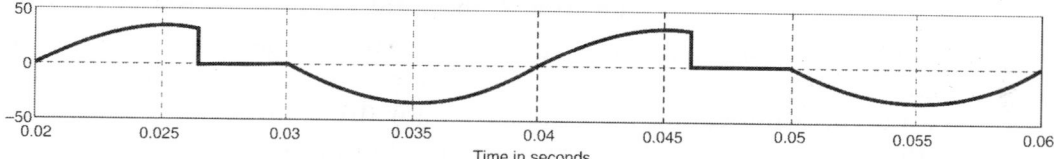

27. Design and set up a triggering circuit for a half-wave thyristors controlled rectifier circuit. The triggering circuit should have an active device. Use a 230/24 V isolation transformer.

B5. Single phase half-wave controlled rectifier

28. Design and set up an experiment to observe the waveforms at firing angle of $\alpha = 60°$ for single phase half-wave controlled rectifier circuit with resistive load. Derive expressions for the average and rms values of the load voltage and predetermine them. Use a 230/24 V isolation transformer.

29. Set up a half-wave rectifier circuit and observe the load voltage wave forms for (a) R load (b) RL Load and (c) RL load with Freewheeling Diode for discontinuous load current. Derive the expression for average and rms values of load voltage in each case with firing angle of 60°. Use a 230/24 V isolation transformer.

B6. Single phase Mid–Point converter

30. Design and set up an experiment to obtain the load voltage waveform for a single phase fully controlled rectifier where a non-conducting thyristor has a peak inverse voltage (PIV) of $2\,V_m$ appearing across it. Use a 230/24 V transformer for isolation.

31. Observe and plot the relevant waveforms for a single phase full-wave rectifier with centre-tapped configuration feeding an RL load under discontinuous current mode of operation. Use a 230/24 V transformer for isolation.

32. Set up a proper thyristor controlled rectifier circuit and obtain the voltage waveform shown in the figure. Use a 230/24 V transformer for isolation.

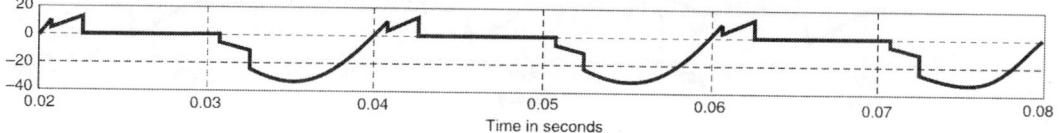

B7. Single phase semi converter

33. Observe and plot the load voltage waveform for a single phase semi converter feeding (a) R load (b) RL load and (c) RL load with freewheeling Diode, under discontinuous current mode operation. Derive the expression of RMS value of the load voltage and pre-determine its value for a firing angle α = 45° in each case. Comment on your observations. Use a 230/24 V transformer for isolation.

34. Observe the Load voltage waveform by setting up a suitable rectifier circuit such that, it can operate only in one quadrant. Take the load as RL load. Use a 230/24 V transformer for isolation.

35. Set up a proper rectifier circuit to obtain the voltage waveform shown in the figure. Use a 230/24 V transformer for isolation.

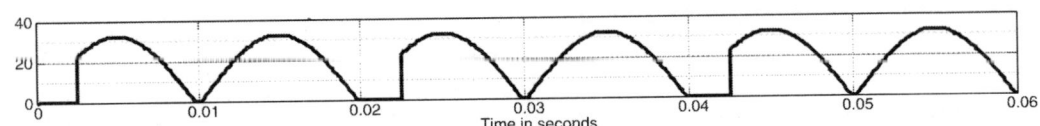

B8. Single phase fully controlled bridge rectifier

36. Design and set up a single phase rectifier circuit which can operate in two quadrants. Use a 230/24 V transformer. Observe the load voltage. Pre-determine the average and rms values of the load voltage for any firing angle.

37. Obtain the following voltage waveform using a controlled rectifier circuit. Use a 230/24 V transformer for isolation.

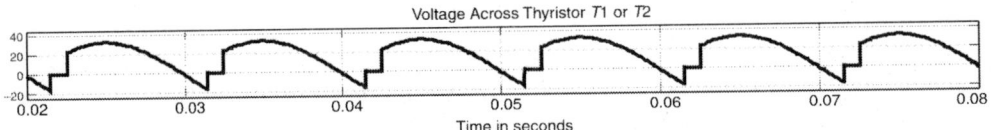

B9. Speed control of a DC motor

38. Design and set up a rectifier circuit to plot the speed of the motor versus armature voltage. The given motor is 24V, 1500 RPM PMDC motor. Use a 230/24 V isolation transformer.

39. Design and set up a circuit to plot the speed of the motor verses armature voltage when 24 V PMDC motor is connected to a single phase half controlled rectifier. Use 230/24V transformer.

B10. AC voltage regulator using thyristors

40. Obtain the following voltage waveform using unidirectional conducting power semiconductor devices. Predetermine and calculate the rms value of the load voltage for a firing angle α = 30°. Given 230/24 V transformer.

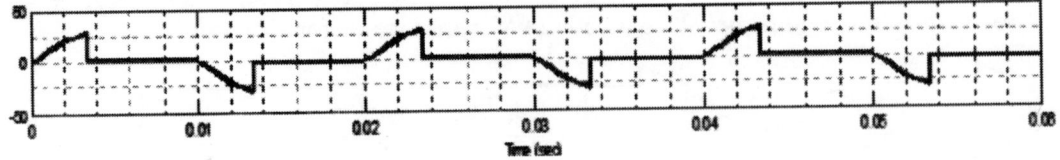

41. Set up a power electronic converter circuit that controls AC voltage using unidirectional controllable devices. Get the following current waveform. Use a 230/24 V transformer for isolation.

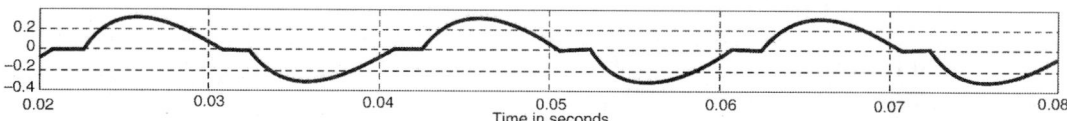

42. Set up a power electronic circuit to obtain the given voltage waveform. Use only one controllable power semiconductor device. Use a 230/24 V transformer for isolation.

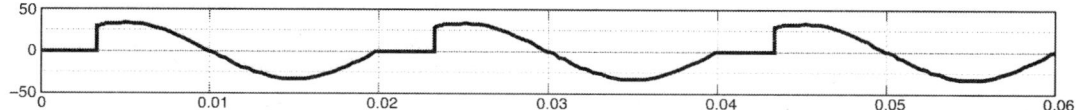

B11. AC Voltage regulator using a triac

43. Design and set up a circuit to obtain the following voltage waveform using a bidirectional conducting power semiconductor device. Predetermine and calculate the rms value of the load voltage for a firing angle α = 45°. Use a 230/24 V transformer for isolation.

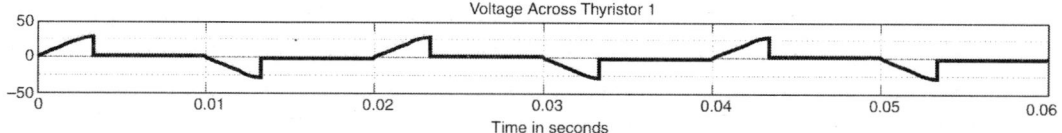

44. Design and set up a circuit to obtain the given voltage waveform using a bidirectional conducting power semiconductor device. Predetermine the rms value of the given waveform for a firing angle α = 45°. Use a 230/24 V transformer for isolation.

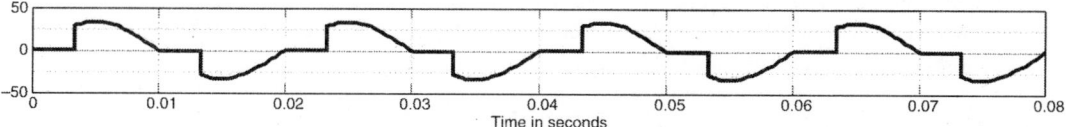

B12. Complementary commutation circuit

45. Identify and set up a circuit to obtain the given voltage waveforms. Plot the observed waveforms on a graph sheet. Supply voltage is 12 V DC.

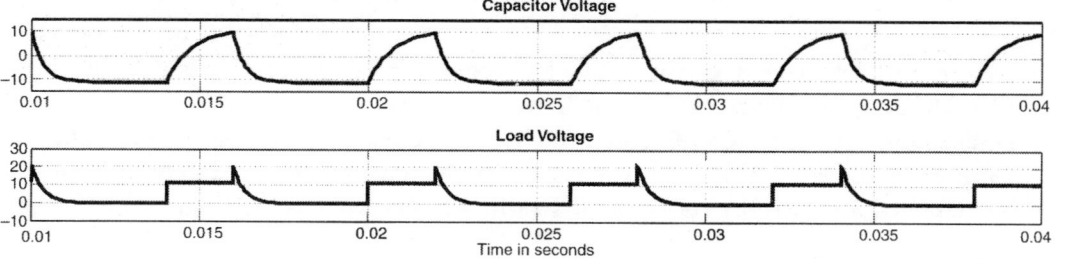

46. Design and setup a commutation circuit to observe the voltage waveforms across its power semiconductor devices shown given below. The supply voltage is 12 V DC.

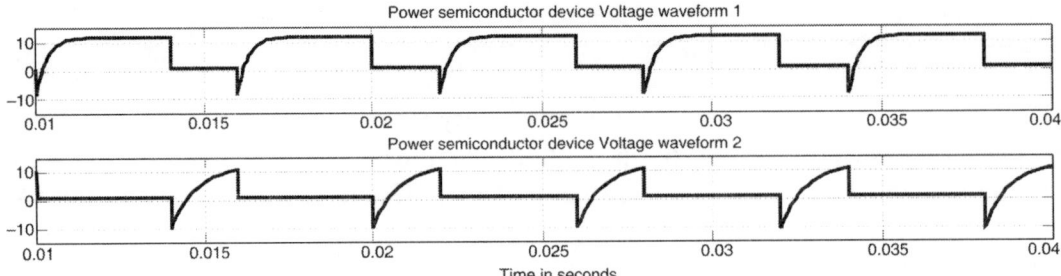

B13. Resonance commutation circuit

47. Design and setup a forced commutation circuit which works on the principle of current commutation technique. Plot the relevant waveforms. The supply voltage is 12 V DC.

48. Design and set up a resonance commutation circuit. Plot the relevant waveforms. Take the supply voltage as 12 V DC.

49. Identify the commutation circuit in which the current waveform in a commutation element is shown below. Design and set up the commutation circuit. Consider a 12 V DC supply.

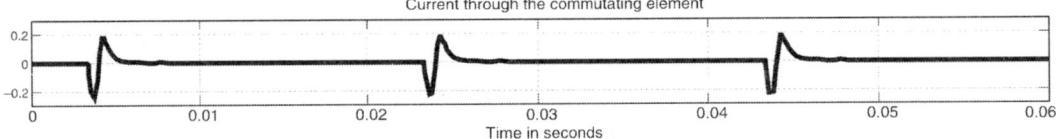

B14. Impulse commutation circuit

50. Design and set up an impulse commutation circuit. Plot the relevant waveforms. Take the supply voltages as 24 V and 12 V DC.

51. Identify the commutation circuit in which the current waveform in a commutation element is shown below. Design and setup the circuit and observe the waveform. Take the supply voltages as 24 V and 12 V DC.

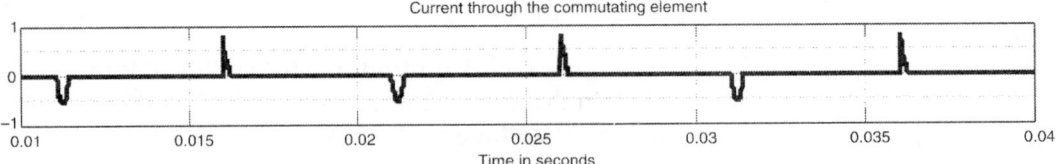

B15. DC–DC Buck regulator

52. Design and set up a step-down DC–DC regulator which produces 6 V DC from 12 V DC supply. Choose appropriate switching frequency.

53. Design and set up a DC–DC converter which performs buck operation. Take the supply voltage as 12 V DC. Observe and plot the load voltage and inductor voltage waveforms. Assume appropriate duty ratio. Choose switching frequency as 5 kHz.

B16. DC–DC boost regulator

54. Design and set up a step-up DC–DC converter which produces 15 V DC from 12 V DC supply. Choose appropriate switching frequency. Plot the load voltage and inductor voltage waveforms.

55. Design and set up a DC–DC converter which performs a boost operation. Take the supply voltage to be 12 V DC. Observe and plot the load voltage and inductor voltage waveforms. Assume appropriate duty ratio. Choose switching frequency as 2 kHz.

B17. Current commutated chopper or oscillation chopper

56. Design and set up an oscillation chopper circuit that operates from a 12 V DC supply. Observe and plot its relevant waveforms. Also perform constant and variable frequency operation.

57. Design and set up a circuit that works on the principle of LC resonance commutation. Observe and obtain the given waveform. Given 12 V DC supply

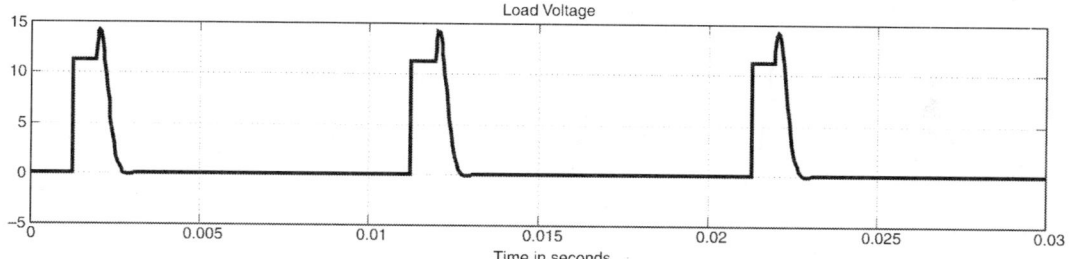

58. Design and set up a circuit that works on the principle of LC resonance commutation to obtain the given waveforms. Identify the control strategy used here. Explain how to achieve the same. Take the supply voltage as 12 V DC.

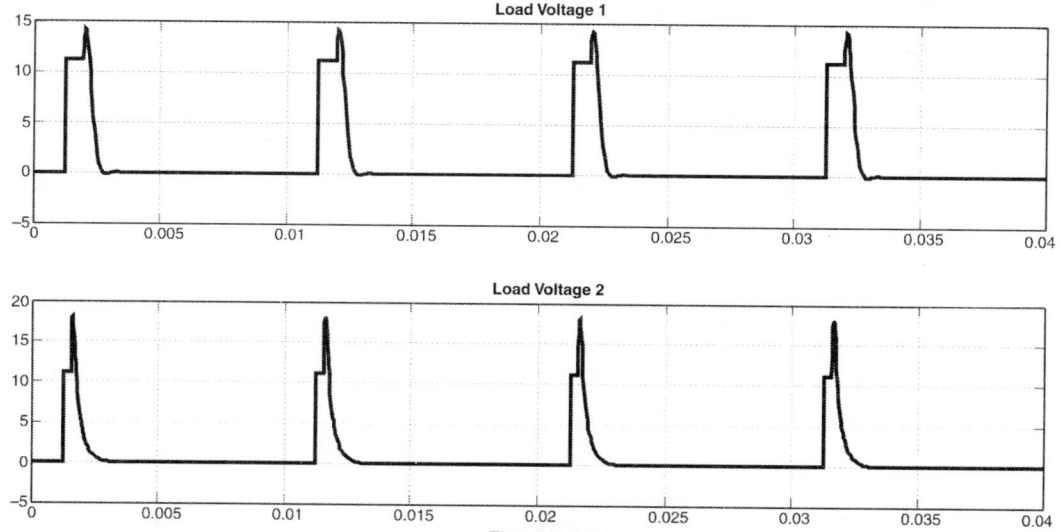

B18. Voltage commutated chopper

59. Design and set up a voltage commutated chopper and plot its waveforms. Take the supply voltages to be 24 V and 12V DC. Obtain experimentally two types of time ratio control (TRC) schemes.

60. Design and set up a circuit that works on the principle of impulse commutation to obtain the given waveforms. Available supply voltages are 24 V and 12 V DC. Identify the control technique used here.

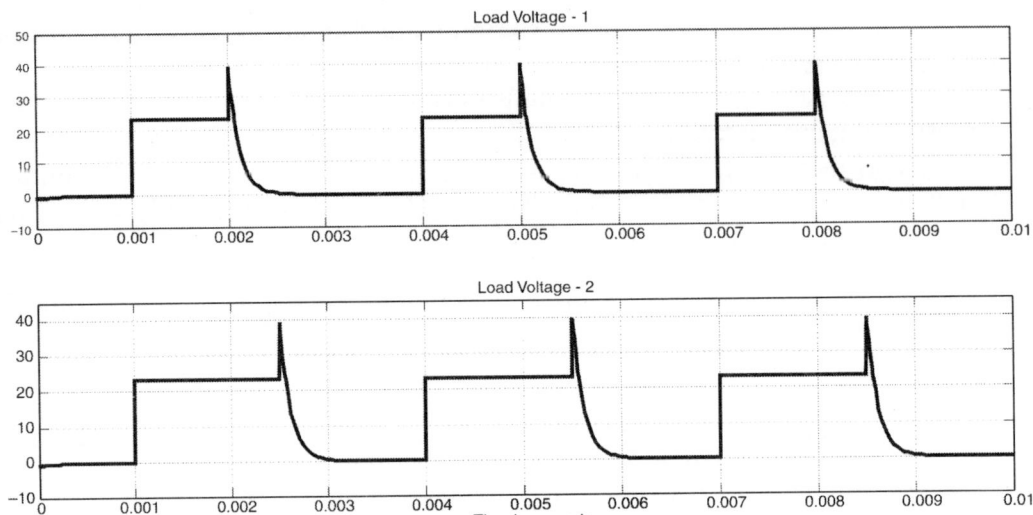

61. Design and set up a circuit to obtain given waveforms in which the main thyristor is turned OFF using a negative voltage. Take the supply voltages as 24 V and 12 V DC. Identify the control technique used here.

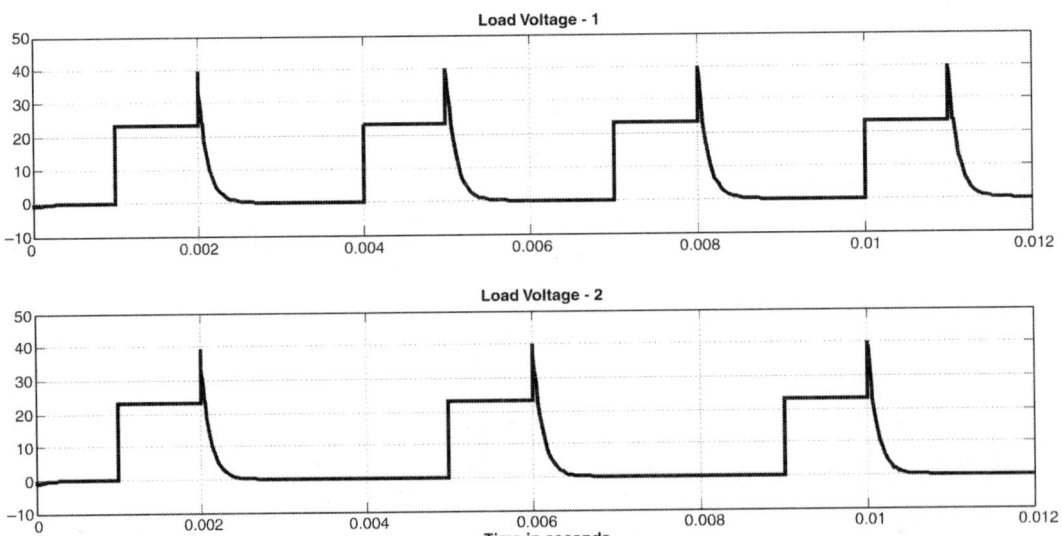

C1. Simulation—R triggering

62. Design and simulate an R triggering circuit for a thyristor, given the following specifications:

    (a) Input voltage of 325 V (max value), 50 Hz
    (b) Firing angle $\alpha = 45°$ and
    (c) $R_{Load} = 100\ \Omega$.

    Observe and note down the source voltage, gate voltage, voltage across the load, and voltage across the thyristor.

63. Design and simulate an R triggering circuit for a thyristor, given the following specifications:

    (a) Input voltage of 200 V (max value), 50 Hz
    (b) Firing angle $\alpha = 30°$ and
    (c) $R_{Load} = 100\ \Omega$.

    Observe and note down the source voltage, gate voltage, voltage across the load, and voltage across the thyristor and comment on result.

C2. Simulation—RC triggering

64. Design and simulate an RC triggering circuit for a thyristor feeding a pure resistor with the following specifications:

    (a) Input voltage of 325 V (max value), 50 Hz
    (b) Firing angle $\alpha = 75°$ and
    (c) $R_{Load} = 100\ \Omega$.

    Load current exists only during positive half cycle of the source voltage. Observe the following waveforms: source voltage, voltage across the load, voltage across the capacitor and voltage across the thyristor.

65. Design and simulate an RC triggering circuit for a full-wave thyristor controlled rectifier with the following specifications:

    (a) Input voltage of 100 V (max value), 50 Hz
    (b) Firing angle $\alpha = 10°$ and
    (c) $R_{Load} = 100\ \Omega$.

    Observe the following waveforms: source voltage, voltage across the load, voltage across the capacitor and voltage across the thyristor.

66. Design and simulate an RC triggering circuit for a half-wave thyristor controlled rectifier with the following specifications:

    (a) Input voltage of 200 V (max value), 50 Hz
    (b) Firing angle $\alpha = 130°$ and
    (c) $R_{Load} = 100\ \Omega$.

    Observe the following waveforms: source voltage, load voltage, load current and source current and comment on the result.

C3. Simulation—half-wave rectifier

67. Design and simulate a single phase half-wave rectifier with the following specifications:

    (a) Input voltage of 325 V (max value), 50 Hz
    (b) Firing angle $\alpha = 45°$ and
    (c) $R_{Load} = 100\ \Omega$.

Simulate the waveforms of source voltage, load voltage, load current and voltage across the thyristor under following load conditions:

(1) Purely resistive load

(2) Resistive-inductive load–Discontinuous conduction mode

(3) Resistive-inductive-back EMF load–Discontinuous conduction mode

(4) Resistive-inductive-back EMF load and with Freewheeling diode-discontinuous conduction mode

(5) Resistive-inductive-back EMF load and with Freewheeling diode-continuous conduction mode

Predetermine the average and RMS value of the load voltage under each load condition and verify the values by simulation.

68. Design and simulate a single phase half-wave rectifier with the following specifications:

(a) Input voltage of 120V (RMS value), 50Hz

(b) Firing angle $\alpha = 30°$ and

(c) $R_{Load} = 100\ \Omega$.

Simulate the waveforms of source voltage, load voltage, load current, freewheeling diode current and voltage across the thyristor under following conditions:

(1) Purely resistive load

(2) Resistive-inductive load–Discontinuous conduction mode

(3) Resistive-inductive-back EMF load–discontinuous conduction mode

(4) Resistive-inductive-back EMF load and with freewheeling diode–discontinuous conduction mode

(5) Resistive-inductive-back EMF load and with freewheeling diode–continuous conduction Mode

Predetermine the average and RMS value of the load voltage under each load conditions and verify the values by simulation.

69. Design and simulate a single phase half-wave rectifier with the following specifications:

(a) Input voltage of 120 V (RMS value), 50 Hz

(b) Firing angle $\alpha = 100°$ and

(c) $R_{Load} = 100\ \Omega$.

Simulate the waveforms of source voltage, load voltage, load current, and source current under following conditions

(1) Resistive inductive load–continuous or discontinuous conduction mode

(2) Resistive inductive load with freewheeling diode–continuous or discontinuous Conduction Mode

How can we say that the input power factor is improved by keeping a freewheeling diode in the circuit? Comment on it by comparing source current waveforms.

C4. Simulation—mid-point converter

70. Simulate a single phase mid-point converter with the following specifications:

(a) Input voltage of 325 V (max value), 50 Hz

(b) Firing angle $\alpha = 30°$ and

(c) $R_{Load} = 100\ \Omega$.

Observe the waveforms of secondary winding voltage, load voltage, load current and voltage across the thyristor under the following load conditions:

(1) Purely resistive load

(2) Resistive-inductive load–Discontinuous conduction mode

(3) Resistive-inductive load–Continuous conduction mode

Predetermine the average and RMS value of the load voltage under each load condition and verify the values by simulation.

71. Simulate a single phase mid-point converter with the following specifications:

(a) Input voltage of 120 V (RMS value) and 50 Hz

(b) Firing angle $\alpha = 90°$ and

(c) $R_{oad} = 100\ \Omega$.

Observes the waveforms of secondary winding voltage, load voltage, load current, freewheeling diode current and voltage across the thyristor under following load conditions

(1) Resistive-inductive load and freewheeling diode–Discontinuous conduction mode

(2) Resistive-inductive load and freewheeling diode–Continuous conduction mode

(3) Resistive-inductive-back EMF load and with freewheeling diode–discontinuous conduction mode

(4) Resistive-inductive-back EMF load and with freewheeling diode–continuous conduction mode

Predetermine the average and RMS value of the load voltage under each load condition and verify the values by simulation.

How can we say that the input power factor is improved by keeping a freewheeling diode in the circuit?

C5. Simulation—semiconverter

72. Design and simulate a single phase semiconverter circuit with the following specifications:

(a) Input voltage of 325 V (max value), 50 Hz

(b) Firing angle $\alpha = 35°$ and

(c) $R_{Load} = 100\ \Omega$.

Observe the waveforms of source voltage, load voltage, load current and voltage across the thyristor under following load conditions:

(1) Purely resistive load

(2) Resistive-inductive load–Discontinuous conduction mode

(3) Resistive-inductive-back EMF load with freewheeling diode–Discontinuous conduction Mode

Predetermine the average and RMS value of the load voltage under each load condition and verify the values by simulation. Assume proper values of inductance L and back emf E.

73. Design and simulate a single phase semiconverter circuit with the following specifications:

(a) Input voltage of 120 V (RMS value), 50 Hz

(b) Firing angle $\alpha = 60°$ and

(c) $R_{Load} = 100\ \Omega$.

Simulate the waveforms of source voltage, voltage across the load, load current, freewheeling diode current and voltage across the thyristor under following conditions:

(1) Purely resistive load
(2) Resistive-inductive load–Discontinuous conduction mode
(3) Resistive-inductive load–Continuous conduction mode
(4) Resistive-inductive-back EMF load with freewheeling diode–Discontinuous conduction mode
(5) Resistive-Inductive–back EMF load with freewheeling diode –Continuous conduction mode

Predetermine the average and RMS value of the load voltage under each load condition and verify the values by simulation. Assume proper values of inductance L and back emf E.

74. Design and simulate a single phase semiconverter circuit with the following specifications:
    (a) Input voltage of 120 V (RMS value), 50 Hz
    (b) Firing angle $\alpha = 100°$ and
    (c) $R_{Load} = 100\ \Omega$.

    Observe the waveforms of source voltage, load voltage, load current, and source current under following conditions:
    (1) Resistive inductive load-Continuous or discontinuous conduction mode
    (2) Resistive inductive load with freewheeling diode–Continuous or discontinuous conduction mode

    How can we say that the input power factor is improved by keeping a freewheeling diode in the circuit?

C6. Simulation—single phase full converter

75. Design and simulate a single phase full converter circuit with the following specifications:
    (a) Input voltage of 325 V (max value), 50 Hz
    (b) Firing angle $\alpha = 35°$ and
    (c) $R_{Load} = 100\ \Omega$.

    Observe and plot the waveforms of source voltage, load voltage, load current and voltage across the thyristor T_1 under following load conditions:
    (1) Purely resistive load
    (2) Resistive-inductive load–Discontinuous conduction mode
    (3) Resistive-inductive-back EMF load with freewheeling diode-Discontinuous conduction mode

    Predetermine the average and RMS value of the load voltage under each load conditions and verify it by simulation. Assume proper values of inductance L and back EMF E.

76. Design and simulate a single phase full converter circuit with the following specifications:
    (a) Input voltage of 120 V (RMS value) and 50 Hz
    (b) Firing angle $\alpha = 60°$ and
    (c) $R_{Load} = 100\ \Omega$.

Observe and plot the waveforms of source voltage, voltage across the load, load current, freewheeling diode current and voltage across the thyristor T_1 under following conditions:

(1) Purely resistive load
(2) Resistive-inductive load–Discontinuous conduction mode
(3) Resistive-inductive load–Continuous conduction mode
(4) Resistive-inductive-back EMF load with freewheeling diode–Discontinuous Conduction Mode
(5) Resistive-inductive-bback EMF load with Freewheeling diode–Continuous conduction mode

Predetermine the average and RMS value of the load voltage under each load conditions and verify the values by simulation. Assume the values of inductance L and back emf E.

77. Design and simulate a single phase full converter circuit with the following specifications:

    (a) Input voltage of 120 V (RMS value) and 50 Hz
    (b) Firing angle $\alpha = 100°$ and $R_{Load} = 100\ \Omega$.

    Simulate the waveforms of source voltage, load voltage, load current, and source current under following conditions:

    (1) Resistive inductive load-Continuous or discontinuous conduction mode
    (2) Resistive inductive load with freewheeling diode–Continuous or discontinuous conduction Mode.

C7. Simulation—speed Control of a DC motor

78. Model and simulate a DC motor speed control technique using any controlled rectifier. A freewheeling diode is connected across the armature. Consider following specifications for the simulation:

    (a) Supply voltage = 24 V RMS, 50 Hz
    (b) Motor is fully loaded
    (c) Firing angle = 60°.

    Assume continuous current conduction. Pre determine and verify by simulation, the armature voltage and speed of the motor in RPM.

79. Model and simulate a DC motor speed control technique using semi-controlled rectifier. A freewheeling diode is connected across the armature. Consider the following specifications for simulation:

    (a) Supply voltage = 24 V RMS, 50 Hz
    (b) Motor is fully loaded
    (c) Armature current is continuous.

    Plot the following graphs:

    (1) Speed of the motor in RPM versus firing angle of converter
    (2) Armature voltage of DC motor versus firing angle of the converter
    (3) Speed of the motor in RPM versus armature voltage

C8. Simulation—single phase AC voltage controller

80. Simulate a single phase AC voltage controller circuit with following specifications:

    (a) AC supply voltage = 24 V RMS and supply frequency = 50 Hz
    (b) $R = 100\ \Omega$
    (c) Firing angle, $\alpha = 45°$

Observe and plot the following waveforms:

(1) Supply voltage

(2) Load voltage

(3) Load current

(4) Voltage across thyristor –1 and

(5) Voltage across thyristor –2

Perform the operation with RL load. Choose the load inductor value L such that the load current is discontinuous.

81. Simulate a single phase AC voltage controller circuit with following specifications:

(a) AC supply voltage = 230 V RMS and Supply frequency = 50 Hz

(b) $R = 100\ \Omega$

(c) Firing angle, $\alpha = 30°$

Observe and plot the following waveforms:

(1) Supply voltage

(2) Load voltage

(3) Load current

(4) Voltage across Thyristor –1 and

(5) Voltage across Thyristor –2

Perform the operation with RL load. Choose the load inductor value L such that the load current is discontinuous.

C9. Simulation—complementary commutation circuit

82. Design and simulate a complementary commutation circuit with following specifications:

(a) DC supply voltage = 24 V

(b) Load Resistor = 100 Ω

(c) Switching period = 8 msec and select appropriate ON time

Observe and plot the following waveforms:

(1) Capacitor voltage

(2) Voltage across main thyristor

(3) Voltage across complementary thyristor

(4) Load voltage.

83. Design and simulate a complementary commutation circuit with following specifications:

(a) DC supply voltage = 12 V

(b) Load resistor = 50 Ω

(c) Switching period = 10 msec and select appropriate ON time

Observe and plot the following waveforms:

(1) Capacitor voltage

(2) Voltage across main thyristor

(3) Voltage across complementary thyristor

(4) Load voltage.

C10. Simulation—resonance commutation circuit

84. Design and simulate the parallel resonance commutation circuit for a thyristor with following specifications:
    (a) DC supply voltage = 12 V
    (b) Load Resistor = 50 Ω
    (c) Switching period = 10 msec

    Observe the following waveforms:
    (1) Capacitor current
    (2) Load current
    (3) Current through the thyristor
    (4) Load voltage
    (5) Voltage across the thyristor

85. Design and simulate a resonance commutation circuit for a thyristor with following specifications:
    (a) DC supply voltage = 24 V
    (b) Load Resistor = 150 Ω
    (c) Switching period = 15 msec

    Observe the following waveforms:
    (1) Capacitor current
    (2) Load current
    (3) Current through the thyristor
    (4) Load voltage

C11. Simulation—impulse commutation circuit

86. Design and simulate a commutation circuit which uses a reverse voltage to turn off a conducting thyristor with following specifications:
    (a) DC supply voltage = 24 V
    (b) Load Resistor= 150 Ω
    (c) Switching period = 15 msec

    Observe and plot the following waveforms:
    (1) Capacitor current
    (2) Voltage across the capacitor
    (3) Current through the main thyristor
    (4) Voltage across main thyristor
    (5) Voltage across auxiliary thyristor
    (6) Load voltage

87. Design and simulate an impulse commutation circuit with following specifications:
    (a) DC supply voltage = 12 V
    (b) Load Resistor = 150 Ω
    (c) Switching period = 20 msec

    Adjust the OFF time of the main thyristor at different values and observe and plot the following waveforms.
    (1) Capacitor current
    (2) Voltage across the capacitor
    (3) Current through the main thyristor
    (4) Voltage across main thyristor
    (5) Load voltage

C12. Simulation—oscillation chopper

88. Design and simulate an oscillation chopper circuit with following specifications:
    (a) DC supply voltage = 24 V
    (b) t_{OFF} = 0.66 msec
    (c) Chopping frequency = 100 Hz

    Observe and plot the following waveforms:
    (1) Capacitor voltage
    (2) Load voltage
    (3) Thyristor voltage and
    (4) Thyristor current.

    Perform constant and variable frequency operations of oscillation chopper. For variable frequency operation choose another chopping frequency as 200 Hz.

89. Design and simulate an oscillation chopper circuit with following specifications:
    (a) DC supply voltage = 12 V
    (b) t_{OFF} = 0.66 msec
    (c) Chopping frequency = 100 Hz

    Observe and plot the following waveforms:
    (1) Capacitor voltage
    (2) Load voltage
    (3) Thyristor voltage and
    (4) Thyristor current

    Perform constant and variable frequency operations of oscillation chopper. For variable frequency operation choose another chopping frequency as 200 Hz.

C13. Simulation—voltage commutated chopper

90. Design and simulate a voltage commutated chopper circuit with following specifications:
    (a) DC supply voltage = 24 V
    (b) t_{OFF} = 0.66 msec
    (c) Chopping frequency = 100 Hz

    Observe and plot the following waveforms:
    (1) Capacitor voltage
    (2) Load voltage
    (3) Thyristor voltage and
    (4) Thyristor current

    Perform constant and variable frequency operations of oscillation chopper. For variable frequency operation choose another chopping frequency as 200 Hz.

91. Design and simulate a voltage commutated chopper circuit with following specifications:
    (a) DC supply voltage = 12 V
    (b) t_{OFF} = 0.66 msec
    (c) Chopping frequency = 100 Hz

    Observe and plot the following waveforms
    (1) Capacitor voltage
    (2) Load voltage

(3) Thyristor voltage and

(4) Thyristor current.

Perform constant and variable frequency operations of oscillation chopper. For variable frequency operation choose another chopping frequency as 200 Hz.

C14. Simulation—buck converter

92. Design and simulate a buck regulator circuit for the following specifications:

(a) Input voltage = 10 V DC

(b) Switching frequency = 10 kHz

(c) Output voltage = 8V

Observe the waveforms of voltage across the load, inductor voltage, inductor current and capacitor current.

Perform both continuous and discontinuous operations of the circuit.

93. Design and simulate a buck regulator circuit for the following specifications:

(a) Input voltage = 20V DC

(b) Switching frequency = 30 kHz

(c) Output voltage = 15V

Observe and plot the waveforms of output voltage, inductor voltage, inductor current, capacitor current, load current and voltage across the switch.

C15. Simulation—boost converter

94. Design and simulate a boost regulator circuit with the following specifications:

(a) Input voltage = 12 V DC

(b) Switching frequency = 10 kHz

(c) Output voltage = 18 V

Observe and plot the waveforms of voltage across the load, inductor voltage, inductor current, capacitor current, load current and voltage across the switch.

95. Design and simulate a boost regulator circuit with the following specifications:

(a) Input voltage = 24 V DC

(b) Switching frequency = 30 kHz

(c) Output voltage = 35 V

Observe and plot the waveforms of voltage across the load, inductor voltage, capacitor current, load current and voltage across the switch. Consider continuous conduction.

96. Repeat question 95 for discontinuous conduction operation.

C16. Simulation—single phase inverter

97. Simulate a single phase inverter circuit having a DC source voltage of 220 V and with resistance load of 100 Ω. Observe the following waveforms:

(a) Input voltage

(b) Load voltage

(c) Load current

(d) Pulses to the gate of the switches.

Also do FFT analysis on the load voltage waveform and find out THD.

98. Simulate a single phase inverter circuit having a DC source voltage of 220 V and with a resistance-Inductance load. Take the switch as MOSFET and $R = 100\,\Omega$ and $L = 100$ mH.

Observe following waveforms:
(a) Input voltage
(b) Load voltage
(c) Load current
(d) Pulses to the gate of the switches and
(e) Current flowing through a switch.

Also do FFT analysis on the load voltage and the load current waveforms. Find out the THD in each case and comment on harmonic values obtained.

99. Simulate a single phase inverter circuit having a DC source voltage of 220 V with Resistance-Inductance load. Select IGBT as the switch and $R = 100 \ \Omega$ and $L = 100$ mH. Observe following waveforms:
    (a) Input voltage
    (b) Load voltage
    (c) Load current
    (d) Pulses to the gate of the switches and
    (e) current flowing through a switch and its anti-parallel diode.

    Also do FFT analysis on the load voltage and the load current waveforms. Find out THD in each case and comment on harmonic values obtained.

C17. Simulation—three phase inverter

100. Simulate a three phase inverter circuit having a DC voltage source of 220 V with a three phase star connected resistance load of $100 \ \Omega$. Observe following waveforms for 180°.
     (a) Phase voltages
     (b) Line voltages
     (c) Pulses to the gate of the switches
     (d) Current in each phase of the load

     Also perform the FFT analysis on the phase and line voltage waveforms and find out THDs. Comment on THDs obtained.

101. Repeat question 100 for 120° conduction mode.

102. Simulate a three phase inverter circuit having a DC voltage source of 220V and with a three phase star connected Resistance-Inductance load. Select MOSFETs as the switches and $R = 100 \ \Omega$ and $L = 100$ mH.

     Observe following waveforms for 180^0 conduction mode.
     (a) Phase voltages
     (b) Line voltages
     (c) Current in the each phase of the load
     (d) Pulses to the gate of the switches.
     (e) Current flowing through a switch.

     Observe current waveform in any one phase of the load and through a switch and comment on the result.

103. Simulate a three phase inverter circuit having a DC voltage source of 220 V and with a three phase star connected resistance-inductance load. Select IGBT as switches and Observe following waveforms for 120° conduction mode.
     (a) Phase voltages
     (b) Line voltages

(c) Current in the each phase of the load
(d) Pulses to the gate of the switches
(e) Current flowing through a switch and its anti-parallel diode.

Observe current waveform in any one phase of the load and through a switch and its anti parallel diode and comment on the result.

C18. Simulation—power quality analysis

104. Design and simulate a single phase half-wave controlled rectifier with the following specifications:
    (a) Input voltage of 325 V (max value), 50 Hz
    (b) Firing angle $\alpha = 45°$ and
    (c) $R_{Load} = 100\ \Omega$.

Perform the power quality analysis of the system with and without filter capacitor and check whether current THD is within the allowable limit.

Also simulate the circuit having RL load with continuous conduction. Select appropriate values of R and L. Check the power factor and THD. How can the power factor be improved?

[**Hint:** Freewheeling diode]

105. Design and simulate a single phase AC voltage controller with the following specifications:
    (a) Input voltage of 100 V (max value), 50 Hz
    (b) Firing angle $\alpha = 30°$ and
    (c) $R_{Load} = 100\ \Omega$.

Perform the power quality analysis of the system and check whether current THD is within the allowable limit.

106. Repeat question 105 with RL load and with discontinuous conduction mode. Select appropriate values of R and L.

107. Design and simulate a single phase fully controlled bridge rectifier with following specifications:
    (a) Input voltage of 230 V (rms value), 50 Hz,
    (b) Firing angle $\alpha = 60°$ and
    (c) $R_{Load} = 100\ \Omega$.

Perform power quality analysis of the circuit with and without filter capacitor and check whether current THD is within the allowable limit.

108. Design and simulate a single phase fully controlled bridge rectifier with following parameters:
    (a) Input voltage of 230 V (rms value), 50 Hz,
    (b) Firing angle $\alpha = 60°$.

Select appropriate values of R and L.

Perform power quality analysis of the circuit with and without filter capacitor and check whether current THD is within the allowable limit.

APPENDICES

FOURIER ANALYSIS

A.1 INTRODUCTION

Fourier analysis is a powerful tool for both analysis and design of power converters. It helps in the selection of switching frequency of any power electronic circuit. It also gives an indication of the harmonic content and the line spectrum of a voltage or current wave. A continuous –time periodic signal can be represented in a Fourier series.

Any periodic waveform, i.e. one for which $f(t) = f(t + T)$, can be expressed by a Fourier series provided that

(1) if it is discontinuous there are a finite number of discontinuities in the period T
(2) it has a finite average value for the period T
(3) it has a finite number of positive and negative maxima.

These are called *Dirichlet conditions*.

Fourier series can be represented in following three ways:
1. Trigonometric fourier series
2. Compact trigonometric fourier series or polar fourier series
3. Exponential fourier series

A.2 TRIGONOMETRIC FOURIER SERIES

The output voltage (or current) of a converter is generally a periodic function of time and is defined under steady state condition as

$$v(\omega t + T) = v(\omega t)$$

$T = 1/f$ is the period

Angular frequency (in radian/second)

$$\omega = 2\pi f = \frac{2\pi}{T}$$

If a periodic function of any signal is repetitive in every 2π radians, $v(\omega t + 2\pi) = v(\omega t)$, it may be expressed as the sum of a series of sinusoidal and cosinusoidal components. Thus $v(t)$ can be expressed as

$$v(t) = \frac{a_0}{2} + \sum_{n=1}^{\infty} \{a_n \cos(n\omega t) + b_n \sin(n\omega t)\}$$

where,
1. a_0, a_n and b_n are fourier coefficients
2. ω is called the fundamental frequency
3. $n\omega$ = harmonic frequencies
4. n = order of harmonics
5. $(a_0/2)$ is the average value of $v(t)$ or the DC component of $v(t)$.

$$a_n = \frac{2}{T} \int_0^T v(t)\cos(n\omega t)dt = \frac{1}{\pi} \int_0^{2\pi} v(t)\cos(n\omega t)d(\omega t)$$

$$b_n = \frac{2}{T} \int_0^T v(t)\sin(n\omega t)dt = \frac{1}{\pi} \int_0^{2\pi} v(t)\sin(n\omega t)d(\omega t)$$

Also
$$v(t) = \frac{a_0}{2} + \sum_{n=1}^{\infty} C_n \sin(n\omega t + \phi_n)$$

C_n = peak value of nth harmonic and its value is $\sqrt{a_n^2 + b_n^2}$

and ϕ_n = phase delay of nth harmonic component $\tan^{-1}(a_n/b_n)$

A.3 THE EFFECTS OF SYMMETRY ON THE FOURIER COEFFICIENTS

Four types of symmetry used to simplify Fourier analysis are:
1. Even-wave symmetry
2. Odd-wave symmetry
3. Half-wave symmetry
4. Quarter-wave symmetry

1. Even-wave symmetry

If $f(t) = -f(-t)$ then it is said to have even symmetry.
Fourier series will have only cosine terms and all sine terms are absent.

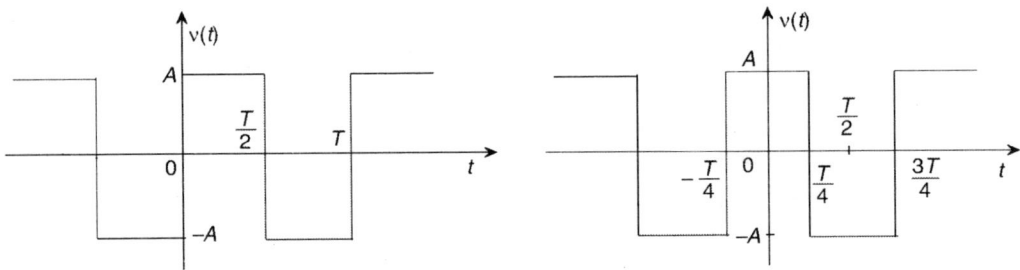

Fig. A1: A waveform with even symmetry   Fig. A2: A waveform with odd symmetry

2. Odd-wave symmetry

If $f(t) = -f(-t)$ then it is said to have odd symmetry.
Fourier series will have only sine terms and all cosine terms are absent.

3. Half-wave symmetry

If the inverted signal, shifted by one half period, is identical to original then it is said to have half-wave symmetry.

$$f(t) = -f\left(t + \frac{T}{2}\right)$$

Fourier series can have sine terms and cosine terms. Also even harmonics are absent.

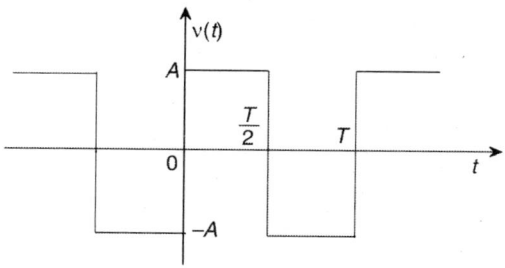

Fig. A3: A waveform with half-wave symmetry

4. Quarter-wave symmetry

A signal that has both half-wave symmetry and that is either even or odd symmetry is said to have quarter wave symmetry.

(a) Half-wave and even symmetry: All even harmonics are absent in the fourier series and it contains only cosine terms of odd order.

(b) Half-wave and odd symmetry: All even harmonics are absent in the fourier series and it contains only sine terms of odd order.

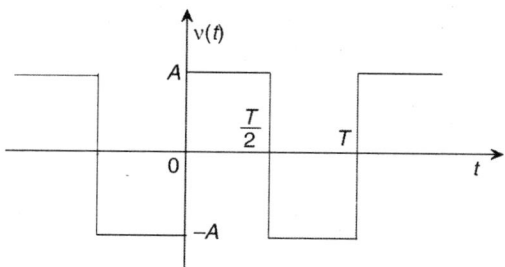

Fig. A4: Waveform with half-wave and even symmetry

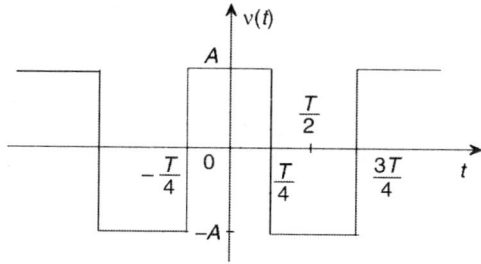

Fig. A5: Waveform with half-wave and Odd symmetry

Line Spectrum

A plot showing each harmonic amplitude in the wave is called the *line spectrum*. The harmonic content decreases rapidly for a convergent series. The line spectra for signals with discontinuities, like saw tooth and square waveform, will be with slowly decreasing amplitudes. An example is shown here.

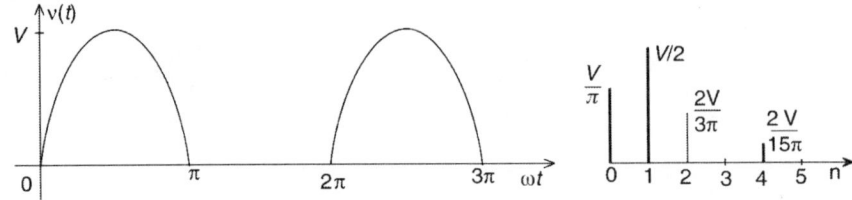

Fig. A6: Periodic signal   **Fig. A7:** Line spectra

Appendix B

B1 HARMONIC/FFT ANALYSIS USING MATLAB/SIMULINK

In MATLAB or in DSP processors, fourier analysis or harmonic analysis is performed with the help of fast fourier transforms (FFT). In fact, FFTs are the algorithms that are used to calculate discrete fourier transform (DFT).

In MATLAB, signals or waveforms are discrete in nature. So we can use either discrete time fourier series (DTFS) or discrete time fourier transform (DTFT). DTFS is used for the frequency domain analysis of the discrete time periodic signal. But its frequency spectrum is not periodic. DTFT is used for the frequency domain analysis of the discrete time non periodic signal and its frequency spectrum is periodic.

So the major difference between the DTFT and DTFS lies in the frequency spectrum. Frequency domain spectral analysis is easy when the frequency spectrum is continuous which is obtained in DTFT.

DTFT can also be applied to discrete time periodic signals.

Since frequency spectrum of DTFT is continuous, it cannot be processed on Digital Signal Processing (DSP) platforms where only discrete values can be processed. So we need to sample continuous frequency spectrum and resulting frequency domain representation of a signal is known as discrete fourier transform (DFT).

The direct computation of DFT involves large number of arithmetic operations and so it is inefficient. Fast fourier transform (FFT) is an algorithm that calculates DFT quite efficiently. Because of FFT algorithms, DFT can be calculated in real time.

DSP processors have special architectural provisions to implement FFT algorithms efficiently.

Table B1: Comparison of DTFS and DTFT

Nature of signal	Fourier representation	Frequency spectrum	When the frequency Spectrum is sampled
Discrete	DTFS	Continuous and aperiodic	—
Discrete	DTFT	Continuous and periodic	DFT (Computation of DFT is fast and easy with FFT)

B2 FREQUENCY SPECTRUM ANALYSIS IN MATLAB/SIMULINK USING POWERGUI BLOCK

The powergui block can be used to obtain the frequency spectrum of any signal directly by clicking the FFT analysis tab. The steps to be followed to perform FFT analysis of a signal is illustrated below with an example.

Consider a square wave and let us do FFT analysis of the same. The steps are as follows:
Step 1: Enter "simulink" in the MATLAB command window to open the simulink Library Browser or clicking Simulink icon in the MATLAB toolbar.
Step 2: Select **File > New > Model** in the simulink library browser to create a new model or press Ctrl+N. An empty model window is opened and it can be saved by selecting **File > Save** in the model window.
Step 3: Drag and place the following building blocks from *Simulink library* and *Sim Power Systems library* to the new simulink model file
  (a) Powergui block [Sim Power Systems]
  (b) Signal generator [Simulink > Sources > Signal Generator]
  (c) Scope [Simulink > Sinks]
Step 4: Connect the signal generator to the scope as shown in Fig. B1.

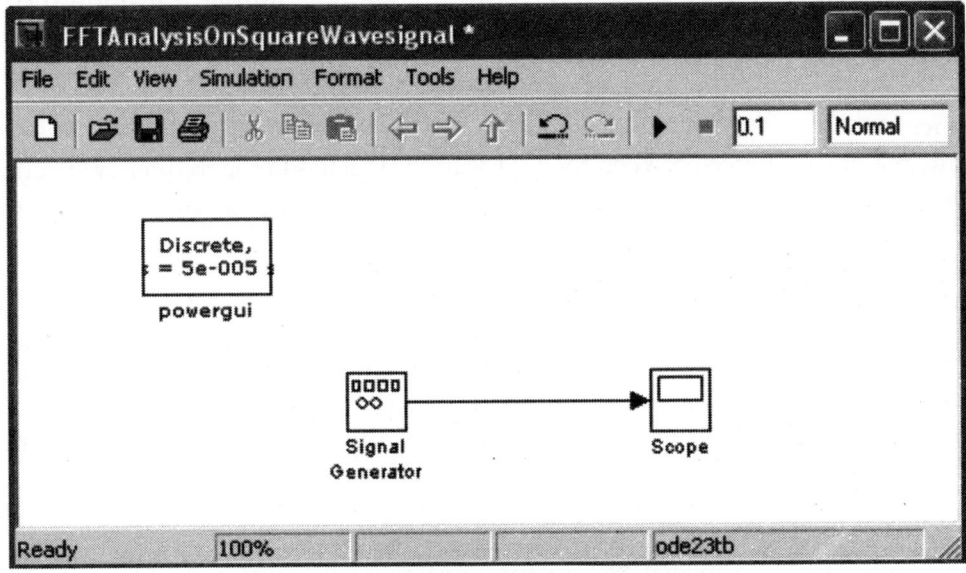

Fig. B1: The model window with building blocks

Step 5: Double click the Signal Generator block in the model window. Select waveform as square and set amplitude as 100 and frequency as 50 Hz.
Step 6: Double click the scope block in the model window. Select the parameters button on the toolbar of the scope block's display.
Select data history in the 'scope' parameters (Refer Fig. B2). Then,
• Deselect the check box "Limit data points to last: 5000".
• Select the check box "Save data to workspace:" and then
• Enter a variable name in the Variable name field.

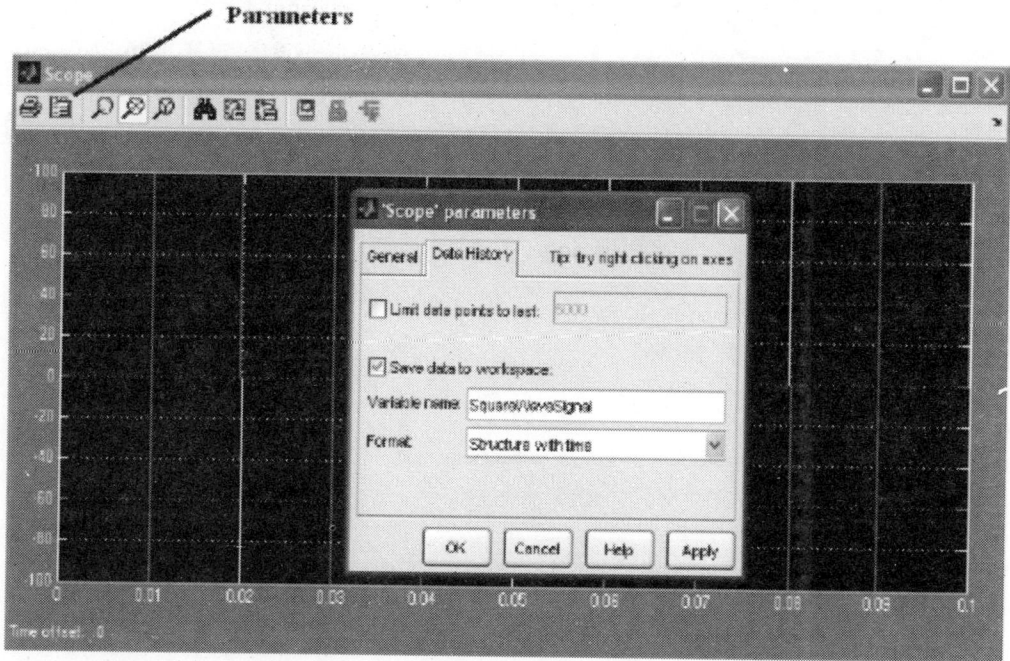

Fig. B2: Scope block with scope parameters

By default, the variable name is assigned as Scope Data which is automatically modified to Scope Data 1, Scope Data 2, and so on, for different scopes placed in the model window. Let us enter variable name as "square wave signal" in the variable name field.

- If the scope has more than one signal, set data format to structure or structure with time.
- Click apply or OK button

Step 7: Select simulation > configuration parameters in the model window. Refer Fig. B3.

The configuration parameters dialog box appears. Then,

- The simulation time can be specified, i.e. start time and stop time is selected. Let us select start time as 0 seconds and stop time as 0.1 seconds.
- In solver options, we can see the maximum step size with default value as auto. Then type 0.00001 in the maximum step size box. This is necessary to make the simulated waveforms smooth.
- From the solver options, solver can be selected. Simulation speed can be increased by selecting proper solver from solver list. Refer to simulink/help for choosing solver options.
- Click apply or OK button.

Step 8: Double click powergui block in the model window and select the simulation type as **discrete** from configure parameters tab.

Step 9: Run the simulation by selecting simulation > start in the model window or use **start button** in the model window toolbar.

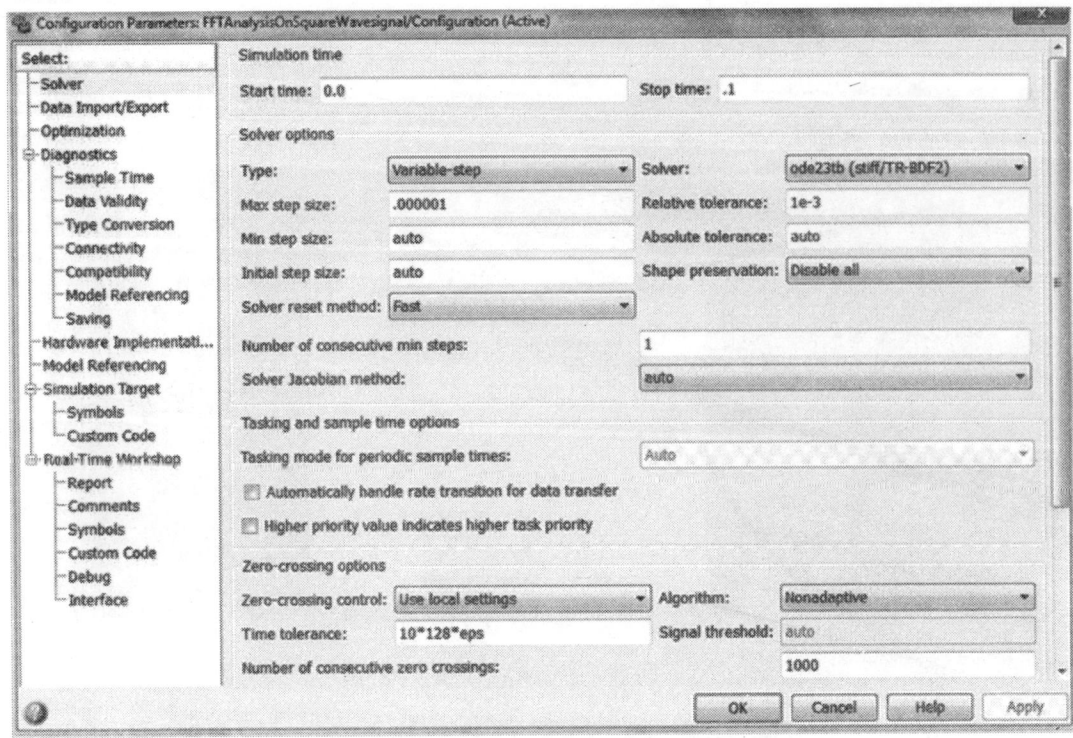

Fig. B3: Configuration parameters of the model window

Step 10: Double click **powergui** block in the model window and click **FFT analysis** tab from the **analysis tools**. A "powergui FFT analysis tool" window will pop up as shown in the Fig. B4.

Step 11: Commonly used settings for FFT analysis in the "powergui FFT analysis tool" window are

- Start time of the waveform and number of cycles of the waveform on which FFT analysis is performed.
- Selection of fundamental frequency which is selected as 50 Hz.
- Display style of FFT analysis can be bar type or list type.

Step 12: Select "Bar (relative to fundamental)". Now FFT analysis can be seen with magnitude (% of fundamental) versus frequency as

- Fundamental (50 Hz) =127.3.
  Equation of fundamental component of square wave is $4V_{dc}/\pi$. (Refer to appendix C). $(4 \times 100/\pi = 127.3 \text{ V})$
- Total harmonic distortion THD = 48.22%
- Magnitude of square wave corresponding to each frequency, i.e. odd order harmonics is shown here as we discussed in Appendix A.

Step 13: Select "List (relative to fundamental)". Now FFT analysis can be seen as (Refer Fig. B5)

- Fundamental = 127.3 peak (90.03 rms)
- Total harmonic distortion, THD = 48.22%

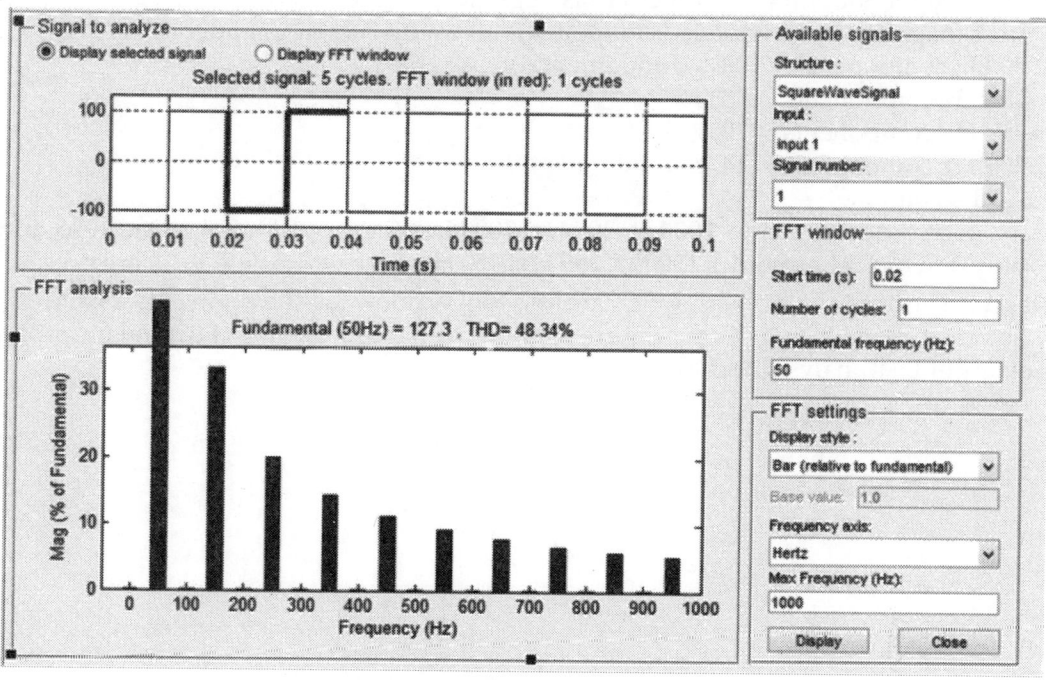

Fig.B4: The powergui FFT analysis tool window (display style: Bar)

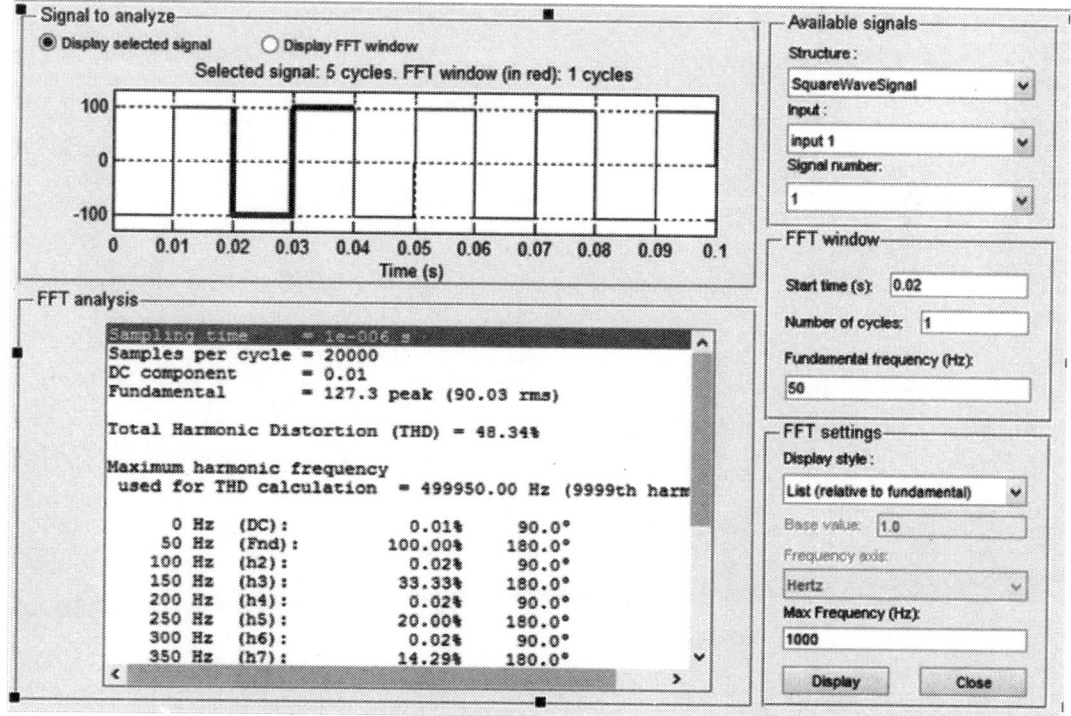

Fig. B5: The powergui FFT analysis tool window (display style: List)

- DC component or 0 Hz component is negligibly small.
- % of magnitude of frequency components as we discussed in appendix A are
- 50 Hz components = 100% (fundamental)
- 150 Hz components = 33%
- 250 Hz components = 20%
- 350 Hz components = 14.29%
- And so on.
- All even order harmonics 100 Hz, 200 Hz, 300 Hz,...(i.e., $n = 2, 4, 6, 8...$) are zero and only odd order harmonics 150 Hz, 250 Hz, 350 Hz,...(i.e., $n = 3, 5, 7, 9,...$) are present.

Step 14: The image of 'powergui FFT analysis tool window'(as shown in Figs B4 and B5) is obtained by choosing '**Copy Figure**' option available from the '**Edit**' menu of the 'powergui FFT analysis tool' window.

Appendix C

HARMONIC ANALYSIS OF SINGLE PHASE FULL BRIDGE INVERTER

The inverter output voltage waveform, which is a square wave, has both *half-wave symmetry* and *odd symmetry*. (Refer to appendix A). Hence it contains only sine terms and will not contain cosine terms. Generally a voltage waveform can be expressed in trigonometric Fourier series as,

$$v(t) = \frac{a_O}{2} + \sum_{n=1}^{\infty} \left(a_n \cos(n\omega_S t) + b_n \sin(n\omega_S t) \right)$$

From the inverter output square waveform, it can be seen that the average value is zero. So it don't have DC term, i.e. $a_0 = 0$. Since the waveform has odd symmetry, $a_n = 0$ for all values of n, (i.e. $n = 1, 2, 3, 4, 5, 6,...$). The switching frequency in rad/sec, $\omega_S = 2\pi/T_S$. Therefore, the square wave output voltage can be expressed in Fourier series as,

$$v(t) = \sum_{n=1,3,5...}^{\infty} b_n \sin(n\omega_S t); \qquad \omega_S = \frac{2\pi}{T_S}$$

The Fourier coefficient,

$$b_n = \frac{2}{T_S} \int_0^{T_S} v(t) \sin(n\omega_S t)\, dt; \qquad n = 1, 3, 5...$$

$$= \frac{2}{T_S} \left(\int_0^{T_S/2} V_S \sin(n\omega_S t)\, dt + \int_{T_S/2}^{T_S} (-V_S) \sin(n\omega_S t)\, dt \right)$$

$$= \frac{2V_S}{T_S} \left[\left(-\frac{\cos n\omega_S t}{n\omega_S} \right)_0^{T_S/2} - \left(-\frac{\cos n\omega_S t}{n\omega_S} \right)_{T_S/2}^{T_S} \right]$$

$$= \frac{2V_S}{n\omega_S T_S} \left[-\cos\left(n\frac{2\pi}{T_S}\frac{T_S}{2} \right) - (-1) + \cos\left(n\frac{2\pi}{T_S}T_S \right) - \cos\left(n\frac{2\pi}{T_S}\frac{T_S}{2} \right) \right]$$

$$= \frac{2V_S}{n\omega_S T_S} \left[-\cos(n\pi) + 1 + \cos(n2\pi) - \cos(n\pi) \right]; \quad n \text{ odd}$$

$$b_n = \frac{2V_S}{(2\pi/T_S)T_S}[1+1+1+1] = \frac{4V_S}{n\pi}$$

$$v(t) = \sum_{n=1,3,5...}^{\infty} \frac{4V_S}{n\pi}\sin(n\omega_S t); \quad \omega_S = \frac{2\pi}{T_S}$$

The load current is

$$i_0(t) = \sum_{n=1,3,5...}^{\infty} \frac{4V_S}{n\pi R}\sin(n\omega_S t); \quad \omega_S = \frac{2\pi}{T_S}$$

The RMS value of fundamental voltage is

$$V_1 = \frac{\left(\frac{4V_S}{1\times\pi}\right)}{\sqrt{2}} = \frac{4V_S}{\sqrt{2}\pi} = 0.9V_S$$

The RMS value of fundamental current is

$$I_1 = \frac{0.9V_S}{R}$$

Average power delivered to the load

$$P = \sum_{n=1,3,5...} P_n = \sum_{n=1,3,5...} I_{rms}^2 R = \sum_{n=1,3,5...} \frac{(4V_S/n\pi R)^2}{\sqrt{2}}R.$$

STANDARD VALUES FOR RESISTORS AND CAPACITORS

Standard resistor values (± 5%)						
1.0	10	100	1.0 K	10 K	100 K	1.0 M
1.1	11	110	1.1 K	11 K	110 K	1.1 M
1.2	12	120	1.2 K	12 K	120 K	1.2 M
1.3	13	130	1.3 K	13 K	130 K	1.3 M
1.5	15	150	1.5 K	15 K	150 K	1.5 M
1.6	16	160	1.6 K	16 K	160 K	1.6 M
1.8	18	180	1.8 K	18 K	180 K	1.8 M
2.0	20	200	2.0 K	20 K	200 K	2.0 M
2.2	22	220	2.2 K	22 K	220 K	2.2 M
2.4	24	240	2.4 K	24 K	240 K	2.4 M
2.7	27	270	2.7 K	27 K	270 K	2.7 M
3.0	30	300	3.0 K	30 K	300 K	3.0 M
3.3	33	330	3.3 K	33 K	330 K	3.3 M
3.6	36	360	3.6 K	36 K	360 K	3.6 M
3.9	39	390	3.9 K	39 K	390 K	3.9 M
4.3	43	430	4.3 K	43 K	430 K	4.3 M
4.7	47	470	4.7 K	47 K	470 K	4.7 M
5.1	51	510	5.1 K	51 K	510 K	5.1 M
5.6	56	560	5.6 K	56 K	560 K	5.6 M
6.2	62	620	6.2 K	62 K	620 K	6.2 M
6.8	68	680	6.8 K	68 K	680 K	6.8 M
7.5	75	750	7.5 K	75 K	750 K	7.5 M
8.2	82	820	8.2 K	82 K	820 K	8.2 M
9.1	91	910	9.1 K	91 K	910 K	9.1 M

Standard Capacitor Values (± 10%)						
10 pF	100 pF	1000 pF	0.001 µF	0.1 µF	1.0 µF	10 µF
12 pF	120 pF	1200 pF	0.012 µF	0.12 µF	1.2 µF	
15 pF	150 pF	1500 pF	0.015 µF	0.15 µF	1.5 µF	
18 pF	180 pF	1800 pF	0.018 µF	0.18 µF	1.8 µF	
22 pF	220 pF	2200 pF	0.022 µF	0.22 µF	2.2 µF	22 µF
27 pF	270 pF	2700 pF	0.027 µF	0.27 µF	2.7 µF	
33 pF	330 pF	3300 pF	0.033 µF	0.33 µF	3.3 µF	33 µF
39 pF	390 pF	3900 pF	0.039 µF	0.39 µF	3.9 µF	
47 pF	470 pF	4700 pF	0.047 µF	0.47 µF	4.7 µF	47 µF
56 pF	560 pF	5600 pF	0.056 µF	0.56 µF	5.6 µF	
68 pF	680 pF	6800 pF	0.068 µF	0.68 µF	6.8 µF	
82 pF	820 pF	8200 pF	0.082 µF	0.82 µF	8.2 µF	

Bibliography

1. Alok Jain. Power Electronics: Devices, Circuits and MATLAB Simulations, Penram International Publishing (India) Pvt. Ltd., 2010.
2. Bhim Singh, Ambrish Chandra, Kamal Al-Haddad. Power Quality Problems and Mitigation Techniques, First edn, John Wiley and Sons Ltd., 2015.
3. DW Hart. Power Electronics, McGraw Hill, 2011.
4. Joseph A Edminister, JE Swann. Schaum's Outline of Theory and Problems of Electric Circuits, McGraw Hill International Book Company, New York, 1972.
5. J Vithayathil. Power Electronics: Principles and Applications, McGraw Hill Inc., New York, 1995.
6. KR Varmah, Chikku Abraham. Power Electronics, Cengage Learning India Pvt. Ltd., 2016.
7. KR Varmah, Chikku Abraham, Ginnes K John. Fundamentals of Electrical Machines and Drives, Cengage Learning India Pvt. Ltd. 2016.
8. MATLAB/Simulink and MATLAB/Program Help.
9. MH Rashid. Power Electronics: Circuits, Devices and Applications, 2nd edn, Prentice Hall Inc., New Jersey, 1993.
10. Ned Mohan, TM Undeland, WP Robbins. Power Electronics: Converters, Applications and Design, 3rd edn, Wiley, New York. 2003.